# PICTURING NATURE

# PICTURING NATURE

❖

## AMERICAN
## NINETEENTH-CENTURY
## ZOOLOGICAL ILLUSTRATION

*Ann Shelby Blum*

PRINCETON UNIVERSITY PRESS · PRINCETON, NEW JERSEY

COPYRIGHT © 1993 BY PRINCETON UNIVERSITY PRESS

PUBLISHED BY PRINCETON UNIVERSITY PRESS, 41 WILLIAM STREET, PRINCETON, NEW JERSEY 08540

IN THE UNITED KINGDOM: PRINCETON UNIVERSITY PRESS, CHICHESTER, WEST SUSSEX

LIBRARY OF CONGRESS CATALOGING-IN-PUBLICATION DATA

BLUM, ANN SHELBY, 1950–

PICTURING NATURE : AMERICAN NINETEENTH-CENTURY ZOOLOGICAL

ILLUSTRATION / ANN SHELBY BLUM.

P.    CM.

INCLUDES BIBLIOGRAPHICAL REFERENCES AND INDEX.

ISBN 0-691-08578-1 (ALK. PAPER)

1. ZOOLOGICAL ILLUSTRATION—UNITED STATES—HISTORY—19TH CENTURY.

2. ANIMALS IN ART—HISTORY—19TH CENTURY.   I. TITLE.

QL46.5.B58     1993

591'.022'2—DC20     92-4681     CIP

PUBLISHED WITH THE ASSISTANCE OF THE GETTY GRANT PROGRAM

THIS BOOK HAS BEEN COMPOSED IN LINOTRON TRUMP MEDIAEVAL WITH MONOTYPE CASTELLAR DISPLAY

PRINCETON UNIVERSITY PRESS BOOKS ARE PRINTED ON ACID-FREE PAPER AND MEET THE

GUIDELINES FOR PERMANENCE AND DURABILITY OF THE COMMITTEE ON

PRODUCTION GUIDELINES FOR BOOK LONGEVITY OF THE COUNCIL

ON LIBRARY RESOURCES

PRINTED IN SINGAPORE

1  3  5  7  9  10  8  6  4  2

*To my mother and father*

*with love and thanks*

# Contents

━━━●❀●━━━

# List of Figures

———————◉———————

IN the measurements of figure dimensions, height precedes width. For many publications plate size varies from copy to copy, and even within a single volume, especially in width, because of the tightness of the binding, and also because of irregularities in trimming the plates.

John Lawrence Leconte, "A descriptive catalogue of the Geodephagous Coleoptera inhabiting the United States east of the Rocky Mountains," *Annals of the Lyceum of Natural History of New-York* 4 (1848): 173–474. Plate 13: 23.5 cm × 14.5 cm. Courtesy of Cornell University Library.

xvii

# List of Plates

— ❄ —

IN the measurements of the plates' dimensions, height precedes width. In many volumes, plate size varies, especially in width, because of the tightness of the binding, and also because of irregularities in trimming the plates.

# *Preface*

———◦❖◦———

THIS study of nineteenth-century American[1] zoological illustration is grounded in studies of American science, biographies of American naturalists, zoologists, illustrators and engravers, social and cultural histories, and histories of printing and painting. An extensive and varied literature on American natural history and zoology has reconstructed, from diverse points of view, the beginnings of those disciplines in the eighteenth and early nineteenth centuries, their professionalization and institutionalization, controversies over theory and values, and role in national life. Until fairly recently, however, studies of illustration have been missing from the historical panorama.

I have focused on how American naturalists and their zoologist successors used pictures, and how, in the changing look of the discipline, its changing practices, as well as its changing place in society, are reflected. My study emphasizes taxonomy because of the close descriptive relationship between zoological taxonomy and illustration, but it differs from much of the related work in this area of the history of science in that I focus on practice rather than ideas.

Why American? After independence, American naturalists asserted that their enterprise was different from European science, just as American painters asserted that theirs would be different from that of their European counterparts. Whether the American difference in natural history illustration carried over into zoological illustration seemed worth investigating. In discussing the early period of American natural history, there is some question about what made a naturalist "American." Why, for example, have I not included the work of naturalists of the early period, such as John Abbot of Georgia, an Englishman, or Nicholas Hentz, a Frenchman, though I did include the work of Alexander Wilson, a Scot, and Charles Alexandre Lesueur, another Frenchman? I hope the reasons will be apparent from my discussion below, but I should say at the outset that my interest centers on published illustration and the interrelationship between zoological practice and printing. In the colonial tradition, Abbott sent his commissioned drawings back to England for publication; Hentz's miniatures of spiders were published posthumously, by which time the interest in his work had become virtually antiquarian. Wilson and Lesueur, on the other hand, participated in establishing a natural history publishing-to-public relationship in the United States, and that gave a particular meaning to their illustration. Indeed, I owe a great debt to historians of science who have recognized the role of publishing in the formation of the literary genres and illustration conventions of science. I should add that my emphasis on published illustration and its major trends has excluded from this book many of the beautiful and fascinating drawings and sketches I encountered in my research, as well as some intriguing idiosyncratic printed illustrations.

I was fortunate to come to this project first during the late 1970s and early 1980s, when the history of American science, building on a largely biographical literature, was moving into studies of institutions and social interactions. I would like to think of this book as a contribution to these ongoing projects. Still, some readers may feel that I have overlooked the contributions of important individuals, institutions, or important aspects of the social scene. I have for the most part emphasized the mainstream of the developing discipline of zoology and its dominant figures who worked at major institutions located in recognized centers of activity. I have put greater emphasis on federal publication programs than on those of the state surveys, even though the state surveys led the way in public funding for description of flora and fauna, and hammered out, at the level of local government, arguments about the utility of scientific surveys to the general public. But because of the role of federal publications in consolidating conventions of illustration, I gave them primary attention. The episodes on which I have concentrated would not be significant, however, unless they had taken place in a context constructed out of the myriad local activities of illustrating natural history.

Other aspects of the social context of natural history, such as the spread of teaching natural history to adults and children and the role of educational and popular literatures of natural history, I have only suggested or sketched. This omission slights the developments that led to women's contributions to teaching, writing, and illustrating natural history, and for their entry into the field of zoology. My focus on what came to be mainstream, professional science gives short shrift to the realm of popular natural history, its printed pictures, and social significance, although the divergence of the professional from the popular is an important aspect of my story. Where the growth of a large general audience for natural history interacted with professional practices, including illustration, I have tried to describe that interaction.

In method and terminology this study falls somewhere between the work by bibliographers of the history of scientific publishing and recent work by sociologists of science investigating scientific representation—literary and pictorial. Both groups write for specialized audiences. I draw on both literatures, and my notes refer to sources where readers may find more detailed documentation of the production of scientific books and the lives of their authors, and analyses of contemporary scientific practices that both corroborate and extend the historical developments I explore here. Historians of science working on the establishment of textual conventions in science may find parallels in illustration. There also exists an extensive literature on the careers and illustrations of the naturalist-illustrators of the early nineteenth century. New books about John James Audubon appear almost every year. My placing natural history illustration of that period in the context of a century of scientific animal illustration should offer some new insights into the early conventions.

In sum, my approach to illustration combines the history of natural history and zoological illustration in the nineteenth-century United States with aspects of the histories of publishing and print technology for the same period, and aspects of the history of nineteenth-century biology, especially classification, that are relevant to the central practices of illustration. The complexity and richness of these issues have attracted the extensive work of the many historians without whose studies I could not have contemplated this one. Informed by their work, I have chosen to emphasize episodes that indicate moments of change in the articulation—tacit and explicit—of issues bearing on the production and use of pictures for science, especially episodes that reveal tensions over changing practices. For all the areas that were beyond the scope of this project I have tried to indicate in the notes my debt to sources providing the foundation of my inquiry, to which the reader may turn for amplification.

The foci of interest of the different episodes, as well as the nature of my sources for this study, called for varying approaches from chapter to chapter. The overall organization is chronological. The introduction paints with a broad brush a survey of the European background up to the first productions of natural history illustration in the United States. The early chapters focus on individuals and their work. Later, as societies, museums, and government surveys became more important, the scene shifts to the principal institutions engaged in producing illustration, although the record attests to the continuing influence of individuals in shaping the look of institutional illustrations. An analysis of the influence of photography and photomechanical reproduction on zoological illustration required an investigation of and comparison among a range of publications and the implications of their reproductive printing techniques.

My captions usually reproduce the original captions; I have not attempted to provide revised taxonomic nomenclature. Throughout, I reproduced spelling and punctuation in direct quotations and avoided the distracting "[*sic*]."

# Acknowledgments

———— ❀ ————

I HAVE been looking forward to thanking in print everyone who contributed to this project. When I began my research into zoological illustration, I was the archivist at the Museum of Comparative Zoology, Harvard University. The museum's director at that time, Alfred W. Crompton, and its librarian, Ruth Hill, encouraged me to present my first efforts in a public exhibition. Ruth Hill's successor, Eva Jonas, gave my work shelf space and other accommodation. My library colleagues contributed many valuable references. The staffs of the Academy of Natural Sciences of Philadelphia, especially Carol Spawn, and the American Philosophical Society, especially Steven Catlett, facilitated my first research forays. Bill Deiss and Susan Glenn of the Smithsonian Institution Archives were generous with time and help during my repeated visits and telephone requests, as was Ellen Wells of Special Collections, Smithsonian Institution Libraries. The rich collections and cheerful staffs of the Biological Sciences Library and The Bancroft Library of the University of California, Berkeley, aided in later revisions. Thanks also to the staffs of collections where I did additional research and which facilitated requests for photographic reproduction: the Rare Books and Graphic Arts collections, Princeton University Library; Cornell University's Albert R. Mann Library, Olin Library, and the History of Science Collections, especially to Sam Demas and staff, and to Laura Linke; and the Workingmen's Institute Library, New Harmony, Indiana.

From the beginning, discussions with biological illustrators Peg Estey, Sara Fink, Sarah Landry, and Laszlo Mezoly taught me which questions to ask. Teaching in the Rhode Island School of Design Program in Scientific and Technical Illustration gave me an opportunity to place American nineteenth-century zoological illustration in the historical panorama. Ruth Turner and Herb Levi of the Museum of Comparative Zoology and their students, particularly Brad Calloway, Jonathan Coddington, and Wayne Maddison, offered insights into how today's systematic zoologists produce and use their own illustrations. Paul Hertz, the late Ken Miyata, and Kent Redford, when students at the museum, took a keen interest in the project and spent many hours discussing the pictures and their biology. Ken garnished his insights with dinners of fresh trout. The Museum of Comparative Zoology's Department of Invertebrates, headed by Herb Levi, deserves my warmest thanks for providing a home away from home and a forum for ongoing discussion of the substance and the trivia of my work. Cecile Villars's hospitality and friendship made my work more agreeable; I know she can visualize it, but I wish she could see the finished book. John Nove, former Curator of Natural History at the Peabody Museum, Salem, Massachusetts gave me another thoroughly enjoyable opportunity to synthesize my research for public exhibition.

To Al Coleman, former photographer at the Museum of Comparative Zoology and master craftsman, I owe a special acknowledgment. Al taught me how to look at and think about photography as a translation of the engravings and lithographs about which I wrote. His skill sets the highest standards for photographic copy work. I think of Al as a present-day Sonrel. John Pachai, of University Photography, Cornell, also produced excellent negatives and prints.

My thanks for editorial and historical advice in the middle and late stages of this project to Sharon Kingsland, Charlotte Porter, Peter Taylor, and Mary P. Winsor. Tom Laqueur helped warm up my thinking about illustration as I prepared to revise the manuscript. Late in the day, Peter Dear read and offered suggestions on the introduction. I owe more than thanks to Jehane Burns Kuhn for prodding my thinking and destabilizing my writing habits. She helped me reshape the manuscript in style and substance in its penultimate and ultimate drafts. Her acute and challenging comments provided a bracing interaction with an uncompromising reader.

Emily Wilkinson, my editor at Princeton University Press, undertook this project with enthusiasm, and efficiently—and patiently—piloted the manuscript through the straits of funding applications and production. Her hospitality made picture research in the Princeton University libraries possible and enjoyable.

A generous grant from the J. Paul Getty Foundation supported the reproduction of illustrations for this book.

Family and friends have listened to me complain about writing, cooked me innumerable dinners, made me have fun, and sustained a better faith than my own that I would finish this book. My family has been generous with encouragement, humor, and material aid. Janet Fesler in Washington, and Lou and Kathy Pollak in Philadelphia, put me up on research trips. Paula Chandoha and Victoria Munroe gave imaginative support to the project's early days and brought informed perspectives to their interest in the pictures. During the final revision, I missed acutely the solidarity of my writing group in Berkeley, especially Cheryl Springer, Sheila Tully, and Libby Wood. Ruth Gilmore has always been my inspiration to think harder and more clearly—and to commit it to paper. Peter Taylor coached me down the home stretch and tolerated chaos in the study. And Vann put the whole thing in right perspective.

# PICTURING NATURE

# Animal Pictures and Natural History

THE FORM OF A REPRESENTATION CANNOT BE DIVORCED

FROM ITS PURPOSE AND THE REQUIREMENTS OF THE SOCIETY

IN WHICH THE GIVEN VISUAL LANGUAGE GAINS CURRENCY.

—E. H. GOMBRICH, *ART AND ILLUSION*

ACCEPTANCE of pictures as conveying authentic information about nature lies at the heart of scientific practice. The consistent reading of graphic and pictorial conventions permits the formation of standards of verisimilitude, or of an acceptable equivalency to nature.[1] The process of translating nature onto paper, however, requires a sequence of steps that imply major conceptual leaps.[2] Every scientific illustration of any period and in any medium, whether made from a drawing or a photograph, represents not only the purported subject of the picture, but also these steps of translation. Plants and animals, alive, in motion and changing in nature, become on paper flat and static. So do ephemeral phenomena like growth and change. Lines, dots, hatching, shading, and color translate volume, contour, and textures. Placement and relative size often denote relationships in time and space. The inbuilt graphic conventions of schemata and diagrams both permit and constrain the representation of certain kinds of scientific thinking. Schemata and diagrams rely for the construction of their meaning largely on the Western historical consensus about the meanings of the placement of lines and marks on a page.[3] They are often constructed entirely out of symbols for direction such as arrows or dotted lines, or of conven-tions such as the meanings we give to high and low, large and small, the supposed superiority of regular over irregular shapes or of symmetry over asymmetry. Since the inventions of the telescope and the microscope, objects at great distances or too small to be seen by the naked eye have been enlarged and then translated into pictures composed of lines and marks, or more recently the gray scale of photography, or today, computer screen pixels. Lens-enhanced seeing and picturing only emphasize what holds true in almost all scientific illustration, that natural objects and organisms must undergo enlargement to be read or reduction to fit onto a page. It may even be true that no single factor has influenced scientific illustration as much as the convention of rectangular paper.

The process of translation seems so obvious, its conventions so central to Western picture making and to the way that people have been taught to read pictures that stating them seems tautological. Given the challenge that these translations pose to scientific claims of authenticity, however, it is surprising that science takes any illustration to represent nature. That science accepts and uses pictures demonstrates both the vital role of pictures in science and that science has accepted at face value pictorial conventions and has adopted them as part of scientific communication. Not only the rationalization of space through linear perspective, but also the twin inventions of type and picture printing, the craft and conventions of picture making, were constitutive of scientific practice.[4]

Reliance on pictorial convention and the development of the special conventions of scientific illustration have a long history, which this book takes up relatively late in the game when in Europe the professionalization of natural history and its reliance on printed illustration were well developed. American natural history illustration did not arise spontaneously. The process of establishing a scientific community in the early republic built on the three-centuries-long history of European perceptions of New World nature, mediated in complex ways by ideal and literary constructions and projections as well as by economic expectations and colonial relations. The establishment of an independent and illustrated American natural history required the simultaneous establishment of communities of practitioners, readers, illustrators, and printers. These communities grew within the larger context of the development of a national identity that contributed to the definition of the difference between American and European natural history and demanded recognition of that difference. My study explores how, beginning in the early nineteenth century, American naturalists on the one hand, and artists, draftsmen, and printers on the other, adapted European models of illustration and combined their changing practices to produce pictures of animals for a scientific context. It also traces how changes in those practices affected the appearance of the pictures.[5]

The story of natural history and zoological illustration in nineteenth-century America revolves around two main issues: (1) how a discipline in the process of consolidating its self-definition as a profession and its position in a changing society also consolidated certain conventions of pictorial representation for its own use; and (2) how, at the same time, independent but related developments in printing exerted pressures on the discipline to adopt new technologies and mediums, and how this changing print industry affected the use and appearance of pictures in science.[6] As natural history developed into separate disciplines, one of them zoology, the boundaries and statuses shifted among the multiple activities involved in the study of animals and in the production of the results. During the nineteenth century, as the interrelated practices of systematic zoology and its illustration developed within the larger social context of divergent definitions of science and art, the relations of production of zoological illustration reflected the division of labor and the general social demotion of technique. This shift contributed to the dramatic alteration of the conventions of zoological illustration, away from realism and toward schematization.

In this introduction I sketch the European developments in natural history illustration on which American naturalists built their practices. In Europe natural history became a discipline and developed its pictorial conventions within the larger context of changing perceptions and social roles of animals and their corresponding pictorial representations. By the end of the eighteenth and the beginning of the nineteenth centuries, European landscape, still-life and portrait painters, anatomists and botanists, and the draftsmen and engravers who illustrated their books had established through experiment and tradition a rich fund of pictorial convention to which the emerging discipline of natural history could refer for illustrating its own texts. During the 1760s, when natural history was reorganizing its practices and goals around the classification system of the Swedish botanist Carl von Linné—Linneaus—(1707–1778) and the descriptive approach of the French naturalist Georges-Louis Leclerc, Count de Buffon (1707–1788), the pictures illustrating the field revealed a hodgepodge of influences reflecting the historical accumulation of attitudes toward animals and their representation.

4

Beginning in sixteenth-century Europe, the emergence of a history and later a systematic study of animals both reflected and contributed to changing attitudes toward animals and their pictorial representation; but the ways that people observed, described, and ordered the natural world took shape from human social relationships. Animals existed not only in nature, but as members of human society and economy, as workers, food, clothing and transportation, as symbols of wealth and status, and as symbolic surrogates for human behavior. Above all, they were arranged in a ranked system, with humans at the top as rulers of the animal kingdom. Animals were thus perceived as separate from human society—a distinctly lower order of nature—and were simultaneously woven into the fabric of social meanings.[7] The social changes of the sixteenth through eighteenth centuries, to which the colonization of the New World was integral, grafted new systems of significance for nonhuman creatures onto the accumulated ancient and medieval meanings and roles of animals. Long-term events such as exploration and economic expansion introduced into Europe new animals and plants which had to be incorporated into nature's scheme.

The various meanings of animals were reflected in different kinds of pictures of them. As they had in religious paintings, animals continued to lead an allegorical life in secular pictures after 1500. New World animals such as parrots and alligators together with the native peoples of the Americas appeared throughout the seventeenth and eighteenth centuries in paintings of triumphal processions as tribute, as symbols of the personified continents and of the colonizing powers. Colonial nature was infused with contradictory meanings. It was assigned both a lower status than European nature and was at the same time considered exotic, the two meanings combining in the connotations of primitive. The figure of the American Indian came to personify American nature—interpreted sometimes as noble and sometimes as savage, in European painting, in the decorative and performing arts, and in political, moral, and natural philosophy. Seventeenth-century Gobelins tapestries populated the Americas with the animals of Asia and Africa at a time when all three continents were yielding their resources to build the wealth of Europe. In the late eighteenth-century murals of the Throne Room of the Royal Palace in Madrid, Giovanni Batista Tiepolo painted New World animals among other kinds of American products.[8] Meanwhile, the more familiar animals of Europe appeared in popular texts and primers illustrating homilies, bringing wild animals in pictorial form into the domestic realm and the education of children.

Animal portraiture, a genre from which natural history illustration borrowed, developed from the growing popularity of keeping exotic and domestic pets, and also in relation to important economic and social changes in the European countryside. Enclosure of formerly common land established privileged access to forests for the hunting of wild food and contributed to the romanticizing of wild animals. As a more intensive agriculture eliminated wild nature and dominated the landscape, people of taste cultivated an appreciation for wild nature, its denizens, and their romantic depictions.[9] Enclosure also meant privileged access to grazing land for domestic animals. Domesticated horses, cattle, and other herds appeared in seventeenth- and eighteenth-century European landscape paintings, in which they were integral to the cultivated scene and to the representation of rural harmony. Eighteenth-century English gentry who sat for their own portraits also began to commission portraits of their prize horses and bulls. These animals, like their owners, were exempt from the agricultural labor that had created the landscape often shown in the background. They were painted instead for their breeding or value and often were depicted in settings of leisure and privileged activities.[10] Pet portraiture, like the portraiture of horses and prize bulls, resonated with the meanings of privilege. Commissions to paint the portraits of pet dogs of the French kings both measured and insured the success of artists such as Jean-Baptiste Oudry, also a painter of the royal hunt and its booty.[11]

The developing genre of the still life, or, in French, the *nature morte*, of the seventeenth and eighteenth centuries featured dead poultry, fish, rabbits, and wild game. Such paintings brought the bodies of animals to the foreground, where elaborate technique highlighted the details of feather, fur, scales, and flesh. Although the dead animals in still-life paintings appeared ostensibly as food, the boundaries between still-life and game painting tended to blur, with the paintings of game on the hoof and of game recently killed evoking the privileges of the hunt. In general, small animals indicated a more humble setting, while large game such as deer, wolves, boars, and the larger birds made claims of another class entirely.

Parallel to and interrelated with the development of animal paintings, which were individual works for individual patrons, a kind of illustration devoted to detailed descriptions of plants and animals began to develop around the mid-fifteenth century. These illustrations emulated other pictures in many respects but also tended to restrict the range of social references permitted in the picture. Natural history illustrations borrowed from the conventions of animal portraiture, still life, and game painting. Some of their characteristics developed from the painter's practice of using study drawings of plant and animal as details for larger compositions.[12] But over time the drawings that were used to study and prepare learned treatises of natural history, as well as the woodcuts and copper engravings illustrating them, came to constitute a distinct pictorial genre. Anatomists and engineers developed their own illustration devices, such as the exploded diagram and cutaway view that satisfied the need to display three-dimensional objects on the flat surface of the page. The emergent conventions of natural history illustration acknowledged the two-dimensional surface more openly than did oil painting. In natural history illustration, the animal usually stood or lay in the foreground of a shallow space. Often, foreshortening the animal itself served as the only definition of depth in the picture. Many natural history illustrations had no background at all, although the blank page could be interpreted as a reference to the cabinet drawers or shelves used for storing and displaying preserved specimens. Frequently, the only feature in the picture besides the animal itself was its shadow. When animals were depicted in a landscape, the space tended to be flattened, and the animals were isolated from the background and from each other by posing them on separate clumps of earth or perching them on branches. To a limited extent, the landscape incorporated into natural history illustration served to exhibit habitat, but it served more as a device to locate the animal in a rational space defined by linear perspective.[13] Within that space, however, the figures were meant to be studied one by one.

At the same time, however, natural history illustrations portrayed their subjects with the full kit of devices for rendering a rounded form and for foreshortening, in common with other kinds of pictures and often in deliberate reference to them. In Renaissance anatomical illustrations, for example, the figures of flayed bodies and dissected torsos conformed to conventional classical poses, with torsos and limbs sometimes depicted "broken" like antique marble statuary. The classical and sculptural references in anatomical illustration served a double function: they placed the body within an implied rational space organized by the laws of linear perspective, and in so doing they argued for the naturalness of the depiction.[14] The context of rational space also referred the subject of the illustrations to the larger context of humanism, the intellectual context shared by Renaissance sculptors, painters, cartographers, anatomists and naturalists, and their patrons.[15]

Unlike panel paintings, murals, and tapestries, natural history illustrations took their meaning from adjacent written descriptions. Of course many paintings and other pictorial genres referred to well-known texts such as the Bible and thus could be considered illustrations; but pictures in scientific books since the Renaissance reflected a text that claimed a descriptive or analytical knowledge rather than a merely narrative one. In instances where the text claimed new knowledge based on direct observation, the reader was initiated only through the text into the meaning and frame of reference of the illustration. Further, unlike paintings, murals, or tapestries, natural history illustrations were made for and viewed in the context of the printed book in which they supplemented and were subordinate to text.

The interrelationship between the forms of the book and knowledge about nature and its representation can hardly be overemphasized. Natural history, of course, constituted only one form of knowledge about and representation of nature. Without the aid of libraries or collections, country people, fishing communities, and sailors had long garnered and passed along rich funds of observation. The literature of natural history is full of the surprise of the erudite, sometimes appreciative and sometimes condescending, at the oral traditions of knowledge about nature among the "self-taught."[16] While natural history illustration must be understood as only one kind of depiction of animals, and natural history must be understood as only one expression of complex social perceptions and meanings of animals, the meanings of the natural history book help explain what made these illustrations different from other kinds of animal pictures.[17]

Since late antiquity, well before the mid-fifteenth-century inventions of movable type and picture printing, books had been primary containers and communicators of traditions of learning. The physical form of the book constituted as much the context of scholarly knowledge about nature as universities, operating theaters, or cabinets of specimens. With the invention of printed text and pictures, books became more powerful as containers of knowledge despite the growing emphasis on direct observation in the emerging practices of natural history. The context of the book validated observation as it had validated scholastic tradition. Printing and its interaction with a growing literate public and the developing discipline of natural history consolidated and confirmed the role of the book in the translation of nature into knowledge: the production of natural philosophy became intimately linked to the production of books. The book, often illustrated, represented the final product of the process of natural history, which consisted of varying degrees of observation of live animals in nature, the collection and study of preserved specimens, correspondence among naturalists and exchange of specimens, and consultation of and reference to prior sources—other books. The role of printing in the development of science confirmed that the study of nature was preeminently a learned tradition, translating observation into text and pictures and addressed to a *readership*. Furthermore, the printed book and the concept of copyright established the principle of priority, that is, scientific ideas or discoveries became a kind of personal property.

The printed book retained aspects of the manuscript format that provided some of the most persistent characteristics of natural history illustration. In thinking about the emergence of a learned observational study of nature and its interaction with the twin innovations of movable type and picture printing, we must keep in mind that the new ways of study and of printed pictorial representation built upon the historical continuum of the format of manuscript books and their illustrations, upon centuries of practice of using text and pictures to record and communicate knowledge about the natural world. Printed books, like the manuscript codex, retained a rectangular shape. Their range of sizes fell within certain limits, some technologically constrained, others defined by conventions of use or readership. Printed books inherited from their manuscript predecessors the features of regular margins and unilinear page sequence. These aspects, like the steps of translating nature into pictures, seem so obvious that they should not need mention. Yet the conservative form of the book has played an important role in the development of the relationship of text to illustration in natural history; and these features have persisted as tacit assumptions built into scientific illustration to the present.

Perhaps the most important of the continuities between the manuscript and the printed book in terms of natural history illustration concerned the relationship of image to text. The variability of hand-copied illustration had limited the didactic use of pictures in favor of text. Naturalists as early as Pliny the Elder (ca. A.D. 23–A.D. 79) lamented that students of nature had to forgo the benefit of pictorial illustration because pictures "are very apt to mislead, and more particularly where such a number of tints is required for the imitation of nature with any success; in addition to which, the diversity of copyists from the original paintings, and their comparative degrees of skill, add very considerably to the chances of losing the necessary degree of resemblance to the originals."[18] Pliny's comments indicate a full understanding of the potential of the descriptive relationship and complementary roles between words and pictures. He regretted that without pictures verbal description degenerated into mere lists of names.[19]

In addition to technical constraints on the use of pictures before the invention of picture printing, the classical epistemological hierarchies—notably Plato's—carried into the Christian era associated pictures with illusion and deception, and the word with truth.[20] Historians of the work of the sixteenth-century anatomist Andreas Vesalius noted that "many of the leading physicians of the day [the 1530s] were actively opposed to illustration of the printed word on the grounds that it had not been done in classical times and would degrade scholarship."[21] Many sixteenth-century naturalists

and other authors, however, took a functional approach to the relationship of text to image and adopted with alacrity new picture-printing techniques for illustrating their works. Furthermore, the writers and printers who plagiarized and pirated illustrated works perceived that the illustrations were the books' greatest appeal to potential buyers.[22]

Despite the importance of picture printing to natural history, however, text remained the primary vehicle of scholarly expression. Pictures illustrated text, not vice versa. Pictures served text, supplemented it, completed it, remained subordinate to it, although pictures would come to play various roles in works of natural history. From the beginning of the use of printed illustrations, as in Vesalius's *Epitome* of 1543, a book designed for use by medical students and consisting of a series of plates of human anatomy, the text explained how and in what sequence the reader should view and interpret the figures. In addition, each part of the body had a label or caption, that is, the pictures took their meaning from two systems of referents, the first outside and the second inside the book: (1) the human body observed through sanctioned techniques—in this instance, dissection; and (2) the terms for the body's parts and verbal explanation of their interrelationships.[23]

Printed natural history illustration carried the authority of the book, the security of mechanical replication, and the author's assurance of direct observation. William M. Ivins, Jr., twentieth-century historian of prints, wrote that the promise of the "exactly repeatable visual statement" provided an incentive to original observation because description could be translated and published in a picture.[24] Natural histories and travel accounts, genres that shared an emphasis on eyewitness, asserted that their illustrations conveyed the authority of direct observation (although to the modern reader this assertion is rendered incongruous by the illustration of mythical along with recognizable animals). Picture printing permitted and encouraged naturalists and anatomists, as well as architects, engineers, and astronomers, to invest care and skill in their illustrations as records of their own observations. Each impression of the woodblock, or later the engraved copperplate, replicated the identical content of the original print, which in many instances had been supervised and approved by the author. Authors' repudiations of pirated and plagiarized editions made clear that control was an issue in the pictorial representation of one's own observations.[25] In short, picture printing seemed to offer a means to control the image; "the exactly repeatable visual statement" acted more like printed text than had the hand-reproduced illustration.

By convention, the writer of a work both "owned" and bore responsibility for its intellectual contents, including the illustrations. Even though illustrations would become integral to natural history, the writer had the least control over their production, unless he or she was adept at drawing, cutting a woodblock, or engraving or etching a copperplate. The writer could exert some control by copying or reproducing previously published illustrations, sanctioned through use. The author could also ask the illustrators to copy aspects of previously published illustrations such as shallow space or perches under birds, an approach that worked to consolidate conventions. Some anatomists and naturalists learned how to draw or prepare prints expressly for use in natural history; the examples are numerous and proliferate in the eighteenth century. Some who could not gave special recognition to their illustrators, recognition that indicated a close collaboration or supervision by the author or, in other words, dependence mixed with control. Authorial assertions of precise measurement or direct supervision of the draftsman expressed the tension between dependence and control.[26]

The European audience for natural history diversified during the seventeenth and eighteenth centuries, especially in northern Europe, with a concomitant diversification of pictures of animals both in and out of books. During the sixteenth century, for example, the French traveler in North America, Jacques le Moyne de Morgues (d. 1588), made detailed, frontal watercolor studies of insects and plants intended as natural history drawings, although they did not illustrate a text.[27] In the following decades many painters and printmakers chose animal subjects and grew skilled at depicting them. In Holland, for example, prints of beached whales enjoyed popularity at the turn of the seventeenth century. In one such engraving by Jan Saenredam (1565–1607) the inscription both attested to the accuracy of the artist's measurement of the animal and evoked the popular belief that stranded whales portended disaster.[28] The Czech artist Wenceslaus Hollar (1607–1677) made numerous etchings of birds and insects, printed as single sheets. He often copied from the work of Dürer and the English artist Francis Barlow, best known for his illustrations of Aesop's fables.[29] Pictures such as these were conceived as independent from the relationship to text that distinguished natural history illustration.

Many popular books of the seventeenth century consisted primarily of plates or included pictures drawn from nature, again minimizing the context of text. Crispijn van de Passe II (ca. 1597–1670) engraved a flower book intended as a coloring book and directed at an audience that shared the fashionable interest in gardening. The engravings bore many of the emerging conventions of botanical illustration—shallow space, an arrangement of the plants for simultaneous display of flowers, the top and the underside of the leaves, and an indication of plant-to-animal relationships in the inclusion of garden insects.[30] Illustrations of birds in Dutch emblem books shared pictorial conventions with illustrations in works of natural history such as the stylized branch or stump for the bird's perch. Similarly, plates illustrating shells and fish included indications of the littoral environment.[31]

Like the audiences for the varieties of pictures of nature, the patronage also overlapped. Skill in depicting detail and texture were used in the service of faithful information about nature, but were also prized in themselves. Hollar, for example, spent the most productive years of his career in the employ of the English Earl of Arundel, a collector of the drawings of Dürer and Leonardo da Vinci.[32] Svetlana Alpers, in *The Art of Describing: Dutch Art in the Seventeenth Century*, has proposed an interaction between north European painting and the initiation of the empirical agenda advocated by Francis Bacon (1561–1626), considered the formulator of the program of experiment and observation initiating and characterizing the "scientific revolution": "It would seem to have been not innovating artistic practices but the established practices of the craft, its age-old recording of light and reflecting surfaces, its commitment to descriptive concerns in maps or to botanical illustrations, in short, its fascination with and trust in the representation of the world, that in the seventeenth century helped spawn new knowledge of nature."[33] The combination of elite patronage and the cultivation of a broad audience through popular prints and books assured the social position and social influence of natural history as well as a wide recognition of the basic conventions of animal illustration in book formats.

Meanwhile, the rising prestige of Newtonian mechanics confirmed the image of a privileged domain of natural knowledge whose power rested on measurement, mechanism, and prediction. The effort to bring the study of plant and animal life within that domain raised an implicit comparison between illustrations of the exact sciences such as physics, mechanics, and optics, and natural history illustration. Naturalists began to distinguish their use of pictures from other pictures that looked similar but appeared in different contexts by establishing consistent practices of description. Language and system provided the frameworks within which natural history illustration became distinguished from other pictures

of animals. An early distinction of representation for scientific practice was linguistic, an attempt to refine rhetoric to attain an unmediated representation of nature.[34] The empirical project taken up and promulgated in the late seventeenth century included purging prose of distracting embellishments and misleading metaphors. Language must be grounded in material nature and strive for a direct correspondence between nature and verbal report. The formulation and introduction of a prose style specifically for natural philosophy became part of the larger mission of the Royal Society of London, founded in 1660 for the cultivation and accumulation of scientific observations. The Society's journal spread the reform of the word. In rejecting "amplifications, digressions, and swellings of style" and seeking a speech of "Mathematical plainness," the proponents of the new observational and experimental philosophy established criteria for verbal representation that would qualify as scientific.[35] Late seventeenth-century arguments about the role of language in scientific practice moved toward distinctions in intellectual and cultural production similar to those we live with today. The quest for transparent representation of nature began to establish the oppositional relationship of science to painting and literature, now the repositories, from the scientific point of view, of metaphor, fancy, embellishment, and individual imagination, while science embodied empiricism, order, system, and measurement. With the characterization of the new language of science as virile in contrast to its feminine opposites, we can begin to recognize familiar dichotomies.[36]

Confidence in the possibility of unmediated representation rested on a belief that the senses gave access to knowledge, with sight the primary sense for knowing a world preeminently visible. Philosophy of the seventeenth and eighteenth centuries became preoccupied with the process and limits of sense apprehension and verification.[37] The privileging of vision would seem to give pictures a leading role in the empirical project. The empirical picture, however, seems to have been considered an adjunct of empirical prose; like prose, illustrations must avoid embellishments or stylistic flourishes that distracted from the depicted object of study. At the same time, however, scientific illustration relied on the same devices as other pictorial genres. Except for the conventions already noted, natural history illustration attained its empirical status largely by the reflected light of purified rhetoric.

The illustrations of the *Micrographia* (1665) of Robert Hooke (1635–1703), one of the first microscopists and a regular contributor of his observations to the meetings of the Royal Society, provide a measure of the extent to which a leading proponent of visual empiricism combined an insistence on the visual apprehension of truth with an acceptance, at the core of his practice, of the conventional techniques of drawing and printing. In his preface Hooke stated that "true Philosophy ... is to *begin* with the Hands and the Eyes." Following the empirical method "with diligence and attention, there is nothing that lyes with the power of human Wit ... which we might not compass." With industry, "we might not only hope for Inventions to equalize those of Copernicus, Galileo, Gilbert, Harvy" but also match such achievements with those of the anonymous "Inventors of Gun-powder, the Seamans Compass, Printing, Etching, Graving, Microscopes, etc."[38] Hooke's list of accomplishments is telling, for it yokes the instruments of European expansion with the tools of his own scientific practice. It also indicates the high value placed on printing and the role of printing in the scientific project. Hooke depended as much on printing, etching, and engraving as he did upon his own eyes and hands to portray his observations through the lens.

Hooke expressed satisfaction in the illustrations he presented to the reader: "What each of the delineated Subjects are, the following descriptions annext to each will inform, of which I shall here, only once for all, add, That in di-

vers of them the Gravers have pretty well follow'd my directions and draughts; and that in making them, I indeavoured (as far as I was able) first to discover the true appearance, and next to make a plain representation of it."[39] The difficulty, according to Hooke, lay in seeing rightly, in differentiating hollows from projections, shadows from stains, reflections from light coloration, in finding the right angle of light to reveal best the true structure. For making drawings and translating them into prints, however, there was no "exactness of Method," only a "sincere" commitment to producing a "faithful" record of the appearance of the thing observed.[40]

The *Micrographia* contained three distinct kinds of pictures whose differences demonstrate the consolidation of certain illustration conventions early in the empirical project. The first kind appeared on the title page, a coat of arms complete with plumage, animals rampant, and a scroll inscribed with a motto: on the page beginning the dedication to the king was a decorative heading in which twined foliage flanked an angel figure, possibly Truth. The second type of plate illustrated microscopical and experimental apparatus and optical principles. These plates portrayed the apparatus in a simplified but naturalistic manner, as if they were elements of a still life. Even three-dimensional objects, however, lay on virtually a single plane, notwithstanding the shadows they cast on indeterminate surfaces. Each part of the apparatus bore a letter label, as did the lines and angles of the optical diagrams; all labels referred to the plate's textual description. Hooke's third use of pictures, the plates illustrating observations through the lens, resembled landscapes or still lifes, with one difference: in many of them the subject filled the field delimited by a circular border, simulating the experience of looking through the eyepiece.

In contrast to the distinct stylistic and compositional devices of the three kinds of plates, the variations in engraving techniques reflected more the different hands of the engravers at work on the project than the differences in content. Throughout the work, the predominant manner of engraving consisted of a combination of parallel lines with cross-hatching that made a diamond pattern. In the now-famous illustration of a magnified flea in profile, regular parallel lines and cross-hatching, massed for darks or dispersed for highlights, defined the contours and textures of the insect, which filled the field of the page. We do not know whether Hooke preferred the plates with the regular parallel lines to those engraved in a scratchier, fuzzier manner. He seems to have accepted such variation based on his faith in the ability of engravers to provide a faithful transcription of his own record of observation. In other words, this pioneering visual empiricist accepted as his own standard of verisimilitude for illustration whatever printing was accustomed to produce. There existed no special etching or engraving technique for scientific subjects. Nor would one develop even as the production of scientific publications increased. Instead, style and composition, and not engraving technique, denoted the various purposes pictures served in a scientific text; plates representing direct observation rejected frills and decorations, which, however, remained appropriate in dedications or acknowledgments of patronage.

Hooke's attitude toward the reproduction of his drawings points to the pragmatism and instrumentality of scientific practice in the use of pictures. The empiricists of the late seventeenth century sought to establish a purified language but mounted no parallel critique of existing graphic media for their own use. In seventeenth-century natural philosophy and its practice, the privileging of language permitted the continued and largely uncritical use of the available technology of print and its predominant style, the legacy of which continues to the present. Language remained the primary representation. Illustrations derived their authority from their linguistic context and from the reputation of the observer. The empiricists'

belief in the possibility of transparent representation combined with the prevalent metaphor of the "book of nature," which suggested that observation and representation would follow established literary—and graphic—forms, enhanced the credibility of illustrated works. The acceptance of available tools not only as sufficient for scientific purposes, but as primary tools of the empirical project permitted the continued intimate interaction between science and picture printing.[41]

A view of scientific illustration that keeps the focus on the tools of production and the context of their use provides access to explanations of change in the appearance of illustration.[42] As E. H. Gombrich has pointed out: "The test of an image is not its lifelikeness, but its efficacy within a given context of action."[43] In the context of the emergent culture of science, action required publication of observations with available tools.[44] Ivins labeled the regular web of cross-hatching and the diamond lozenges created by their intersection the "net of rationality," the predominant engraving style of the seventeenth and eighteenth centuries for subjects from portraits of kings to pictures of fleas.[45] Ivins's term might lead one to insist either that the technique fit scientific rationalism and that authors of scientific books liked that fit, or that the "net of rationality" represented a projection of the Cartesian grid onto nature. Gombrich's model of context and action, however, seems more helpful in understanding the process by which science adapted its expectations of pictorial representation to fit available techniques. Instead of developing a special technique of picture printing to characterize "science," practitioners exercised control over illustration through defining permissible compositional devices, which in turn became the recognizable conventions associated with different disciplines.

From the late seventeenth through the eighteenth centuries a diversity of illustration styles coexisted in the literature of natural history. To some extent the different styles connoted different sources of patronage. Illustrations in the publications of scientific societies tended toward the plain, sometimes diagrammatic. Illustrated works produced under private patronage tended toward a more elaborate realism, replete with references to the other interests of the same class of patrons, such as the collection of paintings and fine prints and books.

System, the natural history equivalent of universal law, provided philosophical validation for natural history as a science. In both institutional and private publications, the animal subjects of natural history illustration increasingly acquired the meaning of excerpts from the Natural System. For a growing number of naturalists, the search for a universal system of nature constituted the central aim of the discipline. Beginning in the mid-eighteenth century the descriptive practices of natural history would consolidate within the framework of debates over what, if anything, constituted the Natural System. The manner in which naturalists described an animal, the very features they included in their descriptions, depended on how they defined system and relationship. Illustration played its key role in natural history as a supplement to description.

In the mid-eighteenth century, however, system and description came to constitute distinct approaches to natural history under the names of their leading and antagonistic proponents, the contemporaries Linnaeus and Buffon. System attempted to organize natural diversity into a single coherent order of relationship. Description catalogued, analyzed, and portrayed natural diversity. Both Linnaeus and Buffon grappled with the questions about animal relationships raised by diversity but arrived at different methods for resolving them.[46] They shared an ultimate goal, nevertheless; the complete catalog of all nature. Linnaeus believed that a logical and uniform system would facilitate description of plants, animals, and minerals. Buffon believed that only extensive description would establish the basis for the

eventual discernment of the Natural System. The two differed further in their definition of what constituted description. For Linnaeus, description of selected morphological characteristics identified and placed the plant or animal in the larger order; in contrast, Buffon attempted to construct a full portrait of each animal, its anatomy, appearance, habits, life cycle, diet, and habitat, and its social uses.

The system that Linnaeus devised divided the natural world by logical principles into categories of membership (now called taxa, the plural for taxon) according to structure. Classical Linnaean taxonomy proceeded from analysis of general to particular, that is, from the top down, each level distinguished by a more detailed diagnosis. Thus at the broadest level of generalization, Linnaeus divided the empire of nature into the plant, animal, and mineral "kingdoms." He based his system for the plant kingdom on sexuality. The number of stamens determined the membership of a plant in its appropriate class; each class was subdivided into orders based on numbers of pistils. Within the animal kingdom, anatomical structure distinguished "classes" of which Linnaeus designated six: mammals, birds, reptiles, fish, insects, and vermes, a catch-all category for noninsect invertebrates. He established "orders" to distinguish differences within classes, dividing mammals into orders on the basis of their teeth, birds by their bills, fish by their fins, insects by their wings. Linnaeus achieved a simplicity of arrangement by not subdividing his categories and by concentrating on the lower taxa. Later taxonomists, however, established not only more kingdoms, but additional levels of higher taxa, such as the phylum, as well as subgroups that recognized the finer degrees of similarity.[47] By about 1800, "families," although not part of the original Linnaean scheme, distinguished related groups within orders.[48] For Linnaeus, and still today, related genera comprise an order, and related species a genus. A Linnaean description followed an established sequence and presented each character in formal, abbreviated Latin phrases. The regularizing and formalizing of descriptive terms by Linnaeus left as its principal legacy binomial nomenclature, the designation of plant or animal by the Latin names for genus and species. The use of Latin for description well into the nineteenth century established the discipline of natural history as an international, and patently erudite, practice.

Critics of the Linnaean system argued that the selection of characteristics for description produced an artificial system. Linnaeus admitted that a system based on isolated characters was indeed artificial, for it gave only a partial account of relationships, or affinities.[49] Previous systems, such as that of the English naturalist John Ray (1627–1705) whose classification system was the most widely used in England before the adoption of the Linnaean, had also taken selected features to distinguish groups, and were likewise considered artificial. They shared with the Linnaean system, however, the emphasis on structural rather than behavioral characteristics for establishing classification.[50] Linnaeus believed that the only two categories in his hierarchy that corresponded to natural divisions were genus and species. He taught that the naturalist "discovers" genera, and that the generic characters should represent the "essence" of the group and its members.[51] While he held an essentialist concept of genera, he defined species by the function of reproduction. Each plant or animal belonged to a species, and members of each species reproduced offspring that were alike. Although Linnaeus initially believed in a fixed and static number of species, as a botanist he recognized that hybridism challenged the ideal of stability. Eventually he concluded that every genus had begun with a single species, which, varying through hybridism, produced the related species that composed the expanding genus.[52]

Buffon, naturalist to the king of France, keeper of the royal collections, and the most influential anti-Linnaean, led the French school of natural history in claiming that the imposition of any system of classification imposed artificial categories on nature.[53] He refused to use Latin binomials and insisted instead on vernacular names. Buffon, like Linnaeus, also began by considering species unchanging or static, and his rejection of taxa higher than the species level reflected that belief; for if species remained constant, then no higher level of relationship could develop. As the recipient of collections from around the world, however, Buffon had to reconcile diversity and resemblance with a static concept of species. He assigned a crucial role to migration and climate to explain the diversity he observed. He acknowledged that climate, diet, and domestication could effect change and create varieties within species. Thus Buffon attributed to migration and climate what Linneaus, working primarily with immobile plants, attributed to hybridization. He came to recognize the existence of species in nature and defined a species by the perpetuation of its characters through reproduction. Sterile hybrids such as the mule indicated that the parents were different species.[54] Buffon's concept of species came to correspond with the Linnaean genus, and varieties with Linnaean species.[55]

But for Buffon classification was subordinate to the real purpose of natural history—to give a full account of each species. He wrote: "The history of one animal should be . . . that of the entire species . . .; it ought to include their procreation, gestation period, the time of birth, number of offspring, the care given by the mother and father, their education, their instincts, their habitats, their diet, the manner in which they procure their food, their habits, their wiles, their hunting methods."[56] Putting this method into practice, Buffon produced forty-four volumes of his *Histoire naturelle*. The first printing of the work sold out in six weeks, and it became one of the most popular books in France during the second half of the eighteenth century, creating a large audience for descriptive and narrative natural history.[57] The volumes were sumptuously illustrated with engravings befitting a work based on royal collections. Jean-Baptiste Oudry, painter of royal dogs and hunts, drew the whole horse for the series of engraved plates illustrating its anatomy. The illustrations of primates—drawn making arresting eye contact—were often taken from live individuals kept in the royal menagerie or as pets of the aristocracy. Each plate depicted a single animal in a landscape adorned with antique ruins. Many of the animals, whether depicted whole or as a skeleton, stood on fragments of classical statuary, or appeared as on a stage, framed or draped with tapestries.

FIGURE
INTRO.1

Buffon's eminently social approach to description—recounting the full life history of each species—reflected the fact that he studied animals, principally mammals and birds, whose behavior seemed to mirror the human. In further recognition of the human and animal relationship, Buffon classified mammals as domestic or wild.[58] He described them in order of interest, beginning with the horse, and he catalogued their social uses. The categories that Linnaeus established for ordering the natural world reflected social structures more abstractly—through ordering nature in a top-down hierarchy, basing plant taxonomy on sexual difference, and emphasizing the interrelations of dominance and subordination that maintained the natural order and harmony within it. Both naturalists placed humans within nature, with Linnaeus including humans in the same group as apes, and Buffon insisting that humans were a single species. The effect of locating humankind within nature, however, was less to reduce people to animals than vice versa, that is, to socialize nature.

The Linnaean system offered simplicity, consistency, and clarity in an age of proliferating systems. After the publication of the tenth edition of the *Systema naturae* in 1758–1759, Linnaean classification and binomial nomenclature for plants and animals became widely adopted in England. The adoption of Linnaean systematics spurred a revival of natural history

in the universities and a spate of society founding. The fashionable rise of natural history received additional support from royal patronage; George III's queen and some of his ministers were avid botanists and collectors.[59]

It was not only the convenience of Linnaean binomials that made the system appeal to eighteenth-century English naturalists. They shared with the Swedish botanist a view of a stable economy of nature mixed with a natural theology that attributed the established natural order to the wisdom of God. One of the founding texts of natural theology, John Ray's *The Wisdom of God Manifested in the Works of Creation* (1691), had popularized a theologically infused scientific cosmology. The combined vision of harmony and system, of the law of dominance and subordination regulating an economy of plenitude, gave rise to a particular justification of the pursuit of natural history that the English would contrast, especially after 1789, to what they considered godless French materialism. In the light of natural theology, natural history offered a pursuit above the political tumult and divisions of the era, for it provided examples of stability that transcended conflict.[60] With God the author of the original book of nature, the study of natural history offered a view into the workings of the divine mind and the glories of creation.

Thus the patient and acute observation of the physical facts of nature could build a sense of a stable and harmonious society in works such as the curate Gilbert White's *Natural History of Selbourne* (1789). White's evocation of natural harmony, which ignored the transformations that commercial agriculture worked on the countryside, made his work one of the most influential books in the literature of natural history, appealing especially to nineteenth-century urban readers nostalgic for the supposed harmony of rural life.[61] In contrast, the illustrations of Thomas Bewick (1753–1828) juxtaposed the natural with the social order. Bewick invented the technique of fine-line engraving on the end grain of boxwood blocks,

Pl. X. Page 366.

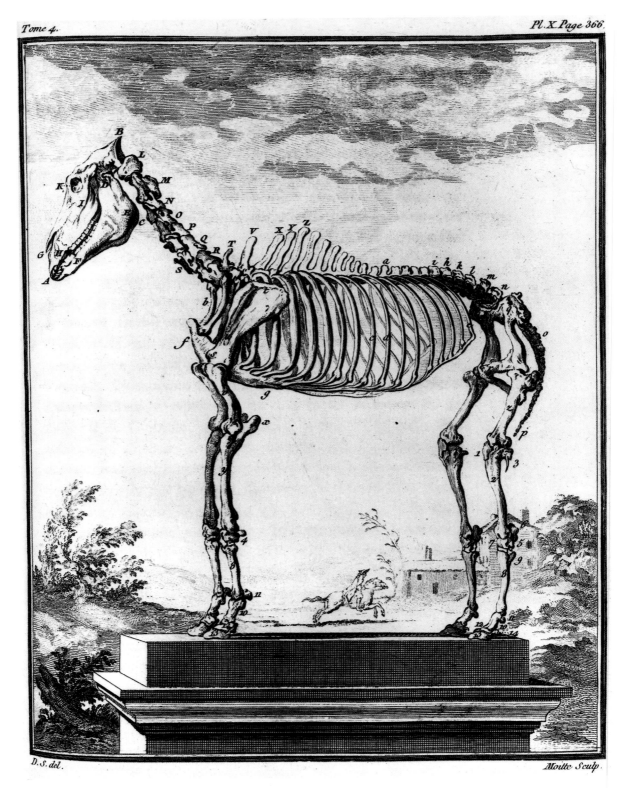

D.S. del.

Moitte Sculp.

Intro.1 Skeleton of the horse, last of the ten plates illustrating the animal's anatomy. Engraving by Moitte from a drawing by De Sève. From Buffon, *Histoire Naturelle*, vol. 4 (1753).

which could be set with the type in the press bed, unlike metal-plate engravings. Bewick's wood engravings of quadrupeds depicted the exotic mammals of Africa and the familiar English bulls and dogs. He illustrated his birds in the hedgerows or by the mill ponds of rural England. The contrast between Bewick's idyllic portrayals of animals and his vignettes depicting human abuse of beasts of burden, peasants suffering in bad weather, and other realities of village life suggested that nature enjoyed a more perfect organization than did society. He addressed his first volume on birds to "men of leisure . . . residing in the country," and looked forward to "times of greater tranquility," when, "undisturbed by public calamities" men could enjoy undisturbed the observation of nature.[62] The economy of printing Bewick's wood engravings with the text made his work available to a wide audience and popularized his combination of the exotic with the domestic. Like White's *Selbourne*, Bewick's *A General History of Quadrupeds* (1791), the even more popular *History of British Birds* (1797) about land birds, and the second volume on water birds (1804) enjoyed wide popularity throughout the nineteenth century.[63]

Exploration and expansion went hand in hand with the rising interest in natural history. The presence of Joseph Banks on Cook's first voyage (1768), organized by the Royal Society, initiated the great age of scientific travel and the popular genre of the scientific travel account.[64] In this developing framework, the colonies played the crucial role of supplying new and exotic species for English collectors. Linnaean taxonomy bestowed coherence on the plants and animals received from the colonies, and transformed cabinets of curiosities into scientific collections. Even before the more general acceptance of Linnaean classification, English collectors, who had special access to North American and Caribbean species, had contributed to the Linnaean project. Many New World species were first described by Linnaeus himself. When he lacked the actual spec-

imens, because they were fragile and perishable, Linnaeus based some of his descriptions on drawings alone, elevating the status of drawing and illustration in the taxonomic project. Linnaeus described some of his New World species from illustrated accounts published by the traveler-naturalist, Mark Catesby (1683–1750), in his *Natural History of Carolina, Florida and the Bahama Islands* (1731–1743), as well as on drawings obtained from the ornithologist George Edwards (1694–1773) and on specimens sent to England by John Bartram (1699–1777) of Pennsylvania, a Quaker farmer and astute observer and collector.[65]

PLATE
1

The role of illustration in English natural history had special relevance to the development of American natural history. The congruence between the Linnaean emphasis on morphology and the emphasis on illustrations in English eighteenth-century natural history had probably helped the general acceptance of the Linnaean system. Richly illustrated works of natural history had enjoyed wide popularity in England during the first half of the eighteenth century.[66] Some English naturalists, notably Catesby and Edwards, considered drawing and even printmaking part of the necessary equipment of a naturalist. Catesby, collector of plants and animals, taught himself etching to be able to produce the plates for his two-volume work. Edwards learned etching from Catesby to illustrate his own publications.[67] Catesby's and Edwards's insistence on producing their own illustrations for publication marked steps toward erosion of the distinction, dominant in English natural history, between the patron-collector who wrote and published, and the traveler-illustrator who supplied his patron with specimens and drawings, often from the colonies. John Abbot (1751–1840), an Englishman who settled in Georgia, sent thousands of drawings and specimens to his patrons back home.[68] Although collectors were eager for Abbot's shipments, rarely did he receive mention for his part in the production process. Catesby and Edwards, by integrating the steps

17

of producing natural history, provided important models for early nineteenth-century American naturalists.

By the late eighteenth century, natural history writing in English had coupled with almost every popular literary genre. The combination of description, narrative, and introduction to taxonomy in many popular works of natural history extended the life and infused new meaning into illustration conventions, some already hundreds of years old. Depicting the animal, usually in profile, in the foreground of the illustration satisfied the requirements of Linnaean taxonomic description. Placing the animal in a landscape, even the abbreviated landscape of a single clod of earth, connoted Buffonian description of the living animal in nature. Pyramids, palm trees, or other climatic or cultural indicators scattered in the background invoked the allure of travel and the exotic. In the context of the increasingly popular illustrated literature of scientific travel, writes Barbara Stafford, "By the end of the 18th century, the word *illustration* had become largely identified with engravings," and its meaning "extended to embrace 'embellishment' as well as 'explanation' or 'intellectual illumination.'"[69]

During the eighteenth century, natural history established itself and developed as a legitimate pursuit among the other branches of natural philosophy. The elaboration of theories of the Natural System gave natural history philosophical justification, while state and private patronage provided financial support for a pursuit that continued to be for most naturalists an expensive hobby. The increase in available facilities for printing and publishing books and pictures in interaction with a growing international readership, both scholarly and popular, cultivated a pool from which to recruit new practitioners. In the newly independent United States, early nineteenth-century naturalists aspired to re-create these conditions and at the same time to break the colonial relations of American natural history with England and Europe.

Before independence, a small group of astute observers and collectors had provided English and European savants with the raw materials for important publications. John Bartram had cultivated native plants in his garden outside Philadelphia and had sent both plant and animal specimens to the avid English collector Peter Collinson (1694–1768), also the sponsor of the collecting expeditions of Catesby through Carolina, Florida, and the Bahamas. Bartram's specimens were described and published in England. Similarly, only a few of Abbot's drawings and observations of butterflies were assembled and published in London by James Smith, while Abbot's drawings and observations of birds remained unpublished.[70] After independence, American naturalists felt keenly the implications of the existing mercantilist scientific relationship with Europe. While natural theology supposedly raised the study of natural history above political divisions, some American naturalists believed that it was intrinsically patriotic to study and publish the natural productions of their national territory. Naturalist-illustrators like Catesby and Edwards had a special importance to Americans, who lacked aristocratic patrons for natural history. Not only did American naturalists work within a different system of support for nature study, they sought to incorporate their own political values with their study of North American natural productions. Identifying with their own political revolution, Americans could perceive change as part of the order of nature. Moreover, local observers noted important differences between European and American species, even though many European authors insisted that American species belonged to established European genera.

Americans adapted English natural history literary genres to accommodate a different readership. Eighteenth-century American contributions to the literature of travel and natural history had drawn on English models; indeed, most books on natural history published in the United States before 1800 were reprints of English titles. The New York wood engraver, Alexander Anderson (1775–1870), one of the early practitioners of the medium in the United States, reproduced Bewick's works for an American audience.[71] But American naturalists placed a growing value on direct observation, which promoted descriptions of local nature and local illustration. The development of a natural history press was central to the development of American natural history. Not only did publication attribute priority of description to the author, but also the specimens described usually remained in the collections of the describers, formerly English and European; publication at home went hand in hand with the development of American collections.

Naturalists of the early republic felt the urgency of producing an American natural history that not only corrected European error, but asserted American parity in contributing to natural history. They adapted and transformed the models they inherited from English and continental natural history, and combined emphasis on the Linnaean system with description and narrative. During the early decades of the nineteenth century, American naturalists intent on establishing a broad public for natural history gave fresh meaning to the European categories of animal study and pictorial representation as they contributed to the worldwide collecting and naming of species.

FIGURE
INTRO.2

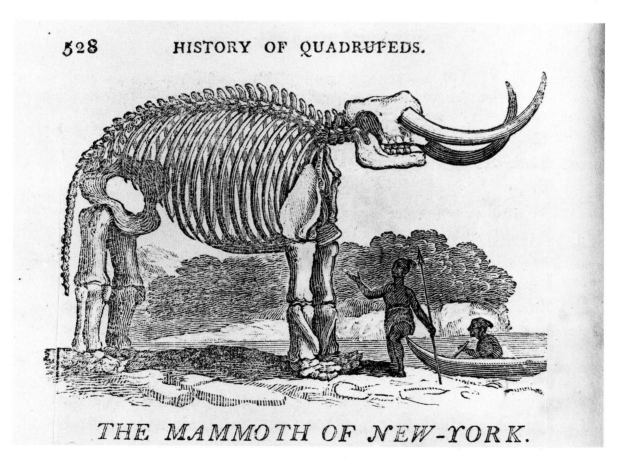

Intro.2 The Mammoth of New York. Wood engraving by Alexander Anderson, from an American edition of Thomas Bewick, *A General History of Quadrupeds*.

19

# 1

# The Naturalist-Illustrator

THE naturalist-illustrator, practicing direct observation and pictorial representation of nature, emerged as the principal practitioner of American natural history in the early nineteenth century. So strongly did the image of the naturalist-illustrator become associated with the founding of an independent American science that it persisted as an ideal, even as a national icon, long after division and specialization of practice had marginalized old-style naturalists from mainstream professional participation. The story of that division of labor and the re-ordering of the status of the different activities in the production of natural history is the story of the institutionalization of science in the United States. Professional scientific institutions of the later nineteenth century would separate the job of illustrator from the role of naturalist and relegate the work of illustration to the work-for-hire draftsman. The later use of the term "artist-naturalist" to refer to the naturalist who produced his own illustration also reflected the nineteenth-century completion of the split between art and science. Throughout the nineteenth century, however, the persistence of the naturalist-illustrator ideal in the growing realm of popular natural history, from which the discipline of zoology recruited members, would periodically impinge on the institutional realm in ways that illuminated changes that were occurring in professional practice.[1]

As central participants in the establishment of the natural history community of Philadelphia during the early decades of the nineteenth century, the Peale family of painter-naturalists gave a visual emphasis to American natural history. Rembrandt Peale (1778–1860), a painter like his father, Charles Willson Peale (1741–1827), the museum founder and portraitist of Revolutionary War patriots, went so far as to insist that the eye of the painter was better trained than that of the anatomist to apprehend natural form. The Peale family's diverse activities, especially those of the elder Peale, included painting, invention, natural history, and the founding of organizations promoting all three.

The Peales seemed to enact the program outlined by Benjamin Franklin in the prospectus of the American Philosophical Society, established in 1743. Franklin's Baconian desiderata had included the study of botany for the improvement of agriculture and medicine; medicine itself; "all new-discovered fossils in different countries, as mines, minerals, and quarries"; mathematics; chemistry and chemistry applied to "distillation, brewing, and assaying ores"; inventions "for saving labour"; "all new arts, trades and manufactures that may be proposed or thought of"; surveying and mapping; the improvement of animal stock; "and all philosophical experiments that let light into the nature of things, . . . to increase the power of man over matter."[2]

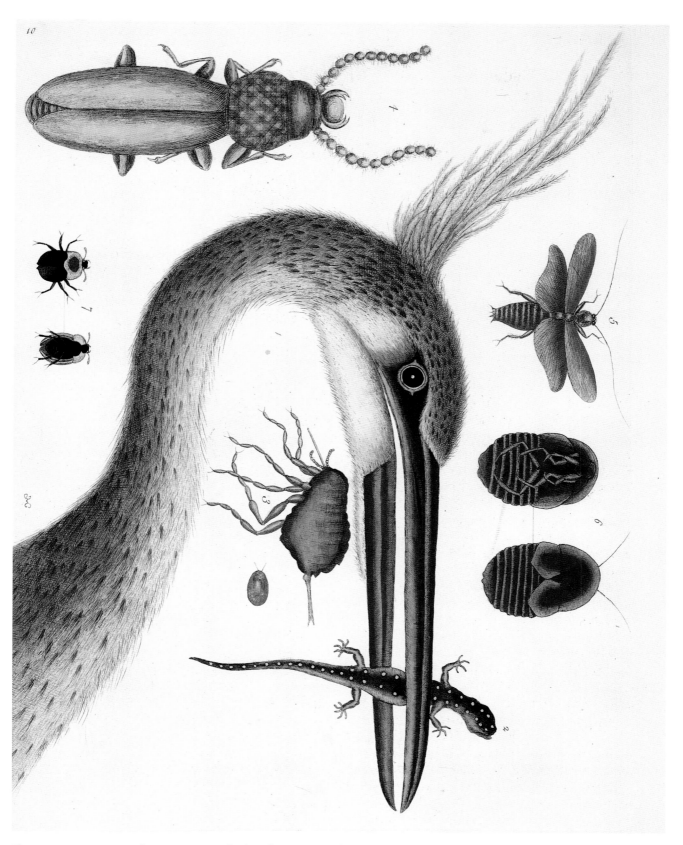

Plate 1. Largest-crested Heron, Spotted Eft, Chegoe, *Scarabaeus capricornus*, Cockroach. Hand-colored engraving by Mark Catesby from his *Natural History of Carolina, Florida, and the Bahama Islands*, vol. 2 (1747).

The Peales' insistence on an American-produced natural history spoke to the widely shared aspiration among newly independent Americans to cultivate arts and letters independently from Europe, and, at the same time, to receive recognition from Europe for the distinctiveness of American cultural production. Charles Willson Peale had a flair for public display and undertook the launching of natural history into the public realm. In contrast to English models of private patronage, Peale also constantly sought support for natural history from the federal government. The family's public approach to natural history contrasted with that of William Bartram (1739–1823), John Bartram's son, who was the first American to devote his life to the study of natural history. As a boy, William Bartram drew and engraved illustrations to send with his father's shipments to British patrons.[3] In his late twenties, Bartram had set out on a four-year wilderness sojourn that he later described in his *Travels Through North and South Carolina, Georgia, East and West Florida*, published in 1791, an account that conveyed a personal, almost mystical communion with nature as well as a wealth of natural history observation.[4] The book became a classic in the genre of travel and natural history for English and European audiences. But Bartram had none of the missionary civic zeal of the Peales and lived the remainder of his life in quiet retirement outside Philadelphia.[5]

The Peales merged the practices of painting with those of natural history and worked to establish institutions for fostering both pursuits that would engage and educate the Philadelphia public. Their museum, founded by Charles Willson Peale in 1790, became the repository for the specimens collected and published by the first generations of nineteenth-century American naturalists, among them Alexander Wilson, Thomas Say, and John Godman. Above the cases displaying the specimens hung portraits of the nation's patriots, statesmen, and naturalists. The successive locations of the museum at Philadelphia's center, first in the hall of the American Philosphical Society, and later in the nearby State House where state and national governments had convened, linked at Philadelphia's center the museum's natural history displays with the seats of two primary cultural and political institutions of the new republic.[6]

The personnel of the national political life and of natural history overlapped to a remarkable extent. In 1797 Thomas Jefferson, author of *Notes on the State of Virginia*, a work published in 1787 that included local natural history observations, arrived in Philadelphia to assume his duties as both vice-president of the United States and president of the American Philosophical Society.[7] Although Jefferson insisted that the federal government should not lend direct support to the work of private individuals, in his twin capacities of statesman and naturalist Jefferson would serve as intellectual patron for the Peales and for Alexander Wilson, author of the first comprehensive work of ornithology published in the United States.[8]

In their museum the Peales arranged specimens in Linnaean order. Their mammal specimens, mounted with mouths agape to display the teeth, demonstrated the principles of the French anatomist Georges Cuvier (1769–1832), who taught that mammals could be characterized by their dentition.[9] The Peales' affinity with Cuvier, founder of the field of comparative anatomy, stemmed in part from Jefferson's interest in fossil mammals and the leading role of Cuvier in analyzing fossil remains. Both Cuvier and Jefferson published descriptions of fossil sloths, Cuvier from engravings of a Paraguayan specimen he called the "megatherium," and Jefferson from bones found in Kentucky.[10] Cuvier's comparison between fossil mastodon bones and those of the elephant would establish the basis for proof of extinction of species.[11] The excitement generated by the descriptions stimulated interest in fossil remains in general. The Peales were the first to assemble an entire mastodon skeleton, one of

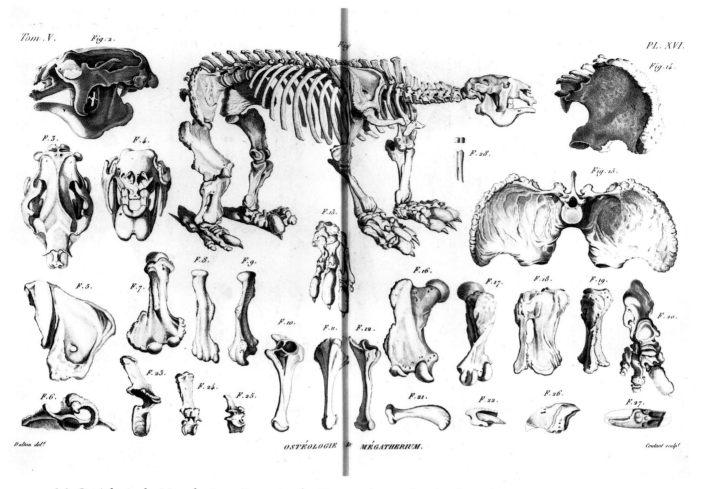

1.1 Ostéologie du Megatherium. Engraving by Coutant from a drawing by Dalton. From Georges Cuvier, *Recherches sur les Ossemens Fossiles*, vol. 5 (1824).

FIGURE
1.1

the two complete skeletons that they had unearthed in New York State in 1801.

Cuvier's principles had a patriotic resonance for the Peales. That so many fossil specimens came from the Americas gave the New World both an ancient history and an important place in natural history. Cuvier's interpretation of the fossil specimens also explicitly challenged that of the Count de Buffon on two points: Buffon's rejection of extinction, and his theory of degeneration of New World species under supposedly harsh climatic conditions. When Buffon had studied fossil mastodon bones, he had pronounced that they were of a living species of elephant, consistent with his belief that extinction violated natural laws. Through a point-by-point comparison of the fossil skeleton with the skeleton of an elephant, Cuvier demonstrated the anatomical differences that supported his claims for extinction. The Cuvierian version of history naturalized revolution. A theory of extinction initiated a view of nature sympathetic to change rather than a divinely ordained stability and gave the static worldview of Buffon decidedly old-regime connotations.[12] Rembrandt Peale celebrated the theory

23

1. *Corvus cristatus*, Blue Jay. 2. *Fringilla Tristis*, Yellow-Bird or Goldfinch.
3. *Oriolus Baltimorus*, Baltimore Bird.

Drawn from Nature by A. Wilson.    1    Engraved by A. Lawson.

Plate 2. Blue Jay, Goldfinch, Baltimore Oriole. Wilson's first plate. Hand-colored engraving. Drawn by Alexander Wilson, engraved by Alexander Lawson, for *American Ornithology*, vol. 1 (1808).

Plate 3. Woodpeckers. The page as a pasted design. Drawn by Alexander Wilson, engraved by Alexander Lawson, for *American Ornithology*, vol. 4 (1811).

of extinction and the concept of a changing nature:[13] "[T]he fanciful chain of nature is not broken! or else a new chain has taken the place of the old. Formerly it was as unphilosophical and impious to say that any thing ceased to exist which had been created, as it is now to say the reverse."[14] For the Peales, the sheer size of the fossil mastodon disproved Buffon's theory of degeneration, which posited that New World species were smaller deformed variants of larger, more perfect Old World species.[15] Like Jefferson, whose writing of the *Notes* had been partly motivated by Buffon's slurs against American nature, the Peales and other naturalists of the period pointed out the error of Buffon's degeneration theory and made it a central theme in their natural history. Rembrandt Peale wrote: "Had the celebrated Buffon attended better" to the truth, "he would have saved himself some needless observations and theoretic fancies, with respect to the old and new world; but we should likewise have lost the able reply of Jefferson."[16]

Rembrandt Peale took the mastodon skeleton to London, where he had it rebuilt for display. To accompany the exhibition, Peale published an account of the discovery and excavation of the bones as well as a description of mastodon anatomy that corroborated Cuvier's findings. Peale's *Disquisition* connected the new principles of Cuvier with a double claim. He insisted on American parity in pronouncing on subjects of natural history and, at the same time, on the central role of the artist—the trained draftsman rather than the savant—in analyzing natural form and its meanings. The eye of the artist, Peale argued, is sufficiently acute to cut through established theory—such as Buffon's—and perceive evidence in natural structure. He wrote:

[T]o judge correctly in osteological comparisons requires not so much the knowledge of the anatomist as the eye of the artist: —and I maintain it as a fact, in which every candid anatomist and every artist will join with me, that the mere artist, by a little attention to the variations of form, will sooner, and with more certainty, establish the characters of skeletons, than the most learned anatomist, whose eye has not been accustomed in an instant to seize on every particularity.[17]

No one would have disagreed with Peale that painting and natural history, along with medicine, shared the study of anatomy, although painting awarded anatomy a different status than did medicine and natural history. In the late eighteenth century, for example, the Royal Academy of Art, placing a higher value on history painting than on anatomical knowledge, did not accept the painter and anatomist George Stubbs as a member.[18] Cuvier had elevated the study of anatomy to a system, adding a functional and systemic dimension to his division of the animal kingdom into four branches, each branch based on morphological distinctions. According to Rembrandt Peale, the artist's visual faculty went beyond system and seized "peculiarity" of form. The "slow anatomist may be sure, but unless he *devotes* himself to the abstracted *study* of his subject"—as by implication does the painter—"he falls short of correct information," Peale asserted. He continued, "It is, therefore, evident, that our hopes for correct knowledge on this subject rest on those in whom the two characters are combined." Peale reiterated his position: "For my part, my decisions are pronounced with no other authority than that of an artist, pretending to very little more knowledge of anatomy than gives me the names and uses of the bones; but, when *forms* and the right comparison of *lines* and *angles* is the subject of investigation, I feel myself, as every artist must, perfectly confident in the assertion of the truth."[19] In the visual approach to natural history that Peale advocated, the painter enjoyed a perceptual advantage over the anatomist. The painter also had the advantage of being able to represent firsthand what he or she had seen.

Alexander Wilson's *American Ornithology*, a nine-volume study appearing between 1808

FIGURES
1.2–4

PLATES
2–8

26

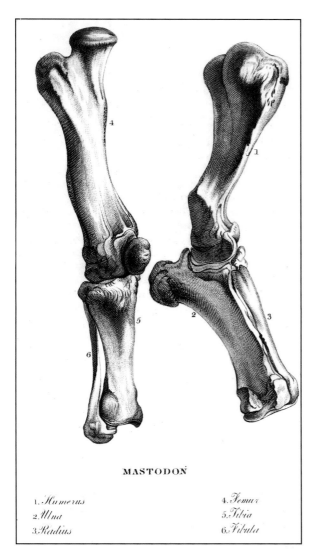

MASTODON

1.2 Mastodon bones. *Left*, unsigned; *below*, engraved by F. Kearny from a drawing by Charles Alexandre Lesueur. From John A. Godman, *American Natural History*, vol. 2 (1826).

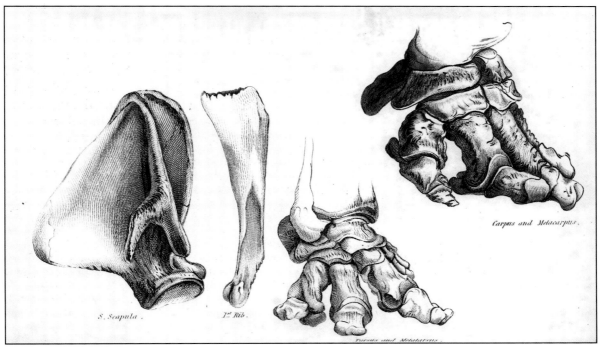

S. *Scapula*.    1.ª *Rib*.

*Tarsus and Metatarsus.*

*Carpus and Metacarpus.*

*1.Rice Bunting. 2.Female. 3.Red-eyed Flycatcher. 4.Marsh Wren. 5.Great Carolina Wren. 6.Yellow-throat Warbler.*

Plate 4.  Rice Buntings, Red-eyed Flycatcher, Marsh Wren and Great Carolina Wren, Yellow-throated Warbler. Hand-colored engraving. Drawn by A. Wilson, engraved by A. Lawson, for *American Ornithology*, vol. 2 (1810).

31

1. *American Crofsbill.* 2. *Female.* 3. *Whitewinged Crofsbill.* 4. *White-crown'd Bunting.* 5. *Baywinged B.*

Plate 5. Cross-bills on a hemlock branch. Drawn by A. Wilson, engraved by A. Lawson, for *American Ornithology*, vol. 4 (1811).

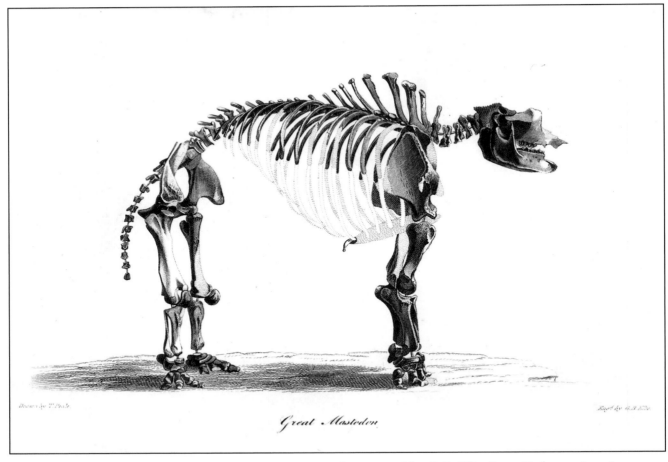

Great Mastodon

1.3 Reconstructed mastodon. Drawn by Titian R. Peale, engraved by G. B. Ellis. From Godman, *American Natural History*, vol. 2 (1826).

and 1814 and the first comprehensive work of American natural history, took its stand firmly on values congenial to the young republic's self-image: outdoor practice as opposed to "speculations of mere closet theory," knowledge defined so as to include vernacular experience, and illustrations made from the living animal.[20] Wilson (1766–1813), a weaver and poet, had emigrated from Paisley, Scotland, to the United States in 1794, settled near Philadelphia, and supported himself teaching school while he devoted his spare time to studying birds.[21] In the preface to his third volume, Wilson summarized the way in which American naturalists, in his view, would make their distinctive contribution to natural history: "Well authenticated facts deduced from careful observation, precise descriptions, and faithfully portrayed representations drawn from living nature, are the only true and substantial materials on which we can hope to erect and complete the great superstructure of science."[22] Wilson's own account and illustrations were to be "a transcript from living nature."[23] Introducing his first volume, he wrote: "It is only by personal intimacy that we can truly ascertain the character" of the birds, "noting their particular haunts, modes of constructing their nests, manner of flight, seasons of migration, favourite food," aspects of North

1.4 Weasels showing teeth. Graphite drawing by Titian R. Peale from a Peale museum exhibit, n.d.

American birds' life histories that were absent from most European literature to date.[24]

Wilson's emphasis on eyewitness of living birds in their natural habitats, his insistence that he base his illustrations on live or freshly killed specimens and draw from preserved specimens only as a last resort, echoed the Peales' visual approach to natural history. Wilson's natural history was intensely visual, the combination of observation of the living bird and its representation. He believed that drawing was an integral part of observation. Wilson placed illustration, both as an observational process and as a means of communication, on a par with written description. He promised that his text and illustrations would "mutually corroborate each other." Like Rembrandt Peale, Wilson privileged vision over verbal description, "so infinitely more rapidly do ideas reach us through the medium of the eye."[25]

Like the Peales, Wilson worked an ocean away from the recognized collections, institutions, and publishers of natural history, which had no counterparts in the United States. His residence near Philadelphia, however, placed him close to some of the most experienced naturalists in the country. He taught himself the skills necessary to study birds and publish on them, but he did so under the tutelage of his aging friend and mentor, William Bartram.[26]

31

1. Belted Kingsfisher. 2. Black and yellow Warbler.
3. Blackburnian W. 4. Autumnal W. 5. Water Thrush.

Drawn from Nature by A. Wilson          Engraved by A. Lawson

33

Plate 6. Kingfisher and Warblers, with pond and mill in distance. Drawn by A. Wilson, engraved by A. Lawson, for *American Ornithology*, vol. 3 (1811).

Plate 7. Great-footed Hawk. Drawn by A. Wilson, engraved by A. Lawson, for *American Ornithology*, vol. 9 (1814).

Wilson had available the resources of Bartram's library and the Peale museum. He was also well situated for undertaking the publication of an illustrated work. Philadelphia, home of the nation's mint and a center of publishing, attracted many engravers and supported them with the work of designing bank notes, printing illustrations for Bibles and encyclopedias, and engraving portraits of American statesmen, Revolutionary War battle scenes, scenic views, and famous paintings.

Fascinated by birds from the moment of his arrival in the United States, Wilson observed them in his spare time from teaching. Although he was a man of solitary habits, he became known for this interest, and neighbors and students brought him live birds, nests, and other natural history curiosities. Around 1803 Wilson began to draw birds seriously and to participate in the natural history community. He met and became friends with Bartram, to whom he sent drawings for criticism, and with Meriwether Lewis, the explorer and a student

of botany, with whom he frequented the intellectual circle that gathered around Alexander Lawson (1772–1846), an engraver and fellow Scot.[27]

Wilson's early attempts at drawing were a first step in embarking on a systematic study of birds. Catesby and Edwards, whose books he studied in Bartram's library, were his principal models, although from his own observations he could already correct errors in their descriptions and illustrations. In a letter to William Bartram of August 1804, written after Wilson had decided to attempt to publish on birds, he indicated that drawing was for him a method of distinguishing species: "I have been drawing Woodpeckers this sometime. Pray be so good as to inform me if there is not 4 different species besides the Fliccer in these parts."[28]

Wilson, for whom drawing, like dissection and marksmanship, was an essential tool for the study of ornithology, found his own early efforts discouraging. At first, intent upon accuracy, he counted everthing—feathers, mark-

1.5 Owl. Drawn by Alexander Wilson, 1804.

than he could create a lifelike portrait from a mummy.[29]

Wilson's correspondence with Bartram traces the shift from frustration to an emerging confidence in his results. In December 1804 Wilson wrote that he considered himself "miserably deficient in many acquirements . . . Botany, mineralogy, and drawing, I most ardently wish to be instructed in, and with these I should fear nothing."[30] Yet several months later, in March 1805, when Wilson wrote Bartram in celebration of the reelection of Jefferson, "enlightened philosopher—the distinguished naturalist—the first statesman on earth, the friend, the ornament of science, . . . the father of our Country," he added as a postscript: "I am at present engaged in drawing the two Birds I brought from the Mohock; which, if I can finish to your approbation, I intend to transmit to our excellent President, . . . as a token of . . . esteem."[31] Wilson described the drawings in a letter to another friend: "I set to work upon a large sheet of fine drawing paper, and in ten days I finished two faithful drawings of them, far superior to any that I had before. In the background, I represented the view of the Falls of Niagara, with the woods wrought as finely as I possibly could do. Mr. Lawson was highly pleased with it, and Mr. Bartram even more so."[32] (see fig. 1.7.) He sent the drawing, of what he believed to be undescribed species, with a cover letter from Bartram that called them "fine and accurate."[33] Jefferson responded, calling the drawings "elegant" and engaging in an exchange of ornithological observations.[34]

During the months that followed, Wilson continued to draw birds in the evenings. In July he sent a package of his work to Bartram for criticism: "I dare say you will smile at my presumption, when I tell you that I have seriously begun to make a collection of the birds to be found in Pennsylvania, or that occasionally pass through it; twenty-eight as the beginning, I send for your opinion. They are, I hope, inferior to what I shall produce, though as close

FIGURES
1.5–6

FIGURE
1.7

ings, scales. The resulting stiff patchwork of lines conveyed nothing of the subtle patterns, soft textures, and contours of the round feathered body. Wilson learned, moreover, that although drawing from a dead or preserved specimen enabled him to count and measure, working from dead models also gave his drawings a dead quality, for which he had already criticized English and European illustrations. In the 1807 prospectus for his book, Wilson would comment on the problem: "Let the abilities of the painter be what they may, if he has nothing but stuffed skins, or dried specimens to draw from, he can no more give the true tints, form and air of the living original"

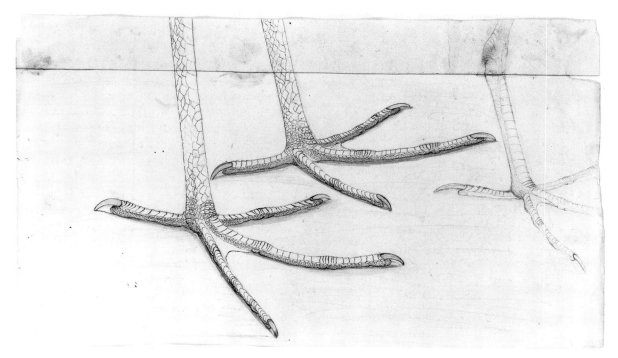

1.6 Bird's feet. Drawn by Alexander Wilson in graphite and ink, n.d.

copies of the originals as I could make." He requested that Bartram "Criticise these, my dear friend, without fear of offending me—this will instruct, but not discourage me. For there is not among all our naturalists one who knows so well what they are, and how they ought to be represented."[35]

That autumn Wilson began to experiment with etching, emulating Catesby and Edwards, who prepared and printed their own plates. Wilson sent the initial "proof-sheet" along with more drawings for criticism to Bartram, who in his youth had also practiced engraving.[36] In January he sent Bartram another proof for his projected "Birds of the United States" and noted that application of watercolor to the print improved it "a little."[37] A month later, Wilson wrote to Jefferson that he had completed nearly one hundred drawings, and had engraved two folio plates of birds, many of which were new to science.[38]

Wilson experimented with etching in part because he believed that the skills needed for making pictures were necessary skills for a naturalist. The process of learning to inform his drawings not only with observed detail but with the quality of life had been integral to his process of learning and observing birds in nature. His readiness to publish, to enter the arena of print, matured when he felt he had mastered all the necessary skills for collecting, observing, and representing birds: wilderness endurance and woodcraft, familiarity with the literature of natural history, and proven ability to produce his own documents of observation—"Upwards of one hundred drawings . . . and two plates in folio already engraved."[39]

By engraving his own plates, Wilson intended to keep the interpretive process in his own hands, but he seems to have reached the conclusion that he must relinquish that control. Alexander Lawson became the principal

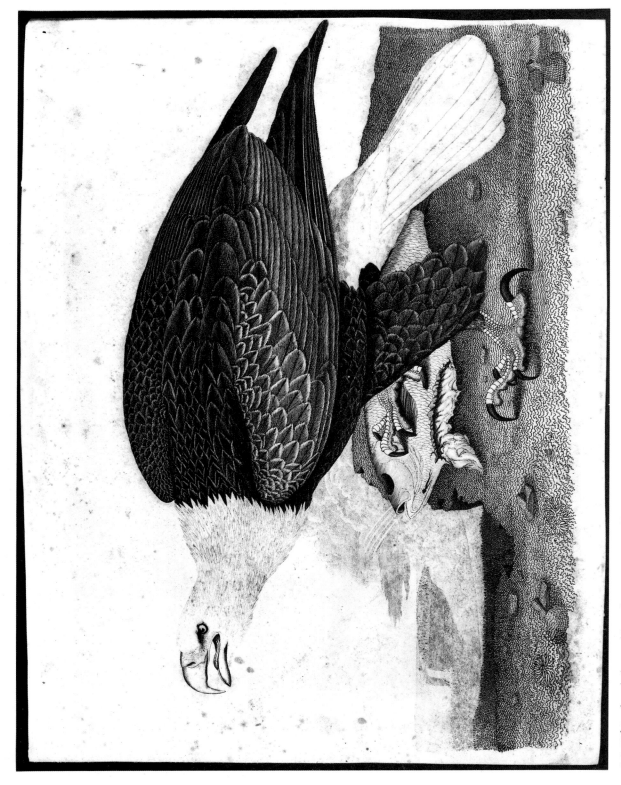

1.7 White-headed or Bald Eagle at Niagara Falls. Drawn by A. Wilson, engraved by A. Lawson. Uncolored proof for *American Ornithology,* vol. 4 (1811).

engraver of the published volumes. Wilson's preface to his first volume suggests at least two reasons for this change. First, by a division of labor he could protect his own time as observer: "Many years of application are necessary to enable a person, whatever may be his talents or diligence, to handle the graver with the facility and effect of the pencil; while the time thus consumed, might be more advantageously employed in finishing drawings, and collecting facts for the descriptive part, which is the proper province of the Ornithologist."[40] In addition, he seems to have found that Lawson's skill and experience could produce an image more faithful to his observations than he could himself: "Every person who is acquainted with the extreme accuracy of eminent engravers, must likewise be sensible to the advantage of having the imperfections of the pencil corrected by the excellence of the graver. Every improvement of this kind the author has studiously availed himself of; and has frequently furnished the artist with the living or newly-killed subject itself to assist his ideas."[41]

Commercial engraving was used primarily to reproduce pictures from other mediums for publication. A commercial engraver could not succeed in business, especially in a competitive engraving center such as Philadelphia, without cultivating an eye for reinterpreting gradations of shade and color into line. As Wilson freely admitted, his principal engraver, Alexander Lawson, did more than copy the ornithologist's drawings; he "corrected" them, and contributed his own "ideas" to the finished picture. Wilson had already interpreted the bird from life onto paper using pencil, ink, and watercolor. His own experience with engraving taught him to emphasize line over shading to conform to the engraving method. The engraver translated the drawing into grids, webs, nets, or dots that would be inked to print a uniform black. The engraving technique itself imposed a significant reinterpretation of visual information. Moreover, the original observation had passed through the mind and hands of an additional person. Everyone who looked at printed pictures, however, including naturalists, accepted the changes wrought by the translation into print.

Wilson considered his printed plates transcripts of living nature. He based his assertions of the authenticity of his observation and its representation on the collaborative relationship between naturalist and printer, yet he differentiated a hierarchy of naturalist's skills and accepted a practical division of labor between the observing and drawing naturalist and the engraver. To the naturalist belonged the role of transcribing observation into words and drawings; to the engravers belonged the role of translating drawings into printed illustration. Each plate produced for Wilson's *American Ornithology* acknowledged the steps of its production, as was customary in published prints. At the bottom of each plate were inscribed, on the left, "Drawn from Nature by A. Wilson," and in the right-hand corner, "Engraved by A. Lawson"—or George Murray or George Warnicke, the other engravers who worked on the *Ornithology*.

For his intial experiment with etching Wilson had borrowed tools from Alexander Lawson. Lawson, orphaned young, had begun to teach himself engraving in England. He came to the United States because revolutionary France, where he wished to participate in political events and learn the refinements of French engraving technique, was closed to British subjects.[42] Lawson produced engravings of engineering projects, architectural views, maps, medical and chemical illustrations—a portfolio reflecting the developing arts and manufactures of the United States. He would apply a regular, ordered engraving style to the bird subjects given to him by Wilson.

Grids and webs of regular lines cut with the graver were the foundation of the commercial engraving style predominant in Europe and the United States throughout the eighteenth and into the nineteenth centuries. A regular linear style had proved more economical than techniques like stipple engraving, mezzotint, etching, or aquatint. Both stipple and mezzotint employed a rocker or roulette, a wheel-like tool with tiny spikes that punctured the copper surface with ink-collecting dots. To etch a plate, one scratched or drew lines through a layer of varnish applied to the copper surface, thereby opening the plate to the aquafortis bath. For aquatinting, the printmaker dusted areas of the plate with resin, then melted it to bond the granules onto the plate. An acid bath bit the pits or spaces between the covered areas to varying depths: the longer the exposure, the deeper the bite, the darker the area in printing. A plate incised with deep lines spaced well apart, however, resisted better the repeated pressure of the press over long printing runs than did those with the shallower pits made by rockers, or irregular ones produced by etching in acid.[43] Lawson, like most engravers, enhanced his engraved lines with etching, with the bite roughening and adding texture and depth to the firm, sharp lines cut with the graver.

For Wilson, publication of an illustrated work presented a steep financial obstacle. In March 1804, when he had first confided his intention to publish his studies, Lawson had discouraged him on the grounds that the project was too expensive. The cost for a single hand-colored plate included copper at $5.66 per plate; engraving anywhere from $50 to $80, depending on the amount of engraving and the engraver's reputation; and finally hand-coloring at 25 cents per sheet. For the project Wilson envisioned, illustrating more than one hundred birds, the copper would cost over $500, the engraving between $5,000 and $8,000, and the coloring of an edition of five hundred would be $12,500. Paper, letter press, and binding would add about $2,000 per volume of ten plates. To cover expenses, the price of each volume would have to be set beyond what most people would pay for a book on birds.[44]

There did exist, however, a proven American audience for natural history, at least in Philadelphia. One city publisher had charged nine

dollars for a four-volume edition of Oliver Goldsmith's *An History of the Earth and Animated Nature*, a work modeled on Buffon's natural history, with engraved but uncolored plates. It had sold three thousand copies.[45] The journals of the few learned societies contained copper-engraved illustrations on a wide range of scientific and technical subjects; they had a more limited audience, and were supported by subscriptions from society members.

Wilson intended his book to entertain and to instruct, to reach both the natural history community and a wider audience interested in more popular publications. He persuaded the publisher Samuel F. Bradford, scion of a long line of Philadelpia printers and publishers, to undertake publication of the *Ornithology*.[46] The publisher insisted that the author finance each volume, at twelve dollars apiece, with advance subscriptions and sales of previous volumes. Bradford further insured himself against loss by linking risk on the scientific book with what he thought would prove a more secure project, and hired Wilson to edit and write for an encyclopedia while working on the less certain *Ornithology*.[47]

The connection between Wilson's plan of an illustrated work intended for a general audience and Bradford's projected encyclopedia was more than coincidental. To date, encyclopedias had been the major American publications illustrated with copper engravings. Thomas Dobson had printed the first American edition of the *Encyclopaedia Britannica* in Philadelphia between 1790 and 1797; it contained 542 copperplates and had employed in their making the cream of the city's corps of engravers.[48] Simultaneously, Dobson had published an edition of *Ree's Encyclopedia*, with 543 copperplate engravings.[49] Bradford's encyclopedia, a new edition of Ree's *Cyclopaedia or Universal Dictionary*, began appearing in 1810. As assistant editor, Wilson adapted the encyclopedia for the American market, writing and rewriting articles on national literature, science, and nature.[50]

Wilson, eager to quit teaching, agreed to an

arrangement that seemed mutually beneficial. The ornithologist met his obligations by combining his wilderness observation and collecting journeys with peddling subscriptions for his own book and, when back in Philadelphia, by working on the encyclopedia while writing his descriptions and overseeing his engravers and colorists. With this hectic and ever-accelerating schedule, he produced eight volumes in six years.

Wilson promised that his text and his plates would "mutually corroborate each other," and, indeed, the text and illustrations of the *Ornithology* should not be treated as separate. The book as a whole was a consistent effort to represent nature in words and pictures. Wilson's illustrations had two referents—the natural object and the text. The plates in Wilson's *American Ornithology* served three main illustrative purposes, corresponding to the three principal narratives that Wilson developed in his book: the first ornithological, the second on the American landscape, and the third on the progress of national arts and manufactures.

The plates illustrated the ornithological text in a number of different but related ways. From his earliest plates Wilson represented each bird, usually in profile, in a characteristic pose and perched on a prop such as a tree stump or leafless branch—both resembling mounts for stuffed display specimens. These perches indicated Wilson's debt both to the European illustrations he had studied and to the mounts in Peale's museum. By deploying these conventions, Wilson made the pictorial assertion that his book belonged to the tradition of natural history books. The leafless branch, like the clod of earth sprouting tufts of undifferentiated grass, dated from the sixteenth century.

In Wilson's first volumes most of the birds sat on such twigs or stumps. Typically, in the early plates, the several birds in profile on their separate perches lay on a flat space, unrelated to one another, filling the page as a pattern. In many plates, scale was not uniform. Such disjointed compositions fell entirely within the conventions of the genre. They also derived

from Wilson's (or Lawson's) practice of cutting out finished drawings and pasting them together onto a single sheet to serve as a model sheet for engraving. Some of Wilson's more idiosyncratic compositions recall Catesby's. The figures are combined and flattened onto a single plane, in a manner similar to the convention in still-life painting of combining disparate objects in a single foreground composition. Catesby, who had felt diffident about his self-taught drawing skills, had sought to make a virtue out of his lack of the techniques of academic drawing: "As I was not bred a Painter I hope some faults in Perspective and other niceties may be more readily excused, for I humbly conceive Plants and other Things done in a Flat tho' exact Manner may serve the Purpose of Natural History, better in some measure than in a more bold and Painter-like Way."[51] In one way, flattening indicates the self-taught status of both Catesby and Wilson. On the other hand, Wilson's emulation of Catesby's "Flat tho' exact Manner" may well have been a deliberate choice. Wilson, like Catesby, "was not bred a Painter" and may have seen no need to venture into deep pictorial space—at any rate after early presentation pieces like his Niagara Falls drawing. Flattening was also entirely consistent with most natural history illustration and connected it to scientific and technical diagrams. Wilson's and Catesby's combination of foreground figures in a flattened arrangement on the page expressed both the literal pasting of separate drawings into a single composition, and the transposition of the bird and plant from nature into the context, or space, of the book.

Wilson's eyewitness experience of nature informed his treatment of these compositional conventions. Here, a bird reached for an insect; there, a lizard blended into the bark of a tree. The lively poses, especially the tilt of the head and firm gazes of the birds, were proof of Wilson's claim to have drawn directly from the living bird, to have produced "a transcript of living nature." In many of his chapters, each opened with the hand-colored engraving figuring the birds under discussion, Wilson told sto-

1. *Winter Falcon.* 2. *Magpie.* 3. *Crow.*

Plate 8. Magpie, Crow, and Falcon, with farm in distance. Drawn by A. Wilson, engraved by A. Lawson, for *American Ornithology*, vol. 4 (1811).

ries of the capture of the live bird, of keeping it as a pet, or drawing it before it died—perhaps in a noble attempt to escape to the wild. These memoirs testified that the illustrations were portraits of individual birds.

FIGURE 1.8

The individual birds, however, also stood as a generalization for all members of that species. Each picture described the form and markings that distinguished one species from another, based on the male bird, just as the text related its characteristic habits, diet, and the requisite taxonomic characters of length, markings, and so on. Wilson took taxonomy seriously, although like other writers directing natural history to a general audience, he believed that a "bare account of scientific names, colour of bills, claws, feathers, &c., would form but a dry detail."[52] The necessity of observing personally each species forced Wilson to publish the birds in the sequence of his own travels and observations instead of in correct systematic order, for which he apologized to his readers.[53] Nevertheless, Wilson's systematic analysis followed that of the English writer John Latham, whose *General Synopsis of Birds* (1781) he considered the closest to the reality he observed: "In particularizing the order, genus, &c. to which each bird belongs, this system, with some necessary exceptions, has been *generally* followed in the present work."[54] Thus each illustration also figured as corroborating evidence for Wilson's placement of the species in an ordered arrangement of nature. Although the plate captions did not always give Latin binomials with the vernacular names, Wilson's chapter headings consistently and properly began with all the synonyms—the names given to the species in the previous literature—and the chapters themselves included discussions of the history of attempts to classify the birds.[55] Some of the plates, notably the one illustrating the orioles, pictured female, juvenile, and mature male, assigned to separate species by European authors who had worked only from preserved specimens.

FIGURE 1.9

The plates also provided visual evidence—size, bright coloration—to support Wilson's ref-

utation of Buffonian error, the catch-all category for mistakes perpetuated about American nature from the distance of Europe. Wilson introduced into his plates what would now be called evidence of the species' ecological relationships. He dissected his specimens and studied their stomach contents to determine diet, and sometimes he pictured the relevant insects. In his illustration of crossbills he perched the birds on a hemlock bough to demonstrate the relationship between the shapes of the bills, which Buffon had considered deformities, and the task of extracting nuts from the cones. In placing crossbills on a hemlock branch instead of a generalized branch or stump, Wilson extended existing European conventions in bird illustration to testify to his eyewitness accounts of living birds.

The plates of the *American Ornithology* also inaugurated in printed American natural history illustration a tradition of landscape that already had its expression in natural history prose—Wilson's own and in William Bartram's before him—and a growing literature with an eager readership.[56] A lyrical celebration of the American landscape and its wild inhabitants ran through the entire work and placed Wilson's text in the literature of nature appreciation as much as in the ornithological tradition. As early as his third volume, published in 1811, a tiny landscape—pond, mill, and horizon—unified the foreground space occupied by the birds on their conventional perches. The introduction of landscape into the space of the ornithological plates illustrated Wilson's prose landscape description in the same way the pictures of birds illustrated his ornithological observations.[57]

The first landscape elements in the plates have a modest quality that recalls Wilson's own early watercolor sketches of his neighborhood schoolhouse or inn. As the work progressed, however, many of the more elaborate landscapes took on a conventionalized or stylized quality, in marked contrast to the minutely observed insects and lizards that had characterized Wilson's early plates and the il-

1.8 Red-tailed Hawk. Drawn by Alexander Wilson in ink, graphite, and watercolor, ca. 1812.

1.9  Wilson's drawing of orioles. Engraved and published as plate 4 in his first volume.

lustration of the crossbills on the hemlock branch. In the later volumes, the more elaborate the landscape, the more it relied on using undifferentiated and unidentifiable foliage, such as Lawson might have used to illustrate other kinds of books. These conventionalized landscape elements resembled many European bird illustrations with generalized backgrounds; they did not bespeak eyewitness or direct observation. Some may have been derived from the painted backgrounds of the exhibits in the Peale museum,[58] but others appear to be Lawson's inventions made on the plate, intended to unify the separate figures.

The landscapes in the plates of the *Ornithology* must be seen as a collaboration between Wilson and Lawson. Wilson's narrative established the context for the gradual elaboration of a full landscape treatment, even if Lawson actually provided the stylized elements. In the later volumes, landscape, both conventional and observed, began to be elaborated into a window onto nature illustrating Wilson's various interwoven narratives. As the landscapes became more elaborate the plates combined the descriptive requirements of natural history with other narrative functions of the work. Whether it was Wilson or Lawson who took the

FIGURE
1.10

1.10  Ducks. Detail of proof engraving from A. Lawson's portfolio. Drawn by A. Wilson, engraved by A. Lawson for *American Ornithology*, vol. 8 (1813).

lead in the change, the plates developed a view of the animal in nature that would reach its culmination in the work of John James Audubon. Over time, natural history illustration with fully elaborated landscape backgrounds that both narrated and described would become associated with the ideal of the naturalist-illustrator, or by then, properly, the artist-naturalist. This association was one of Wilson's most influential and lasting contributions to American ornithological illustration in particular and to natural history illustration in general. The claims he made for firsthand observation, reflected in a text and illustrations that located the living animal in nature, launched a distinctly American narrative of the American naturalist's relationship to American nature.

Finally, Wilson's plates provided material examples of the marshalling of American arts and manufacture to the production of scientific books. Wilson's prefaces to the successive volumes developed a narrative that engaged the reader-subscriber in the production of the book itself. These stories kept the subscribers abreast of their investment not only in the work at hand, but in the larger national project of improving American culture and industry. In his prospectus Wilson promised a superior product, an "Imperial Quarto, on a rich vellum paper": "the type for the letter-press is entirely new, and of singular beauty; and the public may rest assured, that no pains or expense will be spared . . . to render it, in point of accuracy and elegance, worthy of their support."[59] In two subsequent notices promoting the publication, Wilson assured the public that the work would be "in point of elegance not inferior to any of the kind that has appeared in Europe"; that the typography would be "executed in a style" equal to any European production;[60] and that the illustrations would be "engraved in a style superior to any thing of the kind hitherto published."[61] Just as in the finest French ornithological publications, Wilson had some of his plates printed in colored inks, which not only "when judiciously employed in works of this

kind, gives great softness and effect to the plumage," but also avoided the sometimes muddying effect of watercolor over black ink.[62]

All of his materials were manufactured in the United States except the watercolors used to tint the plates. In his second volume, Wilson began to use locally produced paints and notified his audience about it, so that they too could appreciate the progress of national science and manufacture:

The present unexampled spirit . . . for new and valuable manufactures, which are almost every day rising around us; and the exertions of other intelligent and truly patriotic individuals, in the divine science of Chemistry, give the most encouraging hopes, that a short time will render him [the author] completely independent of all foreign aid, and enable him to exhibit the native hues of his subjects in colors of our own, equal in brilliancy, desirability and effect to any others.[63]

The paints, "some beautiful native ochres," were produced in "the laboratory of Messrs Peale and son, of the Museum of this city."[64] Domestic manufacture and supply, always important to the young republic's sense of autonomy, became essential during Jefferson's trade embargo.

Publication proceeded through the war years, although Wilson experienced difficulties with his colorists, whose work he found unsatisfactory. Each issue required forty-five hundred colored plates.[65] "The great precision requisite in this last process, and the difficulty of impressing on the mind of every one whose assistance was found necessary, similar ideas of neatness and accuracy, have been a constant source of anxiety of the author, and of much loss and delay."[66] Alexander Rider, a Swiss painter, was said to have spoiled many impressions by coloring them in opaque paints.[67] Wilson turned the delays into another assurance of quality control: "The correct execution of the plates will be rendered more secure, by the constant superintendence of the author; and by the whole of the coloring being performed in his own room, under his immediate inspection."[68]

Coloring was one stage of book production in which women often worked. The pay was lower than in jobs employing men—twenty-five cents for a completed sheet—and the work could usually be done at home. Wilson hired Nancy Bartram, William's niece, and Mary Leech, a friend of hers; he also paid Anna Peale for coloring plates. Lawson's daughters, Malvina and Helen, would later work as plate colorists for the first reprint edition of Wilson's book.[69] In the winter of 1812–1813, however, Wilson's colorists deserted him, and he was forced to paint all the plates himself.[70]

There were other reasons for delays which Wilson omitted from his narrative. A package of drawings sent to Lawson from Nashville never arrived, and Wilson had to redo the lost work.[71] Moreover, Bradford would not begin a volume's production until subscribers had paid for the previous one.[72] Bradford, who was losing money on the encyclopedia, never paid Wilson his share of the returns on the *Ornithology*.[73] Wilson, in turn, stalled on the encyclopedia until he had been paid for all previous editorial work.[74]

Instead of dwelling on these difficulties, Wilson wove together for his readers the story of his book with that of the developing arts, letters, and natural history in the national life. Wilson's book was not priced to reach every American, but both his intended audience and his actual subscribers—the two groups merged

during the writing and publication process—included artisans and merchants as well as doctors, literary and philosphical societies, and naturalists.[75] He used his introductions to appeal to "gentlemen of leisure, resident in the country" to contribute their ornithological observations to the collective enterprise of natural history. Naturalists Samuel Mitchill of New York and John Abbot of Georgia sent Wilson valuable contributions of observations and specimens. Wilson also promoted in his pages the books and work in progress of other naturalists, such as François Michaux, author of a natural history of North American forest trees. Thus the book not only recorded and documented the progress of natural history, but connected its diverse readership with the active practice of collection and study.

In its scope, the work expressed the broadest national aspirations in both text and plates. In the plate illustrating the magpie, drawn from the specimen that Meriwether Lewis had brought back from the Louisiana Territory, Wilson represented his Jeffersonian hopes for western settlement and cultivation by sketching a farm scene on the distant horizon. Later, he would write: "When the population of this immense western Republic will have diffused itself over every acre of ground fit for the comfortable habitation of man—when farms, villages, towns and glittering cities . . . overspread the face of our beloved country . . . then, not a warbler shall flit through our thickets but its name, its notes and habits will be familiar to all."[76]

# CHAPTER

# 2

# Divergence

THE establishment of new natural history institutions, notably the Academy of Natural Sciences of Philadelphia in 1812, the Linnaean Society of New England in Boston in 1814, and the Lyceum of Natural History of New York founded in 1817 testified to the growing corps of enthusiastic amateur naturalists in the United States. These organizations also began to institutionalize practices quite different from Alexander Wilson's. First, they encouraged the building of libraries and collections of preserved specimens for reference. Natural history in Europe had long since become a historical study based on a literature to which authors were required to refer for clarifying synonyms, acknowledging priority, and correcting error. American naturalists had sorely lacked the historical resources of libraries and study collections of preserved, organized material for comparison and analysis. The Peales and Wilson compensated for the lack of institutions, collections, and books by placing their emphasis on direct observation and representation. So little of American nature was known to naturalists that mere observation and description could make a contribution to knowledge. With the development of American collections, however, natural history study could be differentiated into stages: observation in the field, and study of collections. The location of natural history societies and collections in cities created an urban-rural division of practice, dividing practitioners into those with access to collections and libraries, and those without. Museum study promoted a taxonomic approach, while observation in the field promoted study of the animal's life cycle and relationships to other animals and to plants.

In writing and illustrating the *American Ornithology* Wilson had combined collection and book study with travel and observation. Beginning as a lone outdoorsman and observer in the tradition of Mark Catesby and William Bartram, Wilson had been led into the Philadelphia natural history community. The Peale museum had provided invaluable reference for his work, and he had in turn given the Peales bird skins for mounting and display. But Wilson had founded his motivation and his book on the observations made on his extensive travels, and preferred those times, and his time drawing, to the arduous periods in the city processing observations for publication. Wilson had claimed, moreover, that description of living nature from personal observation would distinguish the American contribution to "the great superstructure of science." Throughout the publication process, Wilson had been able to maintain a balance between his journeys to observe the living bird in nature and his continuing communications with specialist and nonspecialist contributors to the book. That balance was reflected in his text and illustrations, which combined Wilson's narrative of personal observation with taxonomic description. That balance constituted Wilson's meaning of "Drawn from Nature."

Although the *American Ornithology* stood as the exemplar for the next generation of American naturalists, its combination of systematic description and often lyrical narrative proved unstable. Narrative and description pulled in opposite directions during the third and fourth decades of the century. In the initial phase of the process of institutionalization and professionalization of science from the teens through the forties of the century, the place of narrative in the natural history literature became problematic. The emerging category of professional natural history writing differentiated itself by avoiding narrative. Naturalists shifted from strict taxonomic description to a more narrative style when writing for a general readership, because, as in England, the specialized language of taxonomy was generally characterized as exclusive and discouraging to the nonspecialist. Illustration reflected that differentiation. Taxonomic illustration emphasized its difference from narrative by removing the animal from its natural setting—or stylized abbreviations of it—and representing the animal not as a living individual, but as a specimen.

FIGURES
2.1a,b

The early members of the Philadelphia Academy of Natural Sciences were Wilson's immediate successors: Thomas Say (1787–1834), entomologist; Charles Alexandre Lesueur (1778–1846), the self-taught naturalist and illustrator and veteran of the French Australia expedition; the young Titian Ramsay Peale (1799–1885); Richard Harlan, paleontologist; the geologists William Maclure (1763–1840) and Gerard Troost.[1] William Maclure, president of the Academy from 1817 until his death, purchased valuable books and collections for the institution on his extensive European travels and sponsored collecting trips for the Academy's members. The range of illustration styles in their work reflected the tension between narrative and description in natural history publication.[2]

Maclure's support of science and education was animated by a spirit of experiment and by his belief that science and society should both be based on principles of direct participation. His extensive scientific philanthropies reflected his passionate belief in the role of science as an instrument for social change. After emigrating from Scotland at an early age, Maclure made a fortune in commerce, but his life work combined a political evolution from jacobinism to socialism with a consistent dedication to science and education. He believed that universal education would banish superstition and foster a classless society.[3]

In pursuit of these ideals, Maclure and his closest associates played a double role in the initial divergence of description from narrative, or specialist from popular natural history. On the one hand, their work and participation were central to establishing the Academy of Natural Sciences and its specialist values and practices. On the other, their subsequent participation in Robert Owen's community in New Harmony, Indiana, constituted an attempt to integrate specialist values and practice with communitarian social life. Not only was their attempt rejected by their former Academy colleagues, but the demise of the New Harmony scientific and educational community coincided with the rise of a clear distinction between the specialist and popular realms of natural history. Specialist practices were recognized as science while the values and practices associated with natural history were relegated to the popular realm. These distinctions did not constrain the work of Say and Lesueur because they had not yet fully consolidated, nor had the textual and illustration conventions that came to be associated with them. Nevertheless, the range of Say's and Lesueur's publications, their use and nonuse of narrative and narrative illustration conventions, and public response to them, marked the poles that would cohere into distinct realms of nature study. In all their publications, the medium of illustration and method of printing were important expressions of the meaning and of the intended audience.

At the specialist end of the spectrum, this generation of Philadelphia naturalists published many of their new species and observations in the Academy's journal, directed at an audience already initiated into formal taxonomic description. They composed their descriptions in a terse style and format that accommodated little anecdote or metaphor, or the kind of landscape appreciation that had been integral to Wilson's text. Correspondingly, Lesueur's engravings presented the animals positioned to display important characters, that is, the morphological features that identified their genus or species. His plates usually omitted landscape backgrounds, perches, or pedestals and focused on the purely descriptive.

Maclure's scientific and educational philanthropies included active exploration of innovations in printing as well as the establishment of a string of schools and libraries. In the institutions he fostered, Maclure committed his patronage to provide direct access to print. When expenses for the first volume of the *Journal* became too great for the nascent Academy to carry, Maclure purchased a secondhand printing press with which the publication committee completed the volume in Maclure's home.[4] The modest format, small size, and unembellished plates of the Academy journal reflected not only the taxonomic orientation—and shoestring budget—of the institution, but also Maclure's social ideas. He advocated the most economical possible materials for book production so that all classes could afford to buy them.[5] The materials of the *Journal* were economical indeed; Lesueur's engravings, printed on cheap paper, have yellowed and become brittle with time. Maclure's patronage made it possible for the specialized *Journal* in its early years to circumvent the market constraints of commercial publishing and establish a reputation in the scientific community both at home and abroad. Without his support of its publications, the

founding and continued existence of the Academy would have promoted a local culture of science, but would not have projected the work of the Academy and its members into international scientific circles.

Maclure believed that learning, knowledge, and communication should all be grounded in direct experience and placed a high value on pictures in education and science. Understanding through his own geological work the role of mapping, and through his assocation with Say and Lesueur the descriptive relationship between text and illustration, Maclure took a keen interest in Lesueur's experiments with lithography for scientific illustration, and sent Lesueur supplies from Europe. Lithography, a method of printing pictures from crayon drawings on smoothed limestone slabs, was invented in Bavaria by Alois Senefelder in 1798.[6] The technique was hailed as potentially less expensive than engraving both because limestone promised to be cheaper than copper, and because drawing directly onto the stone eliminated a step of interpretation. These economies attracted the interest of the American scientific press.

The first notice concerning the lithographic method in the United States appeared in 1808 in the *Medical Repository* of New York, published by physician and naturalist Samuel L. Mitchill, also a former contributor to Wilson's *American Ornithology*.[7] Nothing immediately came of it, although there were reported experiments with relief etching on stone in Philadelphia, and apparently in Boston, a few years later.[8] When Senefelder's 1818 treatise on the lithographic method was translated into English and French in the next year, however, its appearance seems to have stimulated a fresh round of experiment.

Until exploration had located suitable stone in Kentucky, American printers interested in the technique depended on limestone imported from Paris. In 1819 experiments on Kentucky stone resulted in the first known American

FIGURES

2.2–5

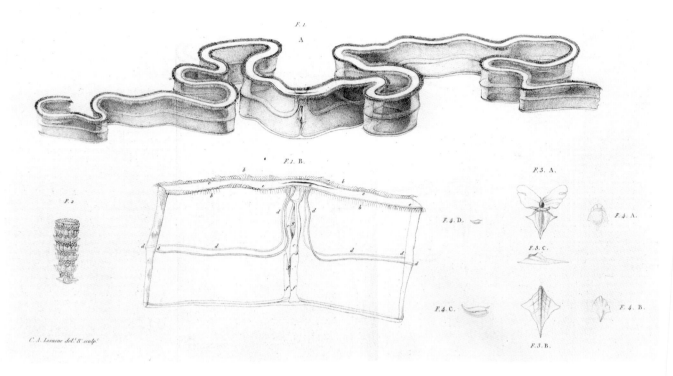

*Noun. Bull. Sc. Tom. III. N 69. Pl. 5.*

2.1a *Cestum veneris, Pyrosoma elegans*, and Hyalaea species. Color engraving by Charles Alexandre Lesueur, *Nouveau Bulletin des Sciences*, Paris (1813).

published lithograph produced by a Philadelphia portrait painter, Bass Otis. Dr. Samuel Brown, a physician and naturalist from Alabama and Otis's sponsor in the experiments, exhibited the trials at a meeting of the Academy of Natural Sciences in May, and "the members present were much pleased with them and expressed a desire to have the illustrations in the next volume of the . . . *Journal* printed from Kentucky stone."[9]

Shortly after the meeting, Lesueur, who engraved with fluency, began experimenting with lithography on imported stone. Accustomed to the crispness of engraving and also to his own ability to produce a varied line, Lesueur may have been frustrated by the soft effect of his crayon lithographs. In 1821, still dissatisfied with his results, he replaced a lithographic il-

lustration for a description of the fish genus *Esoces* with a copper engraving for publication in the *Journal*. Less than a year later, however, he published two lithographic prints to illustrate his descriptions of species of the fish genus *Cichla*.[10] In the intervening year Lesueur had reported on his experiments to Maclure, then traveling in Europe: "I may tell you that all of our experiments in lithography have not been very successful in gratifying our desires, that printing provides an infinity of difficulties which we have not yet been able to overcome."[11]

Meanwhile, a brief notice and description of the process and lithographic illustrations of geological subjects appeared in Benjamin Silliman's *The American Journal of Science and Arts*, indicating the continuing interest in the

FIGURE
2.6

50

Bullet. Sc. Phil. Mai 1815 : Pl. I.

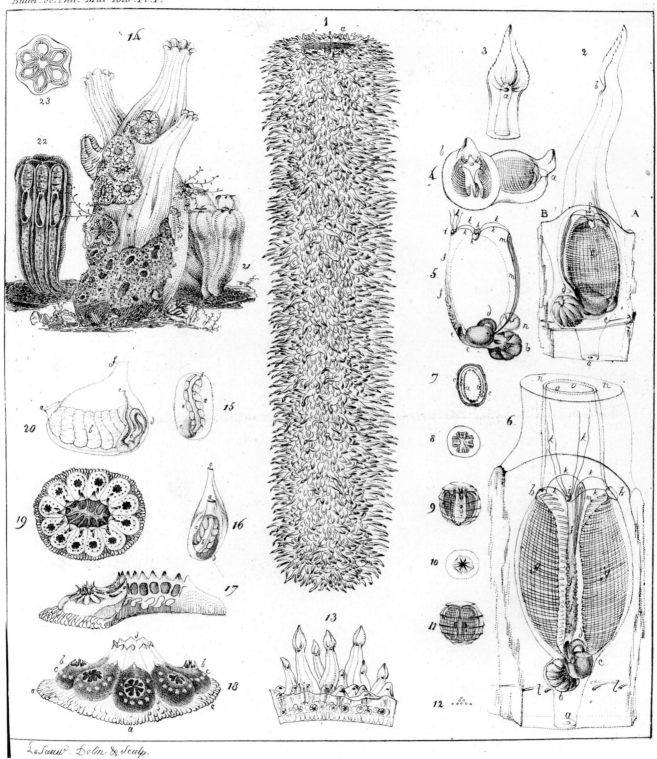

Lesueur. Delin. & Sculp.

2.1b  Pyrosomes. Engraving by C. A. Lesueur from *Bulletin des Sciences*, Paris (1815).

process among the scientific community. Silliman wrote of the lithographic process that "the finest things done in this way are really very beautiful: and they possess a softness which is peculiarly their own. Still Lithography is not *a rival*, it is merely *an auxiliary* to copper plate engraving, which, especially in the higher branches of the art, must still retain the preeminence which it possesses."[12] Despite his strong interest in lithography, Silliman, like others of his contemporaries, could not imagine the new medium displacing a printing technique that had been refined for centuries and whose dominance formed expectations of how a printed picture should look. Moreover, the initial secrecy surrounding lithographic technique meant that experimenters often achieved only crude results.

Lesueur continued to experiment, writing to France for advice on technique. In 1823 he wrote to his former coauthor, zoologist Anselme-Gaetan Demarest:

> If a good lithographer would come to Philadelphia I believe he could prosper by the second year of going into business. Our naturalists feel the need and the importance daily of having such an establishment here, seeing that copper engraving is too expensive and too time consuming. The beautiful impressions which issue from the presses of France are a very strong stimulus for increasing the desire to have something similar.[13]

Lacking an experienced lithographer close at hand, however, Lesueur returned to copper engraving. So did Silliman's journal until 1829, when the Boston firm of Pendleton began print-

2.2–5 Plates engraved by Lesueur from the *Journal of the Academy of Natural Sciences, Philadelphia* 1 and 2 (1817–1822).

2.2 Species of marine mollusc.

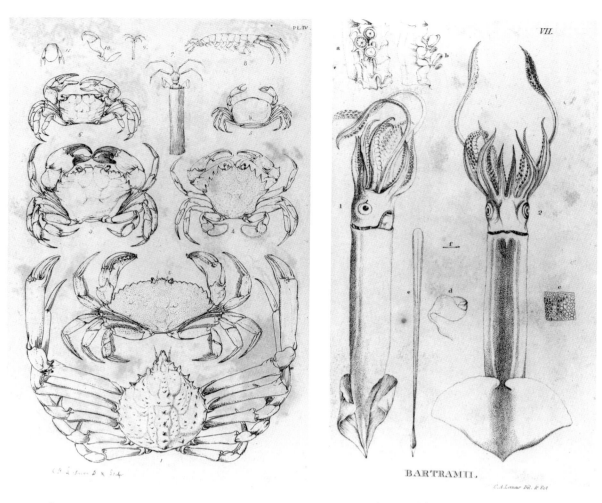

2.3 Crabs.

2.5 Squid named for William Bartram.

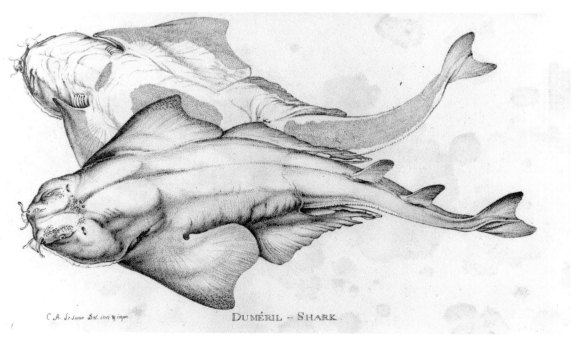

2.4 Shark.

CICHLA. AENEA.                    C.A. Lesueur del:

2.6 *Cichla aenea*. Lithograph by C. A. Lesueur from the *Journal of the Academy of Natural Sciences, Philadelphia* 2 (1822).

ing scientific subjects in lithography. The Academy members continued to use copper engravings to illustrate their books as well as their articles in scientific journals. Lithography could not yet offer the savings that Maclure envisioned would make books available to all at lower cost.

Despite the high cost of engraved illustrations, many of the Academy members sought to follow Wilson in reaching a varied public for their work in natural history. Wilson had been stung by suggestions that the high cost of his illustrated work priced it for an elite readership. He had believed that any work that promoted an appreciation and knowledge of nature contributed to the progress of the nation, and toward that end he had fashioned his text to entertain as well as instruct. And at least for the moment, Philadelphia trade publishers were still willing to undertake scientific books with reasonable expectations for popular sales.[14] The Academy had elected Wilson a member

shortly before his death, and as if taking the *American Ornithology* as the first chapter of the great American volume of natural history, each Academy member undertook the description of an animal group. Say published on insects. Harlan described fossil vertebrates. John Davidson Godman (1794–1830), a physician, wrote about living North American mammals, while Lesueur studied marine invertebrates and fishes, and Maclure surveyed single-handedly the geology of eastern North America. Their debt to Wilson, their commitment to establishing an independent American science, showed clearly in the titles of their books: Thomas Say's *American Entomology*; Lesueur's projected *American Ichthyology*; Godman's *American Natural History*. George Ord (1781–1866) undertook, but never completed, an illustrated work on mammals. He owed his own place in the Philadelphia natural history circles to his part in seeing through publication Wilson's eighth and posthumous volume and

54

his publication of the ninth, largely a biography of the late ornithologist. While for these endeavors the *American Ornithology* established the model, it provided the initial volumes for Charles Lucien Bonaparte's taxonomy of American birds. Bonaparte (1803–1857), a young nephew of Napoleon's, arrived in Philadelphia in 1819. In acknowledgment of Wilson's foundation, Bonaparte entitled his book *American Ornithology* and engaged Lawson to engrave the plates.

When Thomas Say wrote the 1817 prospectus of his *American Entomology,* he seems to have assumed that a general readership would take an interest in the work, for he started with a warning. His personal point of view would intrude as little as possible. He would confine to the preface his account of his lifelong fascination and delight in insects, appreciation of their beauty, and struggles to study them without the established resources of science—all points that Wilson had made about his efforts to study birds. In the book's chapters, each devoted to an insect genus, Say adhered strictly to taxonomic description, with the narrative restricted to the sections labeled "Observation." While Wilson's illustrations had reflected a text that combined description and narrative, the illustrations in Say's works showed clearly how differently the meaning of "Drawn from Nature" could be interpreted and still satisfy the basic requirements of the credo: firsthand observation and faithful representation.

Say focused on taxonomy but he could scarcely be characterized a "closet naturalist." Although he became curator of the Academy's collections and library, he also journeyed to observe and collect insects, both with colleagues from home and as a member of the government explorations of the west that followed the Lewis and Clark expedition of 1804.[15] But much of Say's correspondence with fellow entomologists focused on the exchange and correct identification of specimens. In letters to John F. Melsheimer, a Pennsylvania entomologist and son of Frederick Melsheimer, also an

entomologist, Say wrote as one specialist to another:

1—Your No 51 is the same with 45 (Ammon) that your father sent me it does not agree with the description of that Insect, "Thorax 3-toothed &c"—with the description of *Janus* I am unacquainted it may probably be that insect—the thorax is certainly unarmed . . .

2—Anobium Pertinax differs I find sometimes very much in its depth of colouring, though always brown, yet it is sometimes of a much lighter tint with the eye perfectly black—this may be a sexual difference, my lighter one is I believe a male.[16]

The years between the prospectus and the first volume brought Say considerable experience as a naturalist, especially in the field, and also considerable institutional recognition. He traveled as a member of the scientific corps on government exploring expeditions, one to the Rocky Mountains and another to the Great Lakes. Between 1821 and 1825 he lectured at the Peale museum in the capacity of professor of zoology. From 1821 until 1827 he held the office of curator of the American Philosophical Society, and in 1822 he became professor of natural history at the University of Pennsylvania, both positions, however, providing little income.[17] Neither did he expect to earn much on his book.

Between writing the 1817 prospectus and incorporating its plates into the work's first volume in 1824, Say took an explicit step away from a general readership. In the 1824 preface he addressed, in particular "those whose information is sufficiently comprehensive to enable them duly to appreciate the various departments of human knowledge . . . ; and the author would happily profit by their friendly cooperation in the correction of any errors that may appear, in the enunciation of new facts in the manners and economy of insects, or in the addition of species and localities."[18] While "patrons of science" were Say's primary intended audience, he also fashioned the work as an introduction to the study of insects, for he included a glossary of entomological terms to bridge the gap between initiate and beginner.

The engraved title page designed by Lesueur, alone of the book's illustrations, reflected the more autobiographical preface of the 1817 prospectus. Lesueur's vignette portrayed a wealth of insect activity, including butterflies, ants dragging away a dead beetle, two tumblebugs rolling dung balls in which to lay their eggs, bees hovering outside their nest in a hollow tree on which crawled a termite, and from whose branches hung an orb web in which the spider had caught insect prey. This busy scene was set before a vista more urban than Wilson's landscape backgrounds. Lesueur depicted steamboat traffic on a river with a busy port, possibly Philadelphia, seen through a forest of ships' masts on the far bank. While Say modeled his text on Donovan's *British Insects*, Lesueur's design for both the decorative title page and the front-cover engraving echoed eighteenth-century European conventions in entomological illustration. The cover engraving includes two putti with wings, in allegorical association with Psyche, or butterfly, the classical reference for the study of entomology. The title page resembles a similarly busy scene from a multivolumed natural history of the Bible published in Amsterdam in the 1730s. But although the title page vignette recalls European decorative and allegorical devices, its distant urban and commercial landscape grounds its self-consciously American natural history in an explicitly social and material context.

FIGURES<br>2.7–8 The text of Say's *American Entomology* presented insect descriptions of which the illustrations were a direct mirror, reflecting character by character Say's taxonomic analysis.[19] Say omitted numbers on the text's pages so that the owner of the completed work could arrange the chapters in correct taxonomic order for binding.[20] He himself did not draw. The plates, drawn by Titian Peale, who specialized in the study of butterflies, and Lesueur, among others, showed almost every species with the sprig

2.7 Insect activity in a landscape. Engraved title page designed by C. A. Lesueur for Thomas Say, *American Entomology*, vol. 1 (1824).

of a plant, often in reference to food source or habitat.[21] For Say, "Drawn from Nature" seems to have meant an organism observed and depicted in words and pictures as clearly as possible without elaborations. Discarding Wilson's narrative landscape backgrounds, Say's illustrations acknowledged that scientific observation selected, isolated, and transposed the organism out of nature.

The entomologist strove for impersonality, adopting pared-down prose and pictures to conform as closely as possible to the physical fact

TAB. XII.

GENESIS Cap.I v. 20.
Opus quintæ Diei.

I. Buch Mosts Cap.I.v.20.
Fünffter Tagwerck.

2.8 Lesueur's design for Say's *American Entomology* recalls the illustration of the fifth day of creation, engraved by I. A. Covinus, from J. J. Scheuchzer, *Physique sacrée, ou histoire naturelle de la Bible*, vol. 1 (1732–1737).

American naturalists were discovering. Faithful representation offered the most secure and stable contribution to science under conditions of changing taxonomic criteria.[23] Say's comments concerning his illustration emphasized their fulfillment of taxonomic criteria for the appreciation of specialists: "The graphic execution of the work will exhibit the present state of the arts in this country, as applied to this particular department of natural science, as no attention will be wanting, in this respect, to render the work worthy of the encouragement of the few who have devoted a portion of their attention to animated nature."[24]

Wilson had used stories of his encounters with birds and the behavior of those he kept as pets to attest to his firsthand observation, as well as to place the author in the text in a manner that Say did not. Wilson had also burst into extended lyric verse when mere prose failed to convey his feeling for nature. Say confined verse to his title page. He further minimized the relationship between observer and organism by arranging his chapters into distinct sections. Under *Description* Say presented only physical description of the insect: "Body above black, with a yellow line each side, passing over the origin of the wings, and over the head each side before the eyes," began his description of *Papilio turnus*.[25] He inserted himself only in the sections labeled *Observations*. Wilson's chapters had come to follow a certain pattern, with description and anecdote at the beginning and measurement at the end, but he had never formalized the sequence under separate headings. In contrast, Say seems to have believed that the presence of the naturalist should not intrude in scientific writing. The author appeared in the limited roles of collector and entomologist. Physical detail of the insect alone, translated into text and illustrations, had to authenticate his observation.[26]

Say's avoidance of lyricism and narrative in his text was balanced by special attention to visual elegance. The illustrations brought the book to public attention. The luminous quality of the hand-colored engravings of the *American*

of the insect. "The author's design, in the present work, is to exemplify the genera and species of the insects of the United States, by means of coloured engravings."[22] Say's emphasis on verbal and pictorial reconstruction of the insect compensated for the failure of existing taxonomic systems to accommodate in an orderly fashion the new material that he and other

*Entomology* did much to compensate for the specialized language of the descriptive text and to broaden the work's readership. The quality of the engravings derived in part from the technique of the principal engraver, Cornelius Tiebout (1777–1832).[27] Tiebout specialized in stipple engraving which, like mezzotinting but without the acid etch, employed a rocker or roulette to create areas of tiny dots. When the first volume finally appeared, the *North American Review*, an influential magazine enthusiastic in promoting American letters, published a notice. The reviewer considered it superior to Wilson's *Ornithology* "in the exquisite delicacy of the drawings and beauty of the engraving, as well as the marks of taste indicated in the external attractions of the volume"; the covers were a rose color, and the binding a pale green:

> For beauty and elegance of execution, this work surpasses any other that has been printed in this country. . . . The fanciful and highly emblematic frontispiece was delineated by C. A. LeSueur; the figures of the insects were drawn from nature by T. R. Peale, and engraved by C. Tiebout. The specimens of their labours here furnished are in the highest degree creditable to these artists. The work, as it has come from their hands, affords the most encouraging testimony of the state of the arts in this country, and as such deserves the patronage not more of lovers of science, than of all persons who are disposed to advance the progress of liberal pursuits by rewarding the successful efforts of genius and skill.[28]

As the notice in the *North American Review* attested, an important segment of critical opinion believed in the patronage of natural history by a nonspecialist readership and judged a scientific work on the criterion of elegance. Such expectations helped broaden the audience even for works addressed primarily to specialists. The *North American Review* notice of Say's *Entomology* indicated that promoters of American culture continued to cherish the idea that current advances in science were equally contributions to the nation's developing arts and letters, and should be supported by all who encouraged their progress. The interest of the general reading public in informal accounts of American fauna, flora, and landscape was, if anything, becoming stronger. Wilson's writing had contributed to a taste for the narrative nature essay. An audience for popular natural history flourished, especially in Philadelphia, where an English visitor to the city in the early 1830s would comment with condescension: "In Philadelphia it is the fashion to be scientific, and the young ladies occasionally display the *bas bleu*, in a degree that in other cities would be considered rather alarming."[29] There, an interest in natural history could be pursued in the most congenial settings. Members of the American Philosophical Society continued to hold the soirées initiated during the teens of the century by Dr. Caspar Wistar, where members and guests, often visiting notables, gathered to discuss science and literature and eat supper.[30]

At the same time, as naturalists themselves sharpened their criteria for description and turned increasingly toward a more technical prose and specialized graphic repertoire, they correspondingly restricted their readership. The literature that met popular demand was increasingly shaped by its position outside or on the margins of the consolidating discipline of natural history, even when it was produced by naturalists. Until the distinctions between specialist and general natural history literature became generally acknowledged, however, the meanings conveyed by conventions of narrative and description in natural history prose and illustration remained ambiguous. Narrative elements in natural history illustration did not always mean personal observation of the living animal in nature but could be deployed to lend the authenticity of firsthand observation to secondhand work. The dichotomy between prose and pictorial narrative as a convention and narrative as testimony to personal observation revealed itself in the differences among several publications directed at the growing audience for popular natural history.

In terms of prose style and intended audience, one of Wilson's inheritors in the next generation was John Godman, a physician, naturalist, and Academy member who had married into the Peale family. Godman's book, *American Natural History*, the first volume appearing in 1826, combined the scientific sophistication of his Academy colleagues with an urbane essay style.[31] Godman followed the precepts of Cuvierian classification of mammals, he included in his book discoveries brought back by fellow Academy members from the latest government explorations from beyond the Mississippi, and he discussed the definition of species, a topic emerging as an important issue in American natural history.[32] But a crucial difference distinguished Godman's book from, for example, Say's *American Entomology*. For the most part, Godman reported on the observations of others. His illustrations, largely drawn from the Peale exhibits, reflect the secondhand character of the work. Nevertheless, or perhaps because the material was familiar to local readers, the illustrations insured that Godman's book received critical notice and reached a large readership.[33]

Lesueur was one of several illustrators whose work appeared in Godman's volumes.[34] The plates drawn by Lesueur were doubly secondhand. Originally, he had drawn from Peale's specimens the black bear, wolf, coyote, fox, and raccoon for George Ord's study of North American mammals, which Ord never published.[35] In this work, Lesueur's "magic pencil" failed to animate his drawings and transcribed only what he saw: the quadrupeds in the Godman plates convey more taxidermy than life.[36] A stiffness about the feet and legs betrayed that Lesueur, like the other illustrators for the book, had not observed the animals in motion.[37] Various props from the Peale exhibit backgrounds substituted for living nature. The specimens grazed, or lay outside their lairs. Each animal or animal group stood on a detached piece of ground, which, like the stumps and perches in ornithological illustration, had been conventional for centuries. The floating clods of earth hovered, sometimes several to a page, in a horizonless space. The tiers of little landscape settings, each bearing its own mammal, recalled popular editions of Buffon such as Goldsmith's *Animated Nature*, in which Buffon's original full-page illustrations, abbreviated and reduced, were ranged one above the other. In such works as Godman's—an Americanized Goldsmith, which in turn was an Anglicized Buffon—the conventional landscape setting substituted for firsthand observation.

FIGURES
2.9–11

The increasing formalization of projects that defined themselves as scientific defined by contrast a genre of explictly popular natural history. Readers and authors who considered taxonomic description dull and dry could invoke the ever-popular Buffon to add respectability to a more palatable narrative approach, rich accounts of life histories and habits, often of exotic animals and their strange behaviors. Dr. Reynell Coates of Philadelphia complained in 1833: "It is much to be regretted that many minds capable of enjoying . . . those pleasures that may be drawn from . . . natural history, are arrested on the threshold by the dry and technical systems . . . which are generally regarded as the science itself."[38] Coates liked to dwell on the "beauties and manners" of animals, in which he saw reflections of human behavior and moral values. Describing a "desperate conflict" between fish and crabs, he concluded: "What moral might the observer extract from the high daring and noble prowess of these little aquatics. . . . What exquisite similes might be drawn from such a fertile source to embellish the pages of history."[39] For Coates, natural history remained an embellishment of human history.

While Godman had not claimed that his illustrations were drawn directly from nature, another popular publication, which listed its illustrations under the heading "Embellishments," offered the public pictures supposedly drawn from nature. Doughty's *Cabinet of Natural History and American Rural Sports*, a popular monthly magazine that began publication in Philadelphia in 1830, included natural history articles and pictures of animals in the context of the magazine's principal focus, hunting and sport.[40] The magazine editors stated their purpose clearly; first to amuse and incidentally to instruct. The plates, almost all of them lithographs, resembled illustrations like Wilson's only in that the animals occupied the center foreground of the picture. Otherwise they acted out scenes from a scrapbook of sentimental views. Swans glided through flowering iris; birds sang from berry-laden boughs; ducks paddled in a tranquil millpond. Many of the plates depicted hunting and featured pointing spaniels, unsuspecting quail, or dramatic scenes of wild horses frenzied by a roundup.

In this context the publishers of the *Cabinet* reproduced a drawing by Titian Peale from the 1819 government expedition to the Rocky Mountains led by Major Stephen Long, on which Peale had traveled officially as naturalist and draftsman. Peale's original sketch, one of a series illustrating the life of the Plains Indians, pictured an Indian mounted on a galloping horse, his arrow pointed at a charging buffalo. It is difficult for a twentieth-century viewer to attribute originality to Peale's picture because such scenes, initially popularized in vehicles such as this magazine, became shorthand for

2.9–11 Wolves, Raccoon and Badger, and Bears. Drawn by C. A. Lesueur, engraved by F. Kearny and G. B. Ellis for Godman, *American Natural History*, vols. 1 and 2 (1826).

2.9 Wolves.

2.10 Raccoon and Badger.

2.11 Bears.

frontier adventure during the decades of westward migration. The Long expedition, however, was the first government exploration to follow Lewis and Clark in mapping the trans-Mississippian west. Sponsored by the War Department, the mission was charged specifically with evaluating and reporting on the region's prospects for settlement.[41] Peale's pictures, with those of Samuel Seymour, the official landscape painter of the expedition, were the first published pictures of the Rocky Mountains but they appeared in different venues. Seymour's drawings were published in the atlas of the official report. Like the Indian vocabularies transcribed by Say, Peale's sketches and drawings were part of the official natural history documentation of the expedition. Peale's own attitudes toward his subject matter were

61

FIGURE

2.12

unlikely to have approached twentieth-century norms of ethnography; his preconceptions and probable romanticizations of the frontier must surely have informed his treatment of an Indian buffalo hunt.[42] Nevertheless, the popular context in which Peale's picture was reproduced emphasized its dramatic, exotic character. Its documentary intent would have been foremost in the official report of the expedition, a book that became popular with Americans eager for factual information about the territories they intended to settle. That report, published in 1822, did not reproduce Peale's buffalo hunt picture.[43]

The natural history of the Long expedition was published in several contexts, their variety demonstrating the variety of public expectations of natural history. Say had accompanied the expedition as one of the official naturalists. His observations and discoveries appeared as footnotes in the official report.[44] Descriptions of some species were published in journals like the Academy's. Godman used Long expedition material in his *American Natural History*. In each of these publications, nature meant something different to the intended readership. Potential settlers and speculators, concerned about soil, weather, and hostile or benign con-

AMERICAN BUFFALOE

2.12 American Buffalo. Lithograph, on stone, by M.E.D. Brown, from a drawing by Titian R. Peale for Doughty, *Cabinet of Natural History and American Rural Sports* (1832).

ditions, thought of nature as land or profit. Readers of scientific journals received new species into the literature and added them to the corps of organisms comprising the Natural System. Such readers considered nature a domain of systematic knowledge. Godman's audience might well have been gratified to have access to the latest information from the west presented in so accessible and gracious a medium. Many of his readers considered natural history an entertainment, and nature appreciation a sophisticated attainment. Working on their respective and often overlapping audiences, all these vehicles whetted the public appetite for more natural history and contributed to its patriotic sanction as the description of a national, collective possession.

Illustrated natural history texts, then, made up a continuous spectrum, but the claims of one end of the spectrum tended to undermine the claims of the other. A natural history of taxonomic description like Say's coexisted with a sentimental natural history like Doughty's, both claiming that their illustrations were "Drawn from Nature." The application of the claim to both genres confused its meaning and eroded its authority. Even fellow members of the Academy, Say and Godman, used pictures for very different purposes. Say relied on illustration to establish genera and species, and Godman, to sell books and contribute to the cultivation of the general reader. Freighted with these different tasks, the primacy of illustration that Wilson had worked to establish was weakened; so, accordingly, was the authority of the artist or illustrator. To reassert the authority of observation in prose and illustration, the specialist natural history community began to insist that publication context and intended readership validated content. Increasingly, the only context that counted to establish scientific authority in illustration would be the taxonomic description, and that counted only if published in the specialist journal or monograph directed at a specialist audience. From within the nascent institutions, despite

the fact that the growing memberships of local natural history societies consisted mostly of amateurs, the few specialist members increasingly set the standard style of description and illustration.[45]

At the same time, rising costs for commissioning original illustrations reduced the willingness of trade publishers to undertake illustrated scientific books like Say's or Bonaparte's, both of which emphasized classification. Publishers preferred popularly oriented natural histories, often with plates traced or copied from other sources,[46] or with pictures like the rural scenes of Thomas Bewick's wood engravings or the sentimental views of Doughty's *Cabinet*, depicting animals in a human landscape. In the coming decades, reduced reproductions of Buffon's and even Wilson's illustrations, flipping orientation from left to right in each successive reengraving, betrayed the tracing of a previous edition. These figures, from which generations of children learned about birds and mammals, along with stylized vignettes of country life, became the hallmarks of popular natural history and of many school texts.

Motivation to incorporate natural history into school curricula came from almost as many directions as there were audiences for nature study. While on the one hand the establishment and growing prestige of a specialist discipline influenced legislatures and educators, on the other hand the popularity of natural history among a general middle-class readership persuaded the consumers of education of the value and interest of natural history.[47] Later in the century, illustrations of school texts would occupy an important intermediate position between specialized and popular natural history. By that time, the role of illustrations in pedagogy had become firmly established. William Maclure, however, was not only among the first educational theorists to introduce natural history into children's schooling, but he gave illustration a central place in his pedagogy for social change.

For Maclure and his associates, the commitment to popular natural history took a form very different from Coates's embellished histories or publishers' reissues of eighteenth-century classics. In early 1826 Maclure, Say, Lesueur, and other members of the Academy circle joined Robert Owen (1771–1858), the Scottish industrialist and reformer, in his experimental community at New Harmony, Indiana.[48] Maclure's scientific philosophy found its way into American life through gradual reforms in education, beginning at the school he founded there on the principles of utility and direct experience.

Maclure was the organizer and spokesman for the group he led westward in 1825. For a number of years, Maclure had studied and espoused the educational theories and visited the continental schools of the Swiss educator Johann Heinrich Pestalozzi (1746–1827) and of Robert Owen himself in New Lanark, Scotland. Pestalozzi's system emphasized teaching in accordance with the learner's psychological development and the development of reason through educating the senses. He and his followers rejected corporal punishment and advocated "benevolent superintendence," the cultivation of freedom of choice and self-discipline.[49] One of Pestalozzi's biographers summarized the maxims that laid the foundations of modern pedagogy: "Instruction must be based on the learner's own experience. . . . What the learner experiences must be connected with language. . . . At each point, the instructor shall not go forward till that part of the subject has become the proper intellectual possession of the learner. . . . With knowledge must come power, with information, skill."[50] Robert Owen arrived independently at a similar educational philosophy, but differed from Pestalozzi in rejecting religion as the foundation of moral development. Owen, whose business association with Jeremy Bentham had enabled him to purchase the mills at New Lanark, advanced a theory of pedagogy based on a utilitarian, rational philosophy of social equality.

The formation of the adult, he insisted, must begin before the age at which children usually reached school, for by then, no matter what principles the school espoused, it could not counteract "habits" of mind and behavior learned from the parents. Owen established the first infants' school at New Lanark, where preschool instruction took the form of play and exposure to a variety of experiences rather than the imparting of book knowledge.[51] The children in the higher school studied without texts; their teachers relied instead on the intrinsic interest of useful subjects, including natural history, to develop the powers of the mind rather than fill it with facts.[52]

Maclure similarly rejected rote religious and classical instruction, reliance on punishment to enforce discipline, and unequal education for women. He emphasized the acquisition of knowledge and development of reason through direct observation and through learning useful manual skills. Even the most abstract concepts, Maclure believed, could be presented in simple concrete ways for the comprehension of young children. Maclure's approach to the formation of character through education was more utilitarian than Owen's. They agreed, however, on the placement of the natural sciences at the heart of their pedagogy. Their shared political vision of "raising" the laboring classes to a readiness for participation in social reform was also based on a "scientific" rationale. Science was the model for the apprehension of truth through the senses and the development of reason. Maclure's educational philanthropy was intimately linked with his support of scientific institutions.[53]

Over the course of his travels, Maclure had brought back from Europe a group of teachers—Joseph Neef, Madame Marie Duclos Fretageot,

*On the following pages:*
Plates 9–13. Hand-colored stipple engravings by Cornelius Tiebout from drawings by T. R. Peale, W. W. Wood, and C. A. Lesueur for Thomas Say, *American Entomology*, vols. 2 and 3 (1825–1828).

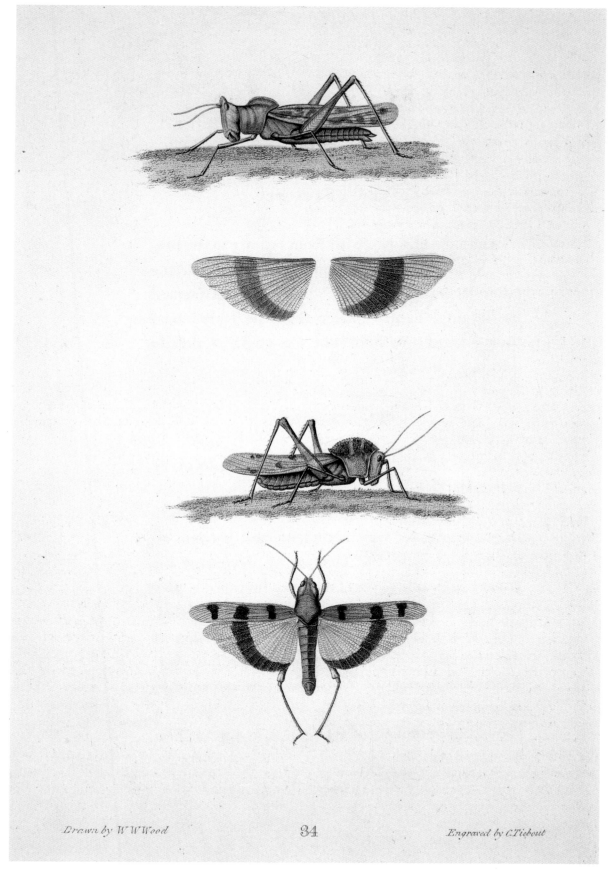

Drawn by W W Wood       34       Engraved by C.Tiebout

Plate 9. Gryllus species.

and Phiquepal d'Arusmont—to introduce his modified Pestalozzian method and administer and staff the schools he founded in the Philadelphia area. Neef, at his school on the Schuylkill River outside Philadelphia, attempted to put into practice his principles of self-government as preparation for adult citizenship.[54] Public complaint against Neef's professed atheism forced him to close the school in 1814. Fretageot's boarding academy for women, however, continued successfully.[55]

At the time that Owen was formulating his plan for the establishment of an American outpost, or "Preliminary Society," of a new social order, Maclure suffered a severe disappointment. The reversal of the liberal constitution in Spain forced him to abandon plans to establish an agricultural and industrial school in Alicante. After the confiscation of Maclure's Spanish properties in 1824, he left Spain and visited New Lanark, where he met Owen. From his inspection of Owen's system, Maclure took fresh inspiration for establishing industrial schools in the United States.[56]

In 1825 Owen carried his mission to the United States, where he believed that the constitutional separation of church and state would permit the foundation of the "New Moral World." He negotiated to purchase the town of Harmony, Indiana, from its founders, Father George Rapp and his religious community, originally from Württemberg, Germany.[57] Owen traveled throughout the eastern states, lecturing on his vision of a rational society and exhibiting the model he had constructed of his ideal city.[58] He made presentations to Thomas Jefferson, in retirement at Monticello, and in Washington, to both the incoming and outgoing presidents, and to the House of Representatives. Notices published in newspapers around the country attracted the attention and interest of a wide public. After Owen visited Philadelphia to publicize his program and exhibit his model, Maclure, still traveling in Europe, received enthusiastic reports about the New Harmony proposal from his associates, especially Mme Fretageot.[59]

2.13  "Experience directing Youth to the contemplation of the works of Art and Nature." Engraving by Alexander Lawson, [1825?].

The community that Owen envisioned appeared to offer a fresh opportunity for educational experiment, and Maclure returned to Philadelphia. He bought partnership in Owen's venture and negotiated for supervision of the educational department. While he was moving to New Harmony, Maclure published his educational theories and proposed curriculum. Printed in Silliman's *American Journal of Science and the Arts* in January 1826, Maclure's statment reached an audience of educated, scientifically inclined Americans.[60] Parents who themselves might be unprepared to join the community nevertheless could send their children to New Harmony to board at the school from the age of two. While boarders paid tui-

FIGURES
2.13–14

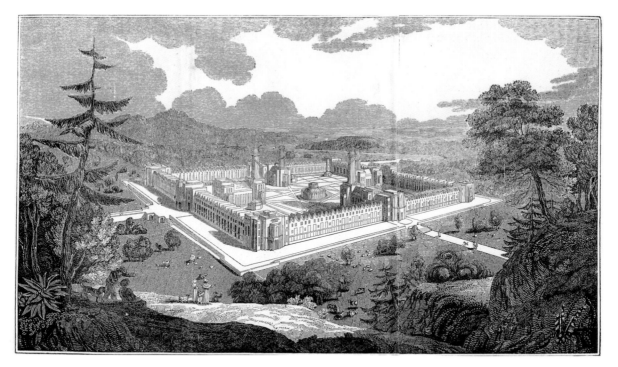

2.14 Lawson may have attended one of Owen's Philadelphia presentations; the structure in the middle ground of his engraving resembles Owen's model for his proposed community, depicted in the frontispiece of *The London Co-operative Magazine and Monthly Herald* (January 1826).

tion, children of the community attended without charge, and the curriculum included training in trades by which, Maclure envisioned, the school would be self-supporting. Describing the educational philosophy and program in the *Journal*, Maclure wrote: "The great and fundamental principle is, never attempt to teach children what they do not comprehend, and teach them in the exact ratio of their understanding, without omitting one link in the chain of ratiocination, proceeding always from the known to the unknown . . . summing up the results free from prejudices, and cautiously avoiding delusions of the imagination."[61]

Maclure's description of his teaching method in many ways recalls the scientific method that Alexander Wilson and other Philadelphia naturalists had advocated as the American approach to natural history. For children, Maclure proposed "a careful examination and inspection of the objects themselves, or of tangible and visi-

ble instruments, calculated to demonstrate their properties and bring them within reach of the senses." Only when lacking the originals should teachers substitute "accurate designs, or representations." Maclure further specified that "natural history in all its branches, is learned by examining the objects in substance, or accurate representations of them, in designs and prints."[62] In a later essay, Maclure elaborated on the relationship between natural history and illustration:

Representation is the only defined language, and is perhaps equal in value and utility to all the languages together; without it, we can have no correct idea of mechanics or natural history; when the objects themselves are absent, descriptions, from the undefined nature of words, must be equally vague and uncertain . . . let any one try to represent to himself a machine, animal, plant, mineral . . . from a description, and he will find

37

Plate 10. *Spectrum femoratum.*

40

Plate 11. *Papilio turnus.*

the great difficulty and in most cases the impossibility of imprinting a correct figure in his mind, without which he cannot have a correct idea of the thing thought of.[63]

The complementary relationship between Say's observations and descriptions and Lesueur's illustrations provided evidence of the practicality of Maclure's proposals. Maclure's educational plan combined Pestalozzian pedagogy with Philadelphia natural history practice, and to insure success Maclure persuaded Say and Lesueur to join him at New Harmony and teach in the school.

At Maclure's school, not only would pictures serve as objects of study second only to specimens and direct experience, but children would learn to draw with facility. "They are taught the elements of writing and designing by the freedom of hand, acquired by constant practice." Maclure included drawing in the curriculum because he believed it was a means of achieving objective understanding. At first, children would "draw right lines, dividing them into equal parts, thereby obtaining an accuracy of eye . . . on which the accurate knowledge of the properties of every species of matter depends."[64] Drawing and music were part of the curriculum because they contributed to useful knowledge and physical skill. "The art of drawing or delineation, which has been placed [(]because its utility was not well understood) amongst the fine arts, must be ranked amongst those which are useful."[65] Maclure wrote that "education ought to be the apprenticeship of life, and children ought to be taught what imperious necessity may force them to practice when men, always preferring the useful to the ornamental." He vehemently opposed the merely decorative arts, associating them with established religion. In his later writings, Maclure described the superior performance in the trades by children who had been educated under the Pestalozzian method.[66]

Maclure's curriculum linked the practice of drawing with the useful skills of printing. "The

boys learn at least one mechanical art; for instance, to set types and print; and for this purpose there is a printing press in each school, by the aid of which are published all their elementary books," although Maclure believed that texts should be kept to a minimum.[67] Because illustrations ranked so high in Maclure's pedagogical system, the skills of printmaking would also be taught:

Lithography being the best, cheapest, and easiest mode of making accurate representations of every thing, and this system requiring so great a number of exact representations, as they are in all possible cases substituted for books or descriptions, the pupils are taught how to design on the stone or cartoons and how to make the proper ink and pencils, as well as all the manipulation of printing and working the press, &c. &c.[68]

Maclure advised: "Obtain a knowledge of the objects of nature and art, and an early habit of receiving pleasure from the examination of them."[69] For Say and Lesueur, who had collaborated in mutually corroborative text and illustrations, Maclure's approach to the acquisition and communication of knowledge was congenial. In the group that Maclure assembled to staff the educational department at New Harmony, the naturalist and illustrator would play leading roles.

From December 1825 through January 1826 Maclure and his party navigated the riverways to New Harmony on a keel boat christened the *Philanthropist*. Besides Say and Lesueur the diverse group included Mme Fretageot, other teachers and students from Maclure's Philadelphia schools, and John Speakman, a friend of Say's and co-founder of the Academy. Others who joined them at New Harmony were Gerard Troost, the Academy geologist, and the engraver Cornelius Tiebout and his children. Along the route the naturalists collected and sketched fish and shells and decorated the cabin with specimens and drawings.[70] The harmony among the passengers during the journey augured well for their new way of life. Robert

Owen, on one of his rare visits to the community, greeted the arrival of the "Boatload of Knowledge" with ceremony and speeches.[71]

Celebration of the party's arrival, however, marked one of the last moments of unity in the struggling community. The New Harmonists had framed a series of constitutions that included common property, community work, group living, and full membership regardless of color or sex. Owen had announced in April 1825 the introduction of "an entire new state of society; to change it from an ignorant, selfish system to an enlightened social system which shall gradually unite all interests into one, and remove all causes for contest between individuals."[72] But the workshops, gardens, and vineyards established by the founders of the town soon fell into ruin. The rigors of frontier life, overcrowding, and the burden of noncontributing opportunists created dissatisfaction and dissension. Some of the disaffected left the parent community to found separate towns, or societies, one named Macluria, another Feiba Peveli, a name based on the assignment of vowels to latitude and consonants to longitude.[73]

Even for those who remained, the ideals of equality proved difficult to put into practice. While most members of the society had to pay for membership or relinquish livestock toward credit for living expenses, the scientific contingent were to receive salaries for their contributions to the community.[74] A foreign aristocrat visiting New Harmony in 1826, Karl Bernhard, Duke of Saxe-Weimar, noted with patent satisfaction: "In spite of the principles of equality which they recognize, it shocks the feelings of people of education, to live on the same footing with every one indiscriminately, and eat with them at the same table." At the school house, "where all the young ladies and gentlemen of *quality* assembled," he observed that despite "the equality so much recommended, this class of persons will not mix with the common sort, and I believe that all the well brought up members are disgusted, and will soon abandon the society."[75]

Maclure saw the splintering of the community as a threat to the success of his school. Owen resisted any suggestions for changing the structure of the community, especially Maclure's proposal for an autonomous educational department. Finally, however, Owen approved a plan to reorganize New Harmony into three units or societies—mechanical, agricultural, and educational—each to lease space from the whole. Having achieved the new arrangement, Maclure considered his partnership with Owen dissolved, and in the ensuing dispute over financial obligation all vestige of the original idea of community property vanished.[76] Owen himself finally abandoned New Harmony in mid-1828. In his farewell speech, he blamed the failure of his society to realize his ideals on the disunity of purpose among the teachers in the schools.[77] Maclure still directed the educational department but moved to Mexico for reasons of health. The teachers remained, however, and the schools and their educational philosophy continued to distinguish New Harmony from other frontier towns.[78]

Thomas Say stayed on, acting as Maclure's agent and teaching as much by example as by instruction. Participating fully in community work, Say gardened, taught without pay, and continued his scientific research. He chose to dress in the special costume, a blousey overall designed for men and women. Say married one of Mme Fretageot's students, Lucy Way Sistaire (1800–1886). He cultivated in his garden the new plants he collected nearby and on exploring excursions with Maclure. Say also edited the town paper, *The Disseminator of Useful Knowledge*, printed by the students as part of their practical training, and there he announced the publication of his continuing natural history.[79] He planned to issue his study of shells, *American Conchology*, in numbers, each with illustrations drawn from nature by Lucy Say and Lesueur and engraved by Tiebout, along with the "necessary letterpress for the description of species represented."[80]

Lucy Say studied drawing with Lesueur and

PLATES
14–16

71

Plate 12. Pompilus species.

Plate 13.  Malachius species.

drew sixty-six of the sixty-eight plates for the *Conchology*.[81] When she had on hand sufficient supplies of paper, colors, and brushes, she also colored thousands of plates, working with Caroline and Henry Tiebout, the children of the engraver. Thomas Say wrote to Mme Fretageot in Paris that Lucy Say had "colored five thousand plates of the Conchology since you left us, more than double the number she ever did before in the same time."[82] As the student population thinned, in part owing to Maclure's incessant reorganization of the school, it seems that more of the coloring fell to Lucy Say. While Caroline Tiebout made two dollars a week for coloring, it is probable that the author's wife worked without pay.[83] On the death of Cornelius Tiebout and the other engraver on the project, Lucy Say aspired to learn engraving to carry on the work.[84]

The object of Thomas Say's work was "to fix the species of our Molluscous animals, by accurate delineation in their appropriate colors, so that they may be readily recognized by those who have not extensive cabinets for comparison," that is, for those who lived outside of scientific centers like Philadelphia.[85] There was considerable local interest in the book. Community members peddled the prints throughout the New Harmony neighborhood and they sold well, although the announced subscription price of a dollar rose by fifty cents on publication.[86] The moderate price, minimum of letterpress, and subordination of text to illustration reflected agreement between Say and Maclure on the purpose of books and the principle of the availability of science for all.[87]

FIGURES
2.15–17

In a series of articles published in the *Disseminator* and later assembled under the title *Opinions on Various Subjects*, Maclure included an essay called "Desirable and probable diminution of the cost of books, for the most general diffusion of knowledge."[88] He expounded: "Book making is like gold beating, the greater the surface . . . the more profit to the man with the hammer, and the more pages a few ideas can be made to occupy, the more

2.15–17 Drawings by Lucy Say for a later edition of Say's *American Conchology*.

profit for the printers, booksellers and authors."[89] Maclure insisted that "it only requires a few correct ideas on utility, to know that gilt edges, hot press, large margins, and larger volumes, add nothing to the knowledge contained in the descriptions."[90] To replace expensive

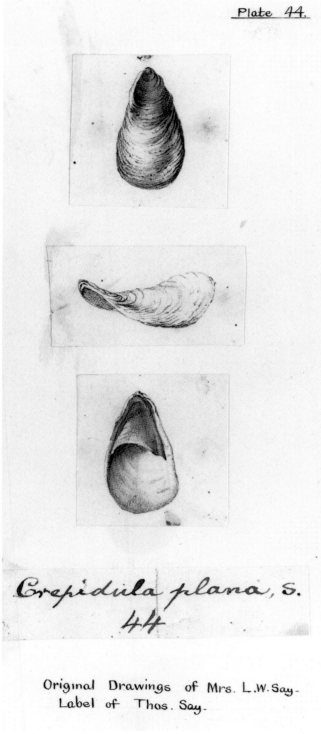

Figure 2.16.

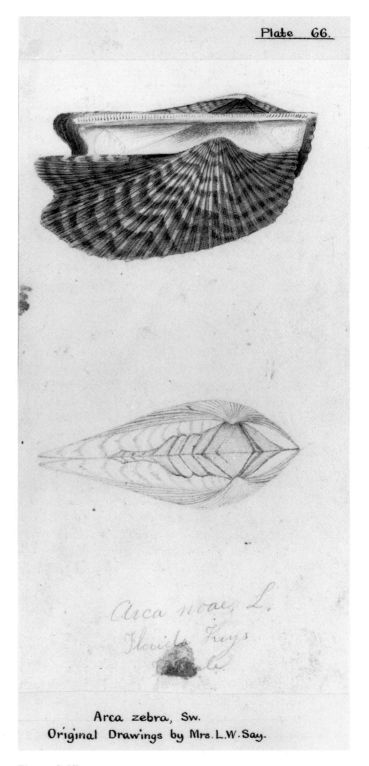

Figure 2.17.

Mrs Say Del                    10                    C.Tiebout Sculp.

Plates 14–16. *Paludina decisa. Unio ventricosus. Arca lienosa* S., *Arca staminea.* Hand-colored engravings by Cornelius Tiebout and L. Lyon, from drawings by Lucy Say, for Thomas Say, *American Conchology* (1834).

Plate 14. *Paludina decisa.*

Plate 15.
*Unio
ventricosus.*

Plate 16.
*Arca lienosa*
S., *Arca
staminea.*

Plate 17.
Black
Cockatoo.
One of
Lesueur's
paintings on
vellum, n.d.

volumes that only those "who can live without labor" could afford, Maclure wanted books designed and manufactured for the "industrious producers."[91] Whereas formerly "elegant and ornamental forms were made to suit . . . luxurious tastes," Maclure foresaw that when "the millions of industrious producers . . . claim and insist on their natural rights," then books would be printed and bound "in the most economical manner" with as few pages "by condensation, containing as much real knowledge and useful knowledge as used to be spread over five hundred; and on cheap paper made of straw or *corn shucks*, printed by steam in hundreds of thousands in an edition, . . . and bring every kind of information within the reach of the poorest."[92] Already he saw that the greater availability of education was working a "gradual" improvement on "the moral . . . man," with "practical effects on society."[93]

Maclure's forecast of public education and inexpensive publishing would prove correct, although indirectly. Neither could quickly eliminate poverty and thereby, as he predicted, eliminate the source of rags and necessitate the use of cornhusks for papermaking. Natural history did become an accepted part of the curriculum of public education. By midcentury, the school text literature filled the gap between the popular and specialized literatures of natural history and their styles of illustration. Shortly after Maclure's death, book production would undergo technological innovations that permitted the expansion of inexpensive mass publishing. By the end of the century natural history would become one of the best-selling subjects available in inexpensive popular editions. Meanwhile, Maclure promoted his cause of the universal availability of books by founding and endowing a system of public libraries in New Harmony and its neighboring communities. And Thomas Say shared his principles and published his own work accordingly.

Lesueur also continued his investigations, drawing and sketching prolifically. Indeed, Lesueur's study of natural history was as visual as Wilson's had been. Lesueur wrote haltingly in French and not at all in English, while he drew and engraved fluently. On his regular river trips to New Orleans to sell New Harmony produce and collect his pension from the French government, he assembled a portfolio of views remarkable for their stark depiction of the raw and isolated river frontier.[94] Traveling overland to visit colleagues in Kentucky and Tennessee, Lesueur also compiled a collection of drawings of fossil shells and sketched cross-sections of fossil-bearing strata.

FIGURES
2.18–24

Lesueur, who had taught drawing in the Philadelphia area, was known throughout scientific circles for the excellence of his instruction.[95] One of his students, Richard Owen (1810–1890), son of the town founder and later professor of natural sciences at the University of Indiana, reminisced that his natural history lessons with Lesueur often took place in the Wabash. There Lesueur instructed him "how to feel with my feet for *Unios* and other shells as we waded sometimes up to our necks in the river . . . searching to add to our collections."[96] Owen recalled that Lesueur, bearded and often barefoot, would pursue his research by exchanging "his fine common fish for the smallest" and give to his companions "the most indifferent-looking, when he recognized some new species or variety." Owen remembered his teacher as "a magnificent artist, good alike at drawing and coloring."[97]

Richard Owen's reminiscences evoke New Harmony as a summer idyll, where natural history was integrated with the community's educational, economic, and leisure pursuits. The actual conditions fell short of ideal. Isolation, perpetual shortness of money, constant difficulty in obtaining recent publications and supplies for printing and drawing, and withdrawal of the community's sustaining members created a discouraging environment for scientific work. Say, overwhelmed by administrative duties and often in poor health, could not or would not translate Lesueur's fish descriptions for publication.[98] In addition, widespread disap-

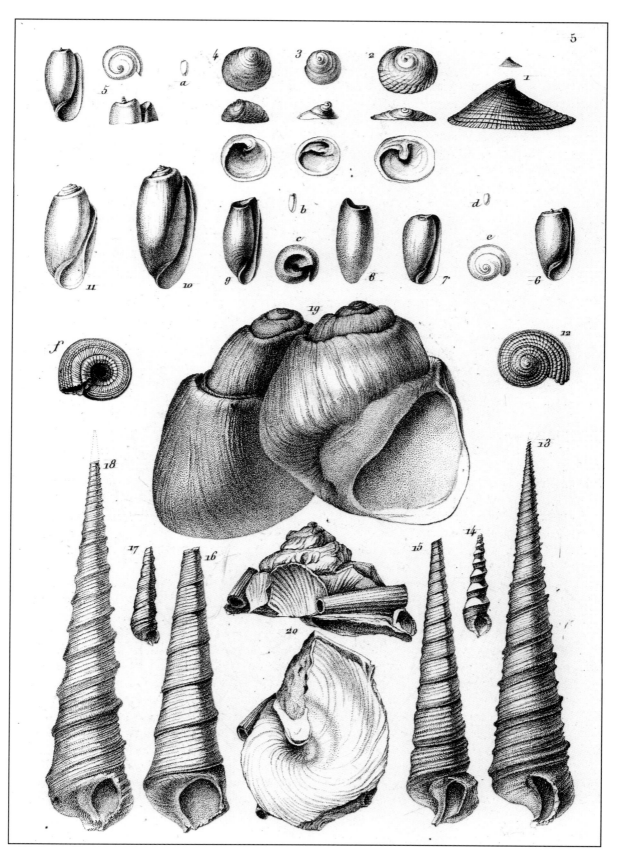

2.18–21 Fossil shells. Proof engravings by C. A. Lesueur, n.d.

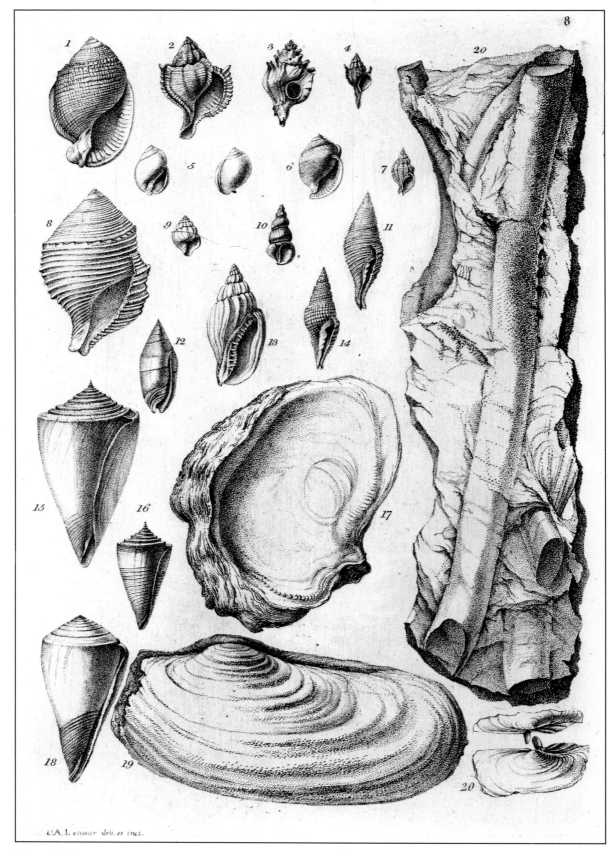

Figure 2.19.

Figure 2.20.

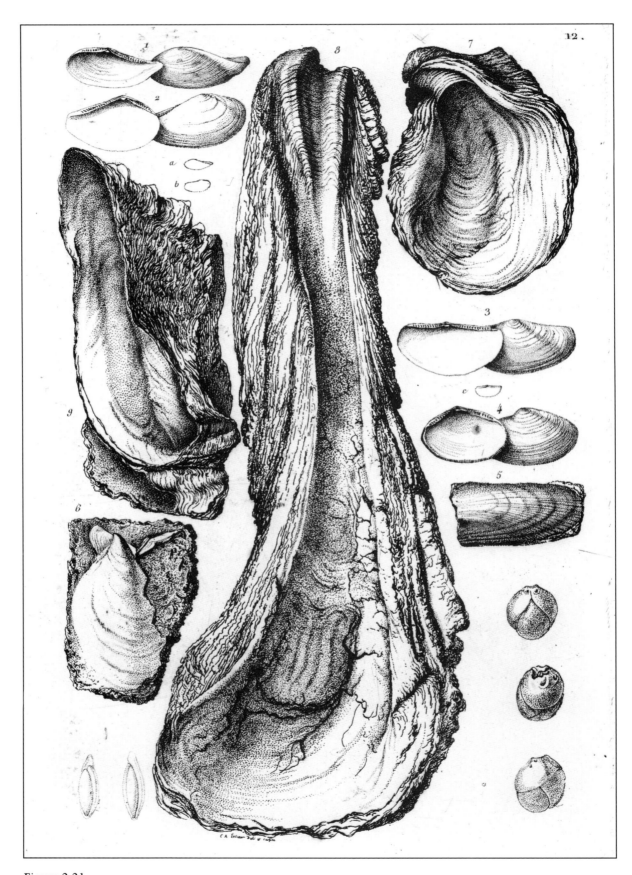

Figure 2.21.

2.22 The Walnut Hills fossil beds. Engraving by C. A. Lesueur, n.d.

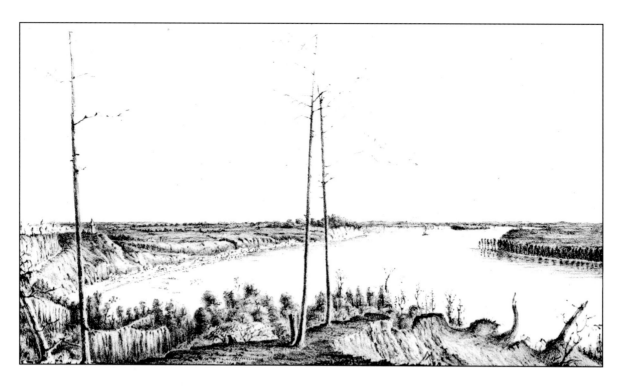

2.23–24  Views of Natchez and Memphis. Lithographs of Lesueur's sketches of the banks of the Mississippi,
n.d.

proval of Robert Owen's socialism and especially his atheism extended by association to Maclure and his friends. To their former colleagues in Philadelphia, participation in the New Harmony experiment seemed a radical and shocking defection, although the Academy retained Maclure, its primary patron, as president in absentia until his death in 1840. One critic writing in the *North American Review* sneered at the shop-worn letterpress and inexpensive paper of the New Harmony publications.[99] Say complained that descriptions he sent to the Academy were denied publication in the journal. George Ord, in his memoir of Thomas Say for the Philadelphia Academy, reviled the *Conchology* as "a disgrace" and "repulsive" and, after Say's death, worked to denigrate his reputation in Philadelphia.[100]

While the cautious Philadelphians may have considered the New Harmony publications a threat to American scientific respectability and international standing, the work of Say and Lesueur in fact commanded such admiration abroad that their presence made New Harmony a requisite stop on the itineraries of visiting European intellectuals. The duke of Saxe-Weimar, whose visit to the community coincided with a rare moment when Owen, Maclure, and Mme Fretageot were all present, passed his evenings in engaging conversation on social and scientific subjects. He "renewed acquaintance . . . with Mr. Say," the "distinguished naturalist from Philadelphia," and discussed geology and ichthyology with Lesueur, who also exhibited for the duke his drawings on vellum from the French Pacific expedition of his youth.[101] Maximilian, prince of Wied-Neuwied, lived in New Harmony during the winter of 1832–1833 while on his grand tour of North America. The prince, a naturalist himself, took a special interest in Lesueur's collections of drawings of fishes, and in the naturalist's projects, among them the exploration of local Indian mounds.[102] Karl Bodmer (1809–1893), the painter who traveled with Maximilian, made a portrait of Lesueur. The style of Bodmer's published views

PLATE
17

and portraits may have been Lesueur's model for finishing and printing his own New World sketches, although Bodmer selected his topics with an eye for the exotic and picturesque, while Lesueur, who never published his sketches, never romanticized the harshness of work and life in the west.

FIGURES
2.25–26

The first episode of science at New Harmony ended when Thomas Say died in 1834, at the age of forty-seven. From Mexico, Maclure continued to send vigorous opinions for publication in the *Disseminator*; but he lost interest in the educational experiment, turning instead to establishing libraries and donating to them his considerable collections of books and prints. Lesueur returned to France in 1837, where he studied lithography and served as curator of the natural history musuem at Le Havre, his birthplace, until he died in 1846. He never published his studies of North American fishes, although his work became well known and highly esteemed by later ichthyologists.[103]

Of the Philadelphia contingent, it was Lucy Say who suffered the greatest bereavement from the disbanding of the New Harmony scientific group. She lost her community and work, as well as her husband. Robert Owen's insistence on the equal education and status for women, and the energetic presence of Frances Wright (1795–1852) during the early days of the community, had made New Harmony an extraordinary outpost of feminism. Although Lucy Say had still performed traditionally female household duties (and although Mme Fretageot accused her of wishing to return to the comforts of New York),[104] she had shared in an atmosphere that encouraged her to think and act as an intellectual worker, a role that translated to outside society only at the considerable emotional cost of isolation. Her abrupt relegation from active participation in the production of science in New Harmony back to her family in New York was painful. She continued to study engraving in the hope of completing publication of her husband's work, but in a poignant letter to friends she wrote: "I am looked

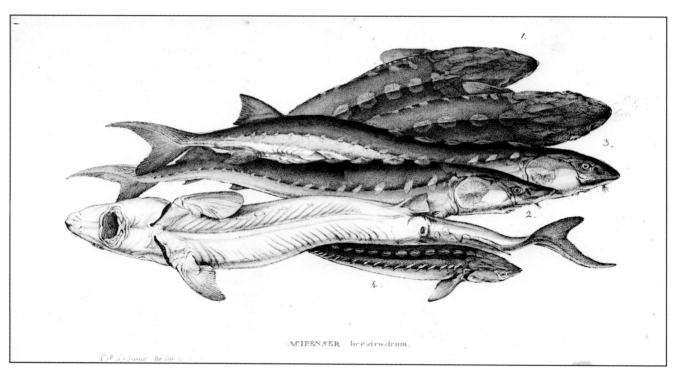

2.25  Sturgeon. Engraving by C. A. Lesueur, n.d.

2.26  A Chinese drawing of sturgeon, probably late eighteenth century. This drawing, collected by Maclure, appears to have informed Lesueur's studies. Both in the manuscript collections of the Academy of Natural Sciences, Philadelphia.

upon as being very singular, particularly since I have commenced Engraving—a gentleman remarked 'well! at what, do you think the ladies will stop?' I replyed, I hoped at nothing short of breaking up the Monopoly so long held by the gentlemen—that we are tired of cramping our genius over the needle and distaff."[105] She found that prejudice against the "Owenites" could still erupt "with a great deal of virulence," especially on the subject of their transgressions against "the mysteries of the holy religion." Her mother took care "that no one in the house, shall know" that her sisters "were ever at New Harmony." She would "seldom sit in the parlor. I am constrained to be quiet or say often what might as well not be said."[106]

Robert Owen's Preliminary Society at New Harmony failed "to remodel the world entirely; to root out all crime; to abolish all punishments; to create similar views and similar wants, and in this manner avoid all dissension and warfare," or even to ignite a transformation of American society.[107] Nevertheless, the scientific work of Maclure's New Harmony educational department had a lasting influence on American science, primarily through Robert Owen's sons, who would carry on the lessons of their New Harmony teachers. Robert Dale Owen (1801–1877), when a member of the Indiana legislature, secured federal money for public education, setting a precedent for the modern public school system that would eventually incorporate many of Maclure's educational innovations. He also introduced into the United States Congress the bill founding the Smithsonian Institution, which would play a central role in training a later generation of American naturalists and in consolidating specialist conventions of scientific publication. His brother, David Dale Owen, led the geological surveys of states including Arkansas, Indiana, Wisconsin, Iowa, Minnesota, and Kentucky. The maps and illustrations of those surveys, many engraved from his own drawings, reflected the strength of his New Harmony associations.[108] Owen, in turn, trained many of the young geologists who worked and drew for later state and federal surveys.[109]

In a limited but important sense, these developments fulfilled the Peales' initial conception of a place for natural history in national life. The institutional gains for natural history were achieved, however, at the cost of the identity of naturalist and illustrator advocated by the Peales and the integration of description and narrative practiced by Wilson. The forms of natural history that gained a prominent place in American social and political life were based on the distinction between specialist and popular conventions. Nevertheless, pictorial representation continued to be the locus where the tension between descriptive and narrative practices was played out, as well as the locus of the argument about what distinguished American natural history from European.

CHAPTER

3

# From Naturalist-Illustrator
# to Artist-Naturalist

JOHN JAMES AUDUBON (1785–1851), who became the best known artist-naturalist, eluded the genre division in natural history between narrative and systematic description by anchoring his enterprises in the domain of fine art. Audubon traced his inspiration, as had Thomas Say, to Wilson's *American Ornithology*. But unlike Say and the other Philadelphia academicians, Audubon staked his lifework on the broad appeal of narrative. The ethos of objectivity, expressed in the technical language of systematic description of generic and specific types, was deeply at odds with Audubon's celebration of the observer as participant and his recording of singular events whose actors were individual creatures. Audubon published his pictures at a time when American specialist natural history practice was separating the activity of systematic description from that of illustration. The distinction between naturalist—or scientist—and draftsman reasserted the primacy of text. Audubon's plates of the *Birds of America*, so large that the sheet size was called double elephant folio, challenged the natural history convention of illustration's subordination to text. Moreover, in the context of European fine arts printing and an aesthetically sophisticated European public, Audubon was able to push the technical limitations of book illustration beyond the established conventions for natural

history, and to invent his own idiom, natural history painting. Audubon would produce a text, but it supplemented or retold the story already narrated by the plates. The artist was author; pictures led text.

Audubon's *Birds of America* elicited, and continues to elicit, controversy over its place in the ornithological literature. Although the scientific status of Audubon's *Birds of America* might be disputed, its visual drama could not be dismissed. Not only the monumental scale of his pictures, but also the immediacy of his observational stance validated his audacious reversal of the relationship of pictures to text. Moreover, his observation of the living bird in nature was undeniably acute. He won the acceptance of American scientific institutions, even if, like so many American painters, he first had to achieve recognition abroad, where he was hailed as both naturalist and painter. Audubon's European acclaim helped American naturalists embrace his romantic view of North American nature as both natural history and nature painting.

Audubon's self-definition as artist-naturalist emerged slowly. During his boyhood in France and his youth in Pennsylvania, Audubon hunted and drew birds.[1] In later life, he would claim to have studied drawing in the studio of Jacques-Louis David (1748–1825), but his early drawings show no influence of such training—

they are stiff profiles of the birds and reveal difficulty in portraying contour.[2] In 1804 Audubon came to the United States to live near Philadelphia on the estate at Mill Grove, given to him by his father. After Audubon's marriage to Lucy Bakewell, the couple headed west to seek their fortune. From 1808 to 1819, while Audubon drifted around the frontier, he hunted birds to escape the responsibilities of his frequent business failures.[3]

In 1810, when Wilson peddled his book at Aubudon's Louisville store, Audubon had as yet no plans to publish his own drawings of birds, and on the advice of his partner he declined to buy the book.[4] Not until 1819, after bankruptcy and imprisonment for debt, did he abandon his half-hearted attempts at commerce, begin to take his drawing seriously, and consider future work in natural history. He consolidated all he had learned about birds and drawing since boyhood and concentrated on improving his skills. With his wife and two sons Audubon moved to Cincinnati, where he worked briefly as the taxidermist in the newly established Western Museum.[5] While he was there, Thomas Say and Titian Peale passed through on their way west with the Long Expedition.[6]

After leaving Cincinnati in 1820, Audubon traveled through the southern states, supporting himself as an itinerant portraitist and drawing master, and observing and drawing birds. Many of his later accounts of bird life and migration were based in Louisiana and provide an interesting regional counterpoint to Wilson's, based in Pennsylvania. During this period, 1820 through 1824, Audubon took the *American Ornithology* for his primary reference, although he did not own a copy. He labeled his drawings with Wilson's identifications, and in contrast to his later work, he composed his drawings as if they were to become book illustrations. But unlike Wilson, who economized by grouping several species in a single plate, Audubon composed each drawing with a single species and a plant chosen to complement the colors of

the bird. The boy Joseph R. Mason, who traveled with Audubon, contributed the meticulous plant illustrations to Audubon's compositions until 1822, when he left Audubon and moved to Philadelphia.[7] By 1824, having mastered a technique that combined watercolor and pastel, Audubon was ready to present his portfolio to the judgment of the scientific community. He journeyed east to Philadelphia to sell his drawings.[8]

Audubon aspired to recognition from the Philadelphia natural history community, but as an outsider from the west he found entrance into that circle blocked by the very work he most admired, Wilson's *American Ornithology*. Since Wilson's death his book had become canonized as a founding document in American scientific ornithology. Alexander Lawson and George Ord in particular, jealously guarding Wilson's achievement, barred Audubon from publishing in Philadelphia. They defended a tradition with which they identified. Not only had Lawson built his subsequent reputation on his work for Wilson, but in 1824, when Audubon arrived in the city with his portfolio, Lawson was engraving the plates for Bonaparte's continuation of Wilson's book. George Ord, who had completed Wilson's ninth and posthumous volume, was issuing at considerable expense the first reprint edition of the *American Ornithology*. Bound up with self-interest was their allegiance to the specialist values of accuracy and priority.

When Audubon arrived in the city in April 1824, he met Charles Lucien Bonaparte, then at work on his continuation of Wilson's *Ornithology*.[9] Enthusiastic about Audubon's drawings, Bonaparte wanted Audubon to contribute to his book and helped him show his work to the Academy and to other Philadelphians of note. Bonaparte, whose father was a noted connoisseur and collector of paintings, may have appreciated in Audubon's work qualities reminiscent of the early still-life paintings of birds and flowers, as well as the later richly textured paintings of dead birds by the

PLATE
18

FIGURE
3.1

No 15. Plate 94.
Published
1829.

3.1 Red-eyed Vireo. The spider detail is invented rather than observed. The spider is identifiably a jumping spider, not an orb weaver, and the angles of the web do not correspond to the angles of actual orb webs. Drawing by J. J. Audubon and Joseph Mason, watercolor and graphite, 1822. (Thanks to Wayne Maddison and Jonathan Coddington for their identification of the spider and critique of the orb web, respectively.)

French eighteenth-century artist Jean-Baptiste Oudry.[10] Audubon's drawings elicited a mixed response. Among the naturalists, it was the Frenchman, Lesueur, who shared Bonaparte's taste for Audubon's work. Painters Rembrandt Peale and Thomas Sully also received Audubon and his work favorably. But Alexander Lawson, whose association with Wilson had given him a position of significant authority in the natural history community, refused to engrave Audubon's drawings for Bonaparte.[11]

William Dunlap, contemporary chronicler of American art to 1834, took a neutral stance on whether Audubon was artist or naturalist, but in comparing Wilson to Audubon, expressed preference for Wilson: "We have seen what Wilson, a modest unpretending man did for the science of Ornithology, and the skill he acquired as a draughtsman, without having his hand guided by *David* and, many masters."[12] After Audubon's triumphant return with his work to the United States, Dunlap would write:

This very enterprising ornithologist and artist has attracted great attention by undertaking to publish from drawings and writings of his own on American ornithology, the figures in which are the size of life. How much science gains by increasing the picture of a bird beyond the size necessary to display all the parts distinctly, is with me questionable; but the work of Mr. Audubon, as far as I have seen it, is honorable to his skill, perseverance and energy.[13]

Dunlap depicted the first encounter between Lawson and Audubon as a hostile one and favored Lawson, "who is undoubtedly biased against the rival of his friend Wilson, but whose character places him above doubt as to the facts he states":[14]

On a certain occasion, a well-known Quaker gentleman of Philadelphia, told his friend Lawson that a wonderful man had arrived in the city, from the backwoods (all wonders come from the backwoods), bringing paintings of birds, beautiful be-

yond all praise, colored with pigments, found out and prepared by himself, of course a self-taught original genius. . . . One morning, very early, Bonaparte roused him [Lawson] from bed—he was accompanied by a rough fellow, bearing a portfolio . . . a number of paintings of birds, executed with crayons, or pastils, . . . were displayed as the work of an untaught wild man from the woods . . . and as such, the engraver thought them very extraordinary. Bonaparte admired them exceedingly, and expatiated upon their merit as originals from nature.[15]

When, however, they came to Audubon's drawing of the horned owl, Lawson perceived and demonstrated to Bonaparte that it was an enlarged and reversed copy of Wilson's plate. The plagiarism only confirmed Lawson's poor impression of the drawings:

Lawson told me that he spoke freely of the pictures, and said that they were ill drawn, not true to nature, and anatomically incorrect. Audubon said nothing. Bonaparte defended them, and said he would buy them, and Lawson should engrave them. "You may buy them," said the Scotchman, "but I will not engrave them. . . . Because ornithology requires truth in forms, and correctness in lines. Here are neither."[16]

Bonaparte bought Audubon's drawing of the male and female grackle. Lawson consented to engrave the drawing "but found it was too large for the book." He refused to reduce it and told Bonaparte: "I will engrave it line for line, but I will not reduce it, or correct it in any part."[17] Audubon returned with the drawing to discuss the engraver's objections. Dunlap re-created their conversation:

"I understand that you object to engraving this." "Yes, it is too large for the book." "And you object to my drawing?" "Yes . . . This leg does not join the body as in nature. This bill is, in the crow, sharp, wedge-like. You have made it crooked and waving. These feathers are too large." "I have seen them twice as large." "Then it is a species of crow I have never seen. I think your painting very extraordinary for one who is self-taught—but we in Philadelphia are accustomed to seeing very correct drawing." "Sir, I have been instructed seven years by the greatest masters in France." "Then you have made dom bad use of your time," said the Scotchman.[18]

Lawson's objections to Audubon's drawings indicated the codification of his working relationship with Wilson into criteria of "correctness." He had become accustomed, for example, to Wilson's characteristic elongating of birds' bodies. Audubon, in contrast, tended to enlarge the head and eyes.[19] In Audubon's drawing to date, the blend of pastel and watercolor conveyed a rich sense of texture of the birds' plumage but sometimes failed to enumerate individual feathers, a necessary part of ornithological description. Audubon's drawing style would have been difficult for Lawson to translate "line for line" in his accustomed technique, especially in contrast to Wilson's pen and pencil sketches. Furthermore, Lawson had called Audubon's pictures "paintings" as if to distinguish between drawing as illustration and drawing in the different tradition of animal painting. The two men had reached a double— and contradictory—impasse. Each claimed more knowledge of birds, Lawson on the basis of his experience engraving the *American Ornithology*, Audubon on his experience observing birds in nature. And each claimed superior judgment about accurate representations, Lawson as an engraver of scientific book illustration, and Audubon, defensively, as a former student of French painting masters.

Certainly Lawson's engravings for Bonaparte adhered to Wilson's model, from the early, pasted compositions to the late and posthumous landscape plates. Yet the engravings for Bonaparte lack the freshness and animation of Lawson's earlier work. The stiff rendering may have come from the hand of Titian Peale, who was preparing the drawings for Bonaparte's first volume. Audubon accused Peale of drawing

birds as if they were seated for their portraits.[20] But Peale had proved himself capable of portraying animation and differentiating alive from dead in his drawings from the Long expedition. Some of Peale's bird sketches, perhaps made in Philadelphia, have a striking immediacy. Bonaparte expressed confidence in his illustrator in the preface of his first volume, writing that he had been "desirous to procure the best representations of birds; in which he hopes he has succeeded, through the happy pencil of Mr. Titian Peale, who has invariably drawn from the recent bird, and not from the preserved specimen; this being the principal advantage of works on Natural History, published in the country where the animals figured are found."[21]

Lawson also made a habit of working "with the bird always before him" and of making measurements and drawings from the living bird when he could.[22] Bonaparte believed that Lawson "transferred our drawings to the copper with his usual unrivalled accuracy and ability. This artist, who acquired so much distinction by the engraving of Wilson's work, has become perfectly master of his art, and so intimately acquainted with the various parts of a bird, that he may be justly styled the first ornithological engraver of our age."[23] The responsibility for the rigidity of the Bonaparte plates may in fact have come from the hand of the aging engraver; it cannot simply be attributed to the lesser skill of Alexander Rider, the illustrator who replaced Peale after the first volume. Note that Bonaparte emphasized "accuracy" over lifelikeness in his praise of Lawson, and that Bonaparte responded with enthusiasm to Audubon's softer, more delicate drawings. In attention to detail "line by line" Lawson seems to have sacrificed interpretation and animation. Not only had he converted his style for Wilson into a formula, but a description of Lawson's working method in one of Bonaparte's later volumes indicates that the engraver had become fixed on "accuracy." While working on the plate illustrating the condor, Lawson made "almost daily

FIGURES
3.2–6

visits" to the captive bird, "for the purpose of measuring and examining accurately every part."[24] The resulting plate conveys little animation. Measurement seems to have cramped Lawson's hand.[25]

Bonaparte had his way, and Lawson engraved the grackles for Bonaparte's book. The male and female, motionless, in profile, nevertheless seem poised for flight. Crossed in an X, they stare intently in the same direction, the male's head turned sharply over the shoulder.[26] The simple composition foretold Audubon's growing preference for dynamic arrangements rather than compositions that squared the birds with the page. He would later create some of his most striking pictures with a similar diagonal device.

FIGURE
3.7

Audubon, meanwhile, was moving further

away from the conventions that constrained bird illustration. Some of the drawings he had brought to Philadelphia already showed birds in motion or in dramatic positions.[27] During the spring of 1824 he began to increase the drama and movement in his pictures, as if deciding that these qualities would distinguish his work from his Philadelphia competition. Audubon portrayed the birds he drew in the city environs engaged in violent activity—a male great-crested flycatcher plucking the tail feathers of a rival, and two tree swallows squabbling on the wing.[28]

After visiting New York, where he found a more encouraging reception, Audubon returned west to complete more drawings before making further attempts to publish. During the period between the Philadelphia visit and

FIGURE
3.8

3.2–4 Field sketches by Titian R. Peale made on the expedition led by Stephen Long to the Rocky Mountains, 1819–1820.

3.2 Squirrel, 1820.

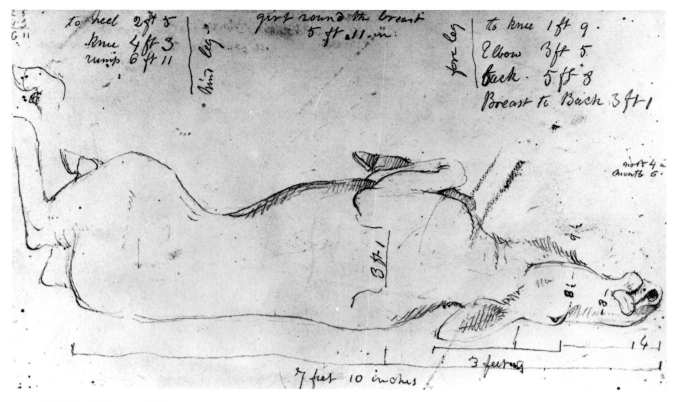

3.3 Dead Moose with measurements, n.d.

PLATES
19–21

his departure for England in 1826, Audubon's drawings became complex compositions that combined several individuals, and sometimes crowds of a single species, with plants as elaborately twined, thorned, or blossoming as the birds were frantically active.[29]

In heightening the movement in his drawings, Audubon abandoned his emulation of natural history book illustration and began instead to explore the possibilities offered by painting. One indication of his shift away from the constraints of the conventions of book illustration for natural history and toward the emotive and active repertoire of painting was his decision to draw and reproduce all his birds life-size, even birds like the wild turkey, the trumpeter swan, and the flamingo. Lawson had already complained that Audubon's drawing was too large for a book. The folio dimensions of Bonaparte's plates were large for scientific books in general,

although standard for engraved atlases and modest compared to some French ornithological plates. (But even French ornithologists were beginning to object to the literature of their discipline being priced for a luxury market.)[30] The larger the book, the more expensive production and the higher the selling price had to be. Moreover, production costs were rising. Wilson's subscribers had paid $120 for ten promised volumes, of which nine were published; Bonaparte's paid $180 for four.[31] The size of Audubon's proposed plates presupposed an audience accustomed to purchasing expensive illustrated books and paintings.

He found that audience in England, where he arrived armed with letters of introduction in 1826. The story of Audubon's successful cultivation of patrons, of his welcome by the natural history circle of Edinburgh, the engraving and publishing of the first ten plates in the Edin-

3.4 Dead squirrel, n.d.

3.5 Ruffed Grouse. Drawing by Titian Peale, graphite, probably made in Philadelphia, n.d.

burgh shop of William Home Lizars and the subsequent move to London and the print shop of Robert Havell and son, the reception of his plates by the great Cuvier and by English royalty alike, and his triumphant return to the United States has been told many times.[32] Here, several points will illuminate why Audubon's work had such a strong impact and

lasting influence on Anglo-American bird illustration, and on the American public imagination, and why the question of its scientific status can still provoke controversy.

In England, Audubon's person—backwoodsman, artist, naturalist—and his pictures were received as authentically American. He was treated as an authority on American nature. In

Plate 18.
Rufous-sided
Towhee.
Drawing by
J. J. Audubon
and Joseph
Mason, water-
color and
graphite,
1822.

Plate 19. Carolina Parokeet. Drawing by J. J. Audubon, watercolor and graphite, Louisiana, 1825.

Wild Turkey. Male and Female.

Meleagris Gallopavo.

3.6 Turkey cock and hen. Drawn by T. R. Peale, engraved by A. Lawson, for Bonaparte, *American Ornithology*, vol. 1 (1825).

3.7 Great Crow-Blackbird. Drawn by J. J. Audubon and A. Rider, engraved by A. Lawson, for Bonaparte, *American Ornithology*, vol. 1 (1825).

the flush of his enjoyment of English patronage, after having been consulted as an expert on American insects, he would write privately: "I would like to know what the entomologist *Thomas Say*, Esq., Academician, &,&,&,&,&,&, would say on such an occasion. American bugs!!! . . . I have seen millions. And Thomas Say has described the same species over and over again, probably hundreds of times."[33] In a letter to Thomas Sully in which Audubon boasted of his reception in Edinburgh and of an invitation to contribute to the journal of the Philosophical Society, he wondered: "What

will my friend Ord say to all this when he sees those journals, the papers, and hears of all those wonderful events?"[34] Audubon recorded in his journal, kept as an extended letter for his wife, the initial enthusiastic response of the engraver, Lizars, to his work, a response that did much to heal the wounds that Lawson, "the Philadelphia brute," had inflicted on Audubon's self-esteem.[35]

In Liverpool, Manchester, and Edinburgh, Audubon arranged public exhibits of his drawings, copies of which he sold. The pictures offered his British audience an unprecedented

Plate 20.  Common Grackle. Drawing by J. J. Audubon, watercolor and graphite, Louisiana, 1825.

Plate 21.  Northern Mockingbird. Drawing by J. J. Audubon, watercolor and graphite, Louisiana, 1825.

3.8 Great-
crested
Flycatcher.
Drawing by
J. J. Audubon,
watercolor
and graphite,
Philadelphia,
1824.

and immediate vision of American nature. They provided a kind of armchair travel similar to the volumes of topographical views popular in England among the same clientele.[36] The meanings that moved Audubon's newfound audience can be read in the variety of drawing styles he had developed. While the delicate compositions of birds and plants evoked the western American woods as a blossoming and fruitful garden, Audubon's more dramatic compositions met other tastes, for Audubon incorporated the conventions of the game paintings that hung in the homes of many of his new patrons. Just as Titian Peale's picture of the mounted Plains Indian resonated with adventure for an urban American audience, Audubon's dramatic depictions of, for example, a red-shouldered hawk attacking a family of bobwhites, peregrine falcons with prey in their talons, or a swallow-tailed kite clutching a snake, provided satisfying evocations of romantic conceptions of wilderness.

In a subsequent meeting with Lizars, the appeal of Audubon's compositions became clear:

Mr. Lizars had not seen one of my largest drawings. He had been enamoured with the "Mocking Bird and Rattlesnake," but Lucy, the "Turkey"—her brood—the "Cock Turkey"—the "Hawk" pouncing on seventeen partridges—the "Whooping Crane" devouring alligators newly born—all were, he said, wonderful productions. According to his say so, I was a most wonderful compositor. He wished to engrave the "Partridges," but when the "Great-footed Hawks" came with bloody rage at the beak's ends and with cruel delight in the glance of their daring eyes, he stopped mute for perhaps an instant . . . then he said, "I will engrave and publish this."[37]

For technical reasons the publication that Audubon envisioned would probably have been impossible in the United States in 1826. The so-called double-elephant folio, 3 feet 3 inches by 2 feet 2 inches, transcended book format. Printing required a huge press bed, such as those used for the maps and nautical charts of

the expanding British empire. Moreover, Audubon's plates required an initial investment in copper that few American engravers could have risked on speculation. The modest pre-Philadelphia compositions were dwarfed by the size of the paper. In contrast, most of the drawings completed after 1825, and especially after the printing had begun, anticipated and took full advantage of the large format, and consequently required more engraving, more coloring—more time and expense. Each number of five plates sold for 2 guineas, or about $10. The first one hundred prints cost 42 pounds, or about $200, bound; the completed work of 435 bound plates would cost $1,050.[38] Furthermore, although American commercial engraving technique could match the European, engraving for fine art reproduction could not. Not only did Audubon produce work the size of paintings and cultivate the same audience that appreciated and patronized painting, but his drawings were transposed into print by a technique developed for publishing reproductions of paintings. English engraving had become distinguished for aquatinting, a technique developed to convey in print the diffuse painting style of the English romantic school, and used to print reproductions of watercolor views for popular albums.[39] Resin particles, shaken over the plate to varying thicknesses and melted to stick, permitted the acid bath to bite only in the tiny crevices between the melted beads. When inked and printed, the irregularly bitten areas created a tonal rather than linear effect. Audubon's engravers, Lizars in Edinburgh and Robert Havell, Jr., in London, added aquatint to etching and engraving to work the large areas of the copperplates that reproduced the watercolor washes of the drawings, although the coloring still had to be applied by hand. (Among other delays, a strike of Lizars's colorists held up completion and distribution of the first issues, prompting Audubon to transfer operations to Havell in London.)[40]

To support his publication enterprise and himself while away from home, Audubon made

Plate 22. Northern Bobwhite and Red-shouldered Hawk. Drawing by J. J. Audubon, watercolor and graphite, with collage elements, 1825.

Plate 23. Tricolored Heron. The background is a Florida key. Drawing by J. J. Audubon and George Lehman, watercolor and graphite, 1831–1832.

Plate 24. Mourning Dove. Drawing by J. J. Audubon, watercolor and graphite, 1825.

oil painting copies of some of his drawings and also painted new compositions not intended for inclusion in his *Birds*.[41] Some of the landscape backgrounds added in oil paint to already-completed drawings probably date from the period when Audubon was exhibiting his portfolio. The subjects he painted in oils emulated English hunting scenes: the bobwhite family attacked by a hawk; an otter in a trap; and a water spaniel attacking pheasants.[42] At the end of 1826 Audubon wrote to Thomas Sully: "Think of my painting oil pieces of eleven wild turkies estimated at 100 guineas, finished in ten days work."[43] He hired the landscape painter Joseph Bartholomew Kidd (1808–1889) to produce oil copies of all the plates; Kidd produced up to ninety-four between 1831 and 1833.[44] The success of Audubon's enterprise enabled him to bring members of his family to England, now his business base, and to make collecting and drawing expeditions to the United States for additional material.

Compositional conventions for painting differed markedly from those of scientific illustration. In illustrations for science the white of the paper, or blank space, indicated a change of subject from one figure to the next. In contrast, the traditions of European painting demanded a unified statement with consistent spatial relation among the parts of the painting. Furthermore, whereas in scientific illustration it was understood that the reader viewed the images in sequence, with an implied comparison both of different figures in the same plate and with figures in other plates, painters assumed that their audience concentrate on each picture as a unique experience, to the momentary exclusion of all other pictures.

Referring to the different sizes of the birds and their relation to page and format, Audubon later claimed, "Chance, and chance alone had divided my drawings into three different classes, depending on the magnitude of the objects to be represented."[45] But the birds' body size did not always determine the composition of the drawing. In his mature work Audubon

often combined the assumptions of natural history illustration with those of painting in a single composition. By elaborating background to fill the picture, Audubon, with the collaboration of George Lehman, a Swiss landscape painter who traveled with him to the Carolinas and Florida in 1831 and 1832, and of his engraver, Robert Havell, Jr., met the expectations for painting.[46] In some pictures, however, Audubon retained the conventionalized foreground tree on which perched several individuals of the same species. In other compositions, Audubon created an active foreground as in history paintings or hunting scenes, but left the page empty behind and around the plant-bird group. One of his favorite devices for charging the foreground with drama, in pictures both with and without fully elaborated backgrounds, was to depict a foreground bird looking directly out at the viewer, creating the startling impression of coming upon the bird and making eye contact with a falcon devouring a teal, or with one of a mob of parakeets. In conventional ornithological illustration, the bird in profile gazed away from the viewer into an unspecific distance, creating and maintaining an emotional and conceptual separation between bird and viewer. The challenging stare of some of Audubon's birds placed the viewer in immediate relationship with the wild bird and provoked an emotional response. Toward the end of the publication process, however, Audubon, pressed for time and also more sensitive to scientific convention, returned to the format consistent with ornithological illustration of many species arranged on the stylized tree. For these later composite plates Audubon drew from preserved skins and often depicted the birds in the conventional profile position.

Audubon's combination of the pictorial conventions for painting and for ornithological illustration explains how his work reached both an audience of art appreciators and of naturalists. By the time the work was completed, British subscribers to the plates included wealthy industrialists of Manchester, a sprinkling of the

PLATE

23

FIGURES

3.9–10

aristocracy, scientific societies and some of their individual members, and the crown. European patrons included French nobility and royal libraries, Baron Cuvier, and the Haarlem Library. Audubon successfully solicited American subscriptions during his return visits to the United States. The list of American subscribers delineated the map of emerging scientific societies and colleges with science instruction, and included the Academy of Natural Sciences in Philadelphia and the Boston Society of Natural History. Legislatures of six states and the Library of Congress, Daniel Webster, Henry Clay, and a number of individual naturalists also subscribed.[47] The high proportion of institutional subscriptions in the United States, compared to the number of private or aristocratic subscriptions in Europe, points to different receptions of the work abroad and at home. Influential spokesmen in England and France pronounced Audubon's production a work of art and recognized that it required private patronage. Although some individual Americans subscribed, the high proportion of public and institutional subscriptions indicated, on the one hand, the need for collective resources to meet Audubon's price and, on the other hand, a public embrace of Audubon's work as a national possession—natural history for a broad audience in the tradition established by Wilson.

The English and European reception of Audubon's work is well characterized by the comments of William Swainson (1789–1855), an English ornithologist and Fellow of the Royal Society, who reviewed the first issues of the plates for a popular English journal of natural history in 1828.[48] Swainson praised Audubon for his "genius and ardour" in pursuing "the study of nature, no less than . . . painting."[49] He characterized both Audubon's work and Wilson's by the "freshness and originality" of their direct observation of nature—"Nature as she really is, not as she is represented in books." While crediting their originality, Swainson distinguished between their practice and his: he was an ornithologist who based his work on

systematics and systematic theory.[50] Audubon was historian, "biographer," and painter of birds:

The observations of such men are the corner stones of every attempt to discover the natural system. Their writings will be consulted when our favourite theories shall have passed into oblivion. Ardently, therefore, do I hope, that M. Audubon will alternately become the historian and the painter of his favourite objects; that he will never be made to convert to any system, but instruct and delight us as a true and unprejudiced biographer of Nature.[51]

He continued: "I am now to speak of M. Audubon more particularly as a painter. I shall, therefore, view the work before me as a speciman of the fine arts, and judge of it by those rules which constitute pictorial criticism."[52] He made the further distinction of calling the work pictures rather than plates, for by implication plates referred to illustrations of a text.

Swainson invoked the names of Rubens and Veronese to describe the pictures. Of the wild turkey hen and chicks he wrote: "The grouping of these little creatures cannnot be surpassed; it would do honour to the pencil of Rubens," although at the same time he found the birds' attitudes somewhat "difficult."[53] Throughout the review, Swainson emphasized that each picture told a story, or offered a portrait. He praised the effects of foreshortening and coloring, and considered the reproduction of the colors accurate. Nevertheless, Swainson found the plumage of the seven Carolina parakeets too gaudy for refined tastes (a statement with Buffonian overtones, only in this instance, depicting an American bird as parvenu).[54] Swainson's descriptions reflected the narrative and romantic qualities of the plates. Of a pair of turtledoves, he wrote: "Their love is in its infancy. The male, seated on the same branch with his intended partner, is eagerly pressing forwards to reach a 'stolen kiss,' but the head of the female is coyly turned."[55]

Summing up his descriptions, Swainson ob-

PLATE
24

107

Plate 25. Golden Eagle. Note the trapper—Audubon—straddling the log. Drawing by J. J. Audubon, water-color and graphite, 1833.

Plate 26. Arctic Tern. Drawing by J. J. Audubon, watercolor and graphite, with collage elements, 1833.

3.9 Hairy Woodpecker and Three-toe Woodpecker. An example of a late composite plate, many of the figures, pasted onto the sheet, were drawn from specimens in London. Drawing by J. J. Audubon, watercolor and graphite, finished ca. 1838.

3.10 Burrowing Owl, Short-eared Owl, Little Owl, Pygmy Owl. Another late composition of pasted drawings. Drawing by J. J. Audubon, watercolor and graphite, ca. 1838.

served that "every object speaks, either to the senses or to the imagination. The examples I have quoted, show that histories are to be narrated. . . . It is this which elevates the character of his paintings, from mere matter of fact portraits, to historical representations." He coined a term for Audubon's new genre—"zoological painting"—which represented "the passions and feelings of birds." Swainson concluded: "It will depend on the powerful and the wealthy, whether Britain shall have the honour of fostering such a magnificent undertaking."[56]

Audubon published his *Birds of America* during a period when the discipline of ornithology on both sides of the Atlantic was consolidating

itself around new standards of professionalism, of which the core was a commitment to system. Even a popular magazine of natural history directed at the improvement of youth, such as the one in which Swainson published his review, emphasized system as the essential foundation for an enlightened enjoyment of nature. "Those of us who know nothing of scientific zoology, still derive much pleasure from observing the great variety of forms, habits, and powers of the animal kingdom," and from projecting "the human virtues to some of the nobler quadrupeds. But what is this interest in animated nature compared with the enjoyments of a scientific zoologist?"[57] Systematic knowl-

111

edge not only increased pleasure in nature's works, but gave the systematist a privileged position from which to view nature: "The man who can trace the powers of the Author of nature . . . from the most minute insect or obscure mollusca . . . and who knows scientifically that man is the most perfectly formed of all animals, lives in a different world from the mere general observer, and enjoys that more exalted pleasure which can be given by scientific knowledge only."[58]

Perhaps the most important contrast between Audubon's *Birds* and conventional scientific illustration was not their scale, technique, or drama, but that Audubon had conceived them, like paintings, to tell their own story without words. Knowing that for his pictures to qualify as natural history they must be associated with a text, Audubon planned a book, the *Ornithological Biography*, to accompany the engravings, but he published it separately, and only after the first one hundred plates had already appeared.[59] He recognized that its strength would be in narrative rather than in systematics. In 1830 Audubon wrote to Swainson of his plans for a first volume comprised of *"one hundred letters addressed to the Reader* referring to the 100 plates forming the first volume of my illustrations. — I will enter even on local descriptions of the country. —Adventures and anecdotes, speak of the trees & the flowers the reptiles or the fishes or insects as far as I know—I wish if possible to make a *pleasing* book as well as an *instructive* one."[60] Although confident in his ability to create a text as descriptive and narrative as his pictures, Audubon sought help with *"the science which I have not!"*[61]

Since writing the review, Swainson had met Audubon and repeatedly offered him corrections on nomenclature. Audubon, however, wanted to publish Swainson's systematic contributions without acknowledging his help on the title page, and Swainson refused to comply. In addition to the issue of credit, Swainson's other reasons for declining to work with Audubon on the text illuminated the disparity be-

tween "professional" practice and Audubon's. First, Swainson and Audubon disagreed over nomenclature. Audubon disliked being corrected, and Swainson clearly disliked having his professional advice rejected. (Later, however, Audubon's nomenclature would reflect a heavy debt to Swainson's influence.) Further, Audubon had distributed the skins of his specimens, making them unavailable for reference and study. Swainson insisted that the systematic work could not be accomplished without the skins and admonished Audubon: "You may not be aware that a new species, deposited in a museum, is of no authority whatsoever, *until its name and its character are* published." Unless Audubon reassembled the collection, "you will, I am fearful, loose the credit of discovering nearly all the new species you possess."[62]

System, collection building, and institutionalized practice were the cornerstones of Swainson's idea of proper scientific conduct. He deplored the admission of amateurs to British natural history societies and advocated institutions strictly for professional initiates.[63] For him, both Audubon and Wilson were amateurs. Indeed, the term "zoology," used by both Swainson and the editor of the popular *Magazine*, indicated how the professional, specialized discipline of the study of animals had begun to distinguish itself from a more generalized, popular natural history, now considered merely narrative and historical.[64] It must have stung Audubon that in declining to collaborate with him, Swainson informed him that he had begun to assist "in bringing out an Octavo of Wilson, by Sir W Jardine which will be arranged according to *my* nomenclature."[65]

Audubon was able to enlist the cooperation of William MacGillivray (1796–1852), a young Edinburgh ornithologist, to assist him with the systematic descriptions for the *Ornithological Biography*, published in five octavo volumes between 1831 and 1838.[66] Audubon wrote the narratives that he had already represented in his pictures, each chapter corresponding to a published plate. MacGillivray supplied the taxonomic requirements. His contribution was credited not on the title page, but in the introduction.

The *Ornithological Biography*, however, developed from emphasizing narrative to introducing more pointedly scientific material. In the first volumes, Audubon's accounts, each ending with a systematic analysis by MacGillivray, were interrupted by occasional chapters of nature description and stories of frontier adventure. Audubon took full advantage of the opportunity to make explicit his persona of artist-naturalist and daring frontiersman, present in his pictures only by implication. Initially, Audubon's text was indeed his own biography, full of stories of his struggles, disappointments, adventures, and triumphs, written in an alternately sentimental and hyperbolic style, like popular novels of the time. But Audubon was producing and publishing the text in an atmosphere of increasing ornithological competition. Much of the pressure came from new editions of Wilson, whose work was being discovered and promoted, especially in Scotland. Jardine's edition of Wilson and Bonaparte, with plates engraved by Lizars, appeared in 1832. It had been preceded in 1831 by an edition of Wilson's with miniature plates. Soon afterwards, in 1835, Thomas Brown published a rewritten and largely redrawn edition of Wilson.[67] Audubon, more anxious now about his priority in describing new species, had his wife copy his manuscript and sent it to the United States to obtain copyright there. In the introduction to the third volume of the *Ornithological Biography*, published in 1835, Audubon pointed out to his readers that he pictured and described species missed by both Wilson and Bonaparte.[68]

The third volume also introduced a discussion of morphology, "Remarks on the Form of the Toes of Birds," with woodcut figures printed with the text to provide the basis for comparison.[69] This addition indicated Audubon's own growing awareness not only of market competition, but of the requirements of the discipline. In the last volumes, the essays of adventure and frontier manners were replaced by descriptions of dissections of esophagi conducted by MacGillivray and illustrated with wood engravings after his sketches. Yet even in presenting the dissections and a technical description of the method of measurement, Audubon maintained the same conversational essay style, addressing the text to his "kind reader." His admission that he himself was learning anatomy from the dissections seemed to solicit the patience of his general readership while also informing his more scientific readership of his professional intent. Audubon's popular audience was not necessarily sympathetic to this shift. In a review of the final volume in the *North American Review*, the writer complained that "such illustrations and descriptions will add to the permanent value of his work; but they take something away from its unity, . . . and lessen its attraction for the general reader; besides that the additional labor, . . . has induced [Audubon] to give up those episodes, descriptive of scenery, life, and manners, which harmonized well with the subject and formed one of the chief attractions of the 'Biography.'"[70]

Even with MacGillivray's taxonomic and anatomical addenda Audubon's text differed fundamentally from its scientific competitors. The early volumes had set the tone of the whole. In recounting and depicting dramatic situations, unusual events and daring exploits, Audubon portrayed as unique his own role of observer-artist. Audubon's picture of a rattlesnake raiding a mockingbird nest drew skepticism from other naturalists. Audubon had shown the mouth open wide enough to see both fangs. Critics, especially Charles Waterton, the English naturalist who led the challenge to Audubon's veracity, protested that rattlesnakes neither climbed trees nor projected their fangs in the way that Audubon had shown. Audubon's observations were later corroborated by other American observers bent on defending national honor as well as Audubon's reputation as a naturalist. But the controversial scene demonstrated his preference for the uncommon or unique event over the repeatable.[71] In principle, professional naturalists—the new "zoologists"—

described observations that others should be able to reproduce; an observation that could not be repeated had little scientific credibility. Audubon created a persona of an individual with a privileged vision of events, but his position differed from the privileged vision of the professional zoologist. As the editor of the *Magazine of Natural History* had indicated, the observer who possessed systematic knowledge was superior to observers who merely enjoyed nature as a mirror of human values. While Audubon placed himself and his reader in nature, the zoologist in contrast extracted or abstracted the object under observation in an attempt to discern its place in the natural order, or system.

Audubon had one impeccable supporter of his scientific claims. Georges Cuvier had praised Audubon's plates to the Académie Royale des Sciences as the greatest monument yet raised to science, and urged that public institutions subscribe to and support Audubon's work.[72] Audubon reprinted Cuvier's praise as an appendix to the *Biography*'s first volume. But in a later review of the *Ornithological Biography*, Frédéric Cuvier (1773–1838), brother of Georges and an ornithologist as well as a student of mammalian behavior, reiterated the distinctions made by Swainson, while bringing to them a fine balance between professional exclusiveness and recognition of its limitations:

> Monsieur Audubon is not . . . a naturalist: he is an able painter and an intelligent observer; but it is perhaps precisely because he is a stranger to the study of nature, that he was led to create an original work of natural history, that no professional naturalist would probably have had the idea of undertaking; for the impetus that science has received, and to which obey more or less all those who are concerned today with the study of natural beings, does not lead to the researches to which Monsieur Audubon devotes himself.[73]

Audubon's plates had a more lasting influence on British ornithology than did his text. One of Audubon's subscribers, John Gould, or-

nithologist and curator of the collections of the Zoological Society of London, would issue volumes of lavishly executed lithographic bird portraits reminiscent of Audubon's style and, like the *Birds of America*, published as a commercial venture.[74] Gould was trained in museum systematics and made substantial contributions to the systematic literature, including a monograph on his own collections of Australian birds and descriptions of the birds that Charles Darwin collected on the voyage of the H.M.S. *Beagle*.[75] Gould stood as a late example of the naturalist-illustrator; tellingly, he adopted the style of the artist-naturalist. He wrote of and illustrated exotic birds during the heyday of Britain's overseas empire; his work anglicized Audubon's elaboration of landscape or habitat and made familiar exotic birds which few among his English audience could know by sight. Similarly for Americans, Audubon's plates elevated the birds of North America to the status of fine art subjects and popularized their habits during the early decades of American westward expansion.

The *Ornithological Biography* reached a wider audience than the expensive full-sized plates and in turn expanded the audience for Audubon's pictorial work. MacGillivary had told Audubon: "Your imperial size and regal price do not answer for radicals, or republicans either. . . . I have often thought that your stories would sell very well by themselves, and I am sure that with your celebrity, knowledge, and enthusiasm, you have it in your power to become more *popular* than your glorious pictures can ever make you of themselves, they being too aristocratic and exclusive."[76] With its mix of description and adventure, celebrating the self-taught naturalist as hero, the text in a sense democratized the plates, which required the patronage of a luxury audience. The popularity of the *Biography* also created a market for smaller, less expensive editions of the plates.

Audubon's work became a widely held model for American expectations of nature, for the way Americans saw birds and saw themselves observing birds. Audubon's ramifying influence on American life reflected the context into which he launched the work. As in England and France, scientific institutions of the United States were beginning to undergo the kinds of changes indicated by Swainson's and the younger Cuvier's criticisms of Audubon— an emphasis on collection building and taxonomy, publication for the specialist, and not general, audience. At the same time, however, a proliferation of local natural history societies continued to rely on amateur membership. From the ranks of amateurs a small but growing number of colleges recruited students for newly established courses of natural history study. The ideal that natural history should be accessible to all continued to inform popular opinion while, at the same time, specialists distinguished their systematic practice from the more general, popular natural history. Nevertheless, the proponents of state funding for botanical, zoological, and geological surveys called upon that commonly held ideal when they argued for the general utility of science.[77]

Having established his own genre of representation, Audubon enjoyed a privileged position among naturalists and field collectors working in the narrative vein. Thomas Nuttall, a distinguished English naturalist best known for his botanical studies, explored and collected widely in the west, and wrote on birds in Wilson's tradition of accessible narrative and description. John Kirk Townsend planned a major work to describe and illustrate his western collections. But the rising generation of specialists characterized the work of Nuttall and Townsend as old-fashioned, and their publications were marginalized. Townsend published only a fragment of his intended work, and Nuttall's writings were relegated to the popular realm. Nuttall was denied access to collections, while he himself had deposited his at the Academy of Natural Sciences. The Academy, meanwhile, gave Audubon priority access to Townsend's bird skins.[78]

The late 1820s and 1830s—the period when the *Birds of America* and the *Ornithological Biography* reached the United States—saw the beginnings of the first flowering of American landscape painting led by Thomas Cole. Audubon's engraver Robert Havell, Jr., would emigrate to the United States to become a landscape painter. George Catlin was painting and describing the American Indian as Audubon had painted and described birds. The improving position of the fine arts and of their practitioners in American society and the increasing popularity of landscape painting enhanced Audubon's reputation at home. At the same time, the growing distinction between naturalist and illustrator in scientific practice created the basis for controversy over defining his true calling. Audubon's supporters insisted that he was an artist, and some that he was both artist and ornithologist. His scientific critics, largely European, characterized him as an artist who lacked the necessary criteria of scientific observation and who failed to conform to the proper conventions of the translation of observation into pictorial representation and text. Audubon's work appeared at a time when most Americans could accept it as both fine art and natural history, if not strictly "science." Soon after the work's completion, however, it would be to the painter that American society assigned the celebration of nature as landscape, and to the scientist its analysis and systematic arrangement.[79] In an important sense, Audubon, man and work, epitomized the transformation of the practice of the naturalist-illustrator into the ethos of the artist-naturalist.

In the nonscientific press, American reviewers embraced Audubon's productions as monuments to national arts and letters. On the occasion of the Philadelphia publication of the *Ornithological Biography*, a writer in the *North American Review* of 1832 expressed the belief that the American public needed pictures to be led into natural history study.[80] Through the work of Audubon, knowledge of the nation's birds and their habits would become widespread: "A few such men as Audubon will soon place the results of their adventurous travels, where men shall see and know them; a taste for their favorite sciences will gradually be created, and they will be sure of general applause."[81] The same reviewer would contribute to the same journal extended essays on Audubon in 1835 and again in 1840, marking the progress of Audubon's publications in the United States. The reviews retold excerpts from the *Biography* of Audubon's travels to Labrador and Florida. These collecting trips, mapping the length of the country, had provided the material to complete the plates and text. For Audubon's audience the combined results offered a sense of geographic expanse through the descriptions of birds and landscape.

American reviewers elaborated the comparison between Wilson and Audubon. They popularized and perpetuated Audubon's efforts to establish priority over Wilson. In defending himself against accusations of having copied Wilson's drawings, Audubon claimed to have lent drawings to Wilson, who then failed to credit him. The writer for the *North American Review* preferred Audubon's drawing to Wilson's. He found Wilson's text and illustrations single-minded and humorless in contrast to Audubon's accounts, which he believed gave a portrait of the country as well as of its birds. His judgment reflected not only a perceived difference between naturalist-illustrator and artist-naturalist, but a change in taste in fine arts and letters. Through the publication of such reviews and of the less expensive editions of plates and text, Audubon's impact on the American popular audience for natural history virtually eclipsed Wilson from popular memory. Wilson became the ornithologist's ornithologist, while in the developing national ideology of nature Audubon became an enduring symbol of the relationship between individual and nature.

Just as the writer for the *North American Review* combined literary with ornithological appreciation, an unattributed review in the *American Journal of Science and Arts* in 1840 could both discuss the fine points of ornithological systematics and credit Audubon for fostering the growing popularity of natural history study. These combinations spoke to the continued linking of natural history to fine arts and literature in the national culture-building project. The writer for Silliman's journal noted that the attractiveness of Audubon's text would spread the improving social influence of natural history, which "must have a necessary and inevitable tendency to impress the mind with the truth of religion, and thereby to improve as well as regulate the moral feelings."[82] In the decades of rising evangelism, temperance, and economic differentiation, the reviewer saw an important social role for natural history:

> How fearfully true is it that nine tenths of the immorality that pervades society, originates in the first place from want of some occupation at least harmless, to fill up vacant time. If, therefore, the study of the natural sciences are as attractive as they have been shown to be beneficial . . . [it] must necessarily exert a beneficial moral influence. . . . Let [the reader], if he still doubts, read the poetic and animated pages of the Wilsons and the Audubons . . . and he cannot fail of being inspired with a love for their pursuits.[83]

Audubon had "probably contributed more toward creating and fostering a taste for nature in this country" than anyone. "His magnificent and unequalled painting created every where a great interest in the subject." The reviewer commended Audubon for a combination of virtues, some appropriate to the frontiersman, some to the rising entrepreneur: "boldness," "perseverence and untiring zeal," "fidelity, industry," and "enterprise."[84] Audubon had succeeded in depicting himself, like his birds, in heroic proportions, "superior to all disasters, surmounting all obstacles, and completing in spite of them, the most magnificent work on natural history the world had ever seen."[85]

Audubon had painted himself permanently into the American landscape. The resonant meanings of Audubon's work and its popular

Red-shouldered Hawk.

FALCO LINEATUS, Gmel.

Male 1. Female 2.

3.11  Red-shouldered Hawk. Uncolored aquatint engraving by R. Havell, Jr., from a drawing by J. J. Audubon, 1826.

interpretation, the cultivation of landscape in painting, and the beginnings of the state scientific surveys converged with the developing ethos of westward expansion into territories whose wealth was represented in part by natural history and landscape, both to be converted through settlement into social wealth. Audubon, the self-taught frontiersman risen to eminence in Europe and at home, was the naturalist par excellence of the Jacksonian era, while at the same time his idealization of nature, and establishment of a privileged vantage point for viewing it, appealed to the Whig elite of New England.[86] Audubon's contemporary culture-hero counterparts opened the Oregon Trail and roamed the best-selling fiction of James Fenimore Cooper.[87] The wilderness celebrated in these mythmaking works and deeds was about to be subjected to speculation, migration, and cultivation. Audubon, like Leatherstocking, symbolized the appropriation for the European population of the intimate knowledge of nature associated with the Indian.

PLATE
25

In 1835 Thomas Cole, like Audubon and Cooper a proponent of the exalted position of the solitary individual in romantic relationship to nature, attempted to reassure himself that nature would continue to assert its moral influence despite rapid settlement: "Nature is still predominant, and there are those who regret that with the improvements of cultivation the

sublimity of the wilderness should pass away; for those scenes of solitude from which the hand of nature has never been lifted, affect the mind with a more deep toned emotion than aught which the hand of man has touched."[88]

Audubon's *Birds of America* gave its viewers access to "those scenes of solitude," and both the work and its creator gained a central place in the cultivation of the artist-naturalist ideal in the realm of popular natural history and nature appreciation. In closing the last volume of the *Biography*, Audubon passed the torch of the frontier artist-naturalist to his readers: "Although I can never entirely relinquish the pleasure of noting new facts in zoology, or of portraying natural objects, whether on canvass or on paper, I shall undertake few journeys, save short rambles for amusement."[89] Many species "yet remain to be added" to the catalog of American nature. "Nay, I look upon the whole range of those magnificent [Rocky] mountains, as being yet unexplored."[90] Evoking the central contradiction inherent in his own role in the popularization of the solitary individual confronting nature, Audubon admonished: pack "your bird's-skin flat in your box" and roll "up your drawing around those previously made," then "repair to Boston, New York, or Baltimore, where you will find means of publishing the results of your journey."[91]

PLATE
26

FIGURE
3.11

# 4

# Scientific Prestige, National Honor:
# Pictures for Federal Science

FIGURES
4.1–3

EVEN as Audubon completed his *Birds of America*, the adventure of field collecting and illustration increasingly passed from the hands of the lone naturalist to appointed members of state-sponsored surveys and federal exploring expeditions. The 1830s witnessed an increase in the publications of state geological and zoological surveys, which received funding on the premise that an inventory of mineral wealth and animal populations, both useful and harmful, had public economic value.[1] Survey publications and their illustrations tended to reflect the modest means of the state programs. Their illustrations contrasted markedly with the refinements displayed in private publications of the same period.[2]

The entry of the federal government into the full-fledged production of illustrated reports of expeditionary science inaugurated a new stage in the process of developing conventions in American zoological illustration. Public debate about federal support of science linked it to national honor and government prestige.[3] Natural history had entered the federal context as an adjunct of discovery, westward expansion, and settlement; it put to sea in the service of navigation and mapping. American vessels were sailing in waters for which there existed no official maps and charts, thus risking considerable investment, particularly in whaling.[4] Rivalry with European countries freighted exploration

with the additional task of establishing the American presence; the illustrations of federal science documented not only knowledge but possession.[5] During the initial stages of the vexed process of publishing the scientific results of the United States Exploring Expedition, which traversed the Pacific during the years 1838 through 1842, the expedition commander, Charles Wilkes (1798–1877), appealed to these concerns to support his complaints of poor treatment: "Contrast our Expedition with those of the French and English engaged in the same service and at the same time, honor and rewards are heaped on all before their return. Examine our results, compare them with theirs, contrast us in every way with them you please, or with Expeditions that have gone before us, and then ask if we have not reason to feel mortified."[6]

The natural history activity of the past three decades had linked the resources for illustration to scientific credibility. Commitment to meeting the expense of engraving and coloring plates was one of the requirements and emerging rules of the production of science.[7] Producing and illustrating government science opened new questions: How was public money being spent? Who decided what kind of science should be done, the scientific community or representative government bodies? Who benefited? Who constituted the audience for govern-

MOUNT KTAADN, from W. BUTTERFIELD'S, (Oct 8th 1836.
Near the GRAND SCHOODIC LAKE.

4.1 Mount Ktaadn, from W. Butterfield's, Oct 8th 1836, Near the Grand Schoodic Lake. Lithographic plate by Moore's of Boston, for the Maine geological survey, from Charles T. Jackson, *Atlas of Plates Illustrating the Geology of the State of Maine* (1837–1839).

ment-published science? What should government science look like? The federal government entered the arena of scientific publication during a period when the American scientific community's answers to these and related questions were in transition. The divergence between the popular and specialized audiences for science continued to widen, particularly as scientists worked to establish their social and intellectual positions independent of arguments for practical utility.[8] A new generation of scientific power brokers worked from in-

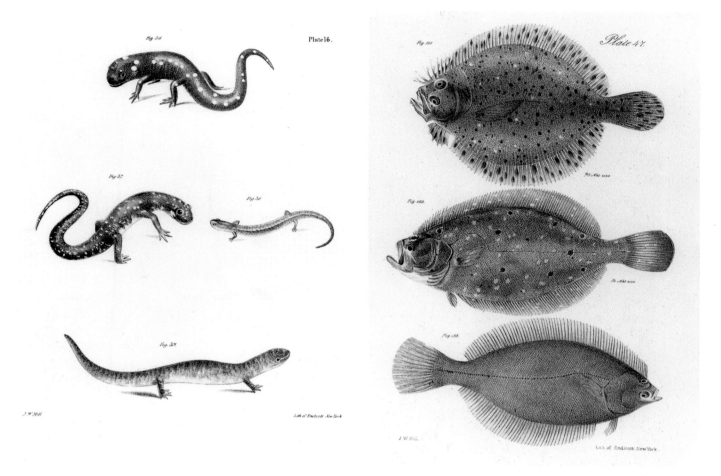

4.2–3 Salamanders (*left*); and a Turbot, a Flounder, and the Spotted Flat-fish (*right*). Lithographs by Endicott, from drawings by J. W. Hill, for De Kay, *Zoology of New-York, or the New-York Fauna*, parts 3 and 4 (1842).

creasingly institutional bases. Benjamin Silliman at Yale still wielded a strong influence through his journal; a well-placed letter to Silliman could exclude a naturalist who failed to conform to standards from publishing in the major journals.[9] Silliman's young protégé, the geologist James Dwight Dana (1813–1895), would play a central role in the next decades of American science. Silliman and Dana, with botanist Asa Gray (1810–1888) and Gray's mentor John Torrey (1796–1873) of the New York Lyceum, helped formulate, and influence their fields to adhere to, a standard of work that

made the aging naturalists of the Philadelphia circles like Coates and Ord seem outmoded.[10]

The look of taxonomic illustration increasingly reflected the institutional base, although the new pictorial conventions would not fully consolidate for several decades. In the period between around 1830 and 1850, the old and the new coexisted in close juxtaposition and to some extent contradicted each other. The ideal of the transcription into words and pictures of the living animal in nature yielded to the institutionally based standard of cataloguing the animal and placing it within a taxonomic system,

121

with the choice of system rather than observation in context the main focus of debate. Plates illustrating the publications of the scientific societies reflected this shift. On the one hand, illustrations still appeared showing a single species in a conventionalized landscape—usually a clod of earth sprouting grasses for mammals, a stylized tree for birds. But plates illustrating several species resting on a single plane, or floating in a space of indeterminate depth, began to predominate. The old style did not always illustrate field-based natural history, nor did the new style of composite plates always illustrate institutionally based taxonomy; recall that Lesueur combined both modes in his illustration. The vestigial landscape conventions for illustrating vertebrates, except snakes, persisted despite changes in taxonomic methods. The smaller invertebrates, mollusks and insects in particular, had long lent themselves to composite arrangements; budget constraints encouraged the practice. Audubon's plates, epitomizing the artist-naturalist ideal of one species—or one dramatic interaction—to a plate, had given new life to the old natural history conventions, but his work hit the United States at the same time as the panics, bank failures, and economic depression of the late 1830s. While Audubon's work was influential, composite plates were more economical, and by their rejection of pictorial space and narrative mode they declared the priority of taxonomy.

Most practicing naturalists still drew, either for study or to illustrate their own work, but there emerged an ethos of division of labor in observation and representation that emulated distinctions well established in Europe.[11] The principal change distinguished scientist from draftsman, with a consequent drop in the status of the draftsman's work. At the turn of the century, natural history had foundations in craft, as in the collaboration of Wilson, a former weaver, and Lawson, an engraver and printer who was also the center of a free-thinking intellectual circle. The rising status of science reflected in part the increasing number of college-trained specialists, when tertiary education was a mark of privilege. Draftsmen, in contrast, often trained in the crafts and trades. Class distinctions, then, had begun to mark the distinction of roles in natural history production.

The division of labor in scientific practice and status distinctions among practitioners also reflected the changing structure of printing and publishing. As the scale of scientific publication grew, drawing and printing scientific illustration moved from family engraving workshops such as Lawson's, whose children Helen, Malvina, and Oscar drew and engraved scientific subjects (or that of Peter Maverick in New York, whose family drew and engraved for authors publishing in the *Annals* of the Lyceum of Natural History), to commercial firms, some large and mechanized.[12] As printing and publishing outgrew the family workshop, the status of draftsman dropped from journeyman to employee. By 1850, for example, 95 percent of employees in the Philadelphia printing industry worked for firms with six or more employees, of whom a striking 43.8 percent worked in firms with fifty or more employees.[13] Moreover, the gradually increasing use of lithography to print zoological illustrations began to interpose the hand of staff draftsmen of commercial lithography ateliers in the production of scientific pictures. During the 1830s and '40s, these changes were in motion, although not yet fully realized; their implications were still open for debate.

The project that marked the federal government's full-scale debut into the production of illustrated scientific reports was the program of publication of zoological, botanical, mineralogical, and ethnographic collections brought back from the Pacific by the United States Exploring Expedition.[14] This unwieldy and unprecedented project highlighted the far-reaching changes in

4.4–6 Three lithographic plates from the *Journal of the Academy of Natural Sciences, Philadelphia* 8 (1839–1842).

122

Lepus Townsendii.

On Stone & Col'd by J. Ackerman.　　　　　　　　　Printed by P.S.Duval, Phil.ᵃ

4.4 *Lepus Townsendii*, on stone and colored by J. Ackerman, printed by P. S. Duval.

Thalamita pulchra.

On Stone & col'd by J. Ackerman.　　　　　　　　Printed by P.S.Duval.

4.5 *Thalamita pulchra*, on stone and colored by J. Ackerman, printed by P. S. Duval.

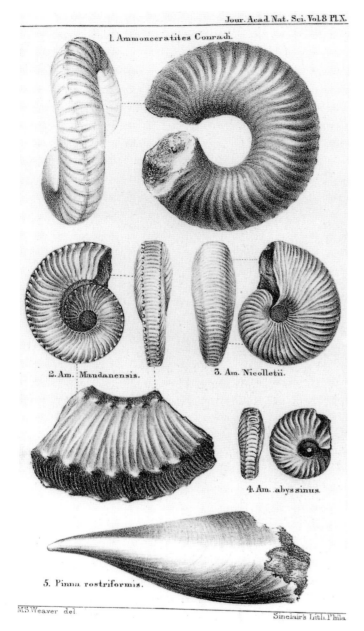

Jour. Acad. Nat. Sci. Vol. 8 Pl. X.

1. Ammonceratites Conradi.

2. Am. Mandanensis.

3. Am. Nicolletii.

4. Am. abyssinus.

5. Pinna rostriformis.

M. S. Weaver del

Sinclair's Lith. Phila

4.6 Ammonite fossils, drawn by M. S. Weaver, printed by the firm of Thomas Sinclair.

the social structure of scientific practice and production, and the corresponding tensions about illustration and its meanings.[15] As a publication program it was a fiasco, especially when compared to the successful production of

other illustrated works of the same period, the 1840s through 1860s. Authors of the expedition reports repeatedly deferred their expedition work while they published under other auspices some of the most elaborate volumes of illustrated science—both governmental and private—of the era.[16] The publications of the Exploring Expedition were buffeted by differences of opinion among its sponsors and proponents about how science should be done, and for whom, and how it should be produced. Much of the tension generated by the publication program of the Exploring Expedition concerned scientific versus navy professionalism. The conflicts between the expedition commander and the scientific corps, backed by the national scientific community, and conflicts between the commander and Congress revealed how assumptions about the practices of natural history and its illustration were fundamentally changing, as well as how many of the former assumptions and practices lingered to confuse the issues at hand. In the main, however, using networks to exert inside pressure, it was the new generation of scientists who won the skirmishes.

Before the expedition sailed, many of the disagreements over its staffing revealed those shifts already in process, for the protracted history of disputes centered on the role and scope of its scientific mission. Before Charles Wilkes's appointment as commander, natural history enjoyed a prominent role in that mission. Jeremiah Reynolds, an ardent proponent of the expedition's scientific purpose and the corresponding secretary for its organization, had elaborated the original plan for the scientific corps in an 1836 letter to President Andrew Jackson.[17] Reynolds considered the natural-historical survey as important as the meteorological and astronomical investigations. He buttressed the pure scientific value of the work by reminding the president of the economic benefits of European voyages. He recommended for the scientific posts men of established standing in their fields, picked from both the

CONTENTS.

PENDLETON'S ESTABLISHMENT.

No. 1. Graphic Court.

ENGRAVINGS

OF ALL KINDS

EXECUTED ON STEEL AND COPPER.

VISITING, PROFESSIONAL, INVITATION, and BUSINESS

CARDS.

**Copperplate Printing and Lithography**

IN ALL THEIR BRANCHES.

Particular attention paid to the execution of

ENGRAVED OR LITHOGRAPHIC COPIES of PORTRAITS
and MINIATURES, VIEWS of PUBLIC and PRIVATE BUILD-
INGS, SURVEYS, MAPS of TOWNS, &c. &c. PLANS of all kinds
of MACHINERY, SPECIFICATION-DRAWINGS for PATENTS,
INVITATION AND BUSINESS CIRCULARS, REPORTS, &c.
in *fac simile* or otherwise. Also every description of EMBELLISH-
MENTS for BOOK-WORK and at every variety of price and finish.
STONES and all necessaries for LITHOGRAPHIC DRAWING, supplied to *Artists*
and *Amateurs.* INSTRUCTION furnished without any charge.

BEST PLATE MUSIC

on TINTED PAPER, at the lowest prices.

ENGLISH AND FRENCH ENGRAVINGS

AND LITHOGRAPHIC PRINTS,

Always on hand, and all novelties in the ARTS regularly imported.

4.7 Notice advertising Pendleton's Establishment in *The Naturalist* of 1832, a short-lived natural history magazine published in Boston.

older and the younger generation of naturalists: Asa Gray for botany, James Dwight Dana for mineralogy and geology, Charles Pickering for all branches of vertebrate zoology, and Reynell Coates—considered by many a purely political appointment—for the invertebrates.[18] (Few of the original nominees in fact sailed with the expedition.)

Reynolds's comments on illustration are revealing of some of the new standards and statuses involved:

I[ts] high importance will be at once seen and acknowledged. Natural His[tory] drawing, particularly, is of the utmost consequence, as well as by far the most difficult province to fill; requiring as it does a kind of talent little cultivated in this country. Still it is probable, persons maybe found capable of executing this description of drawing, under the direction of naturalists, in a creditable manner. They alone are qualified to dec[ide] on the merits of artists of this class, and it is respectfully suggested, th[at] they be authorized, immediately on their appointment, to make the [ne]cessary inquiries, and that they recommend to the Secretary of the Navy such candidates as they approve.[19]

Reynolds further recommended that the landscape and ethnographic portrait artists selected for the voyage "be skilled in Natural History sketching; so as to be capable of assisting Natural History draughtsmen, when consistent with their regular employment; or occupying their places should the latter be temporarily incapacitated."[20] Reynolds's candidate for portraitist came with the recommendation of the Lyceum of Natural History of New York. And for the landscape artist, Reynolds suggested one ranking "only second to Mr. [Thomas] Cole among the landscape painters of the United States."[21] Note the discrepancy between Reynolds's high opinion of the importance of illustration and his low opinion of its current American practitioners, especially in contrast to his respect for landscape painters. Clearly, the fine arts enjoyed a rising status in American society, comparable to the growing recognition of the value of scientific endeavor. Yet Reynolds considered those who provided the all-important pictures for science explicitly subordinate to the scientists themselves; even the appointed landscape painters and portraitists were, at times, to submit to the direction of the scientific corps. A letter of recommendation for the scientific corps written by a committee at the American Philosophical Society confirmed Reynolds's perceptions of the division of labor and status between scientist and illustrator: "The Zoologists should observe, draw and describe the various animals inhabiting the Countries which may be visited by the Expedition. The Assistants should be qualified to collect, draw or prepare specimens for preservation."[22] One sees two distinct functions of drawing implied here: drawing associated with observation and taxonomic description by the scientist, and drawing as record keeping, an essential step in collecting, by an "assistant."

With the surprise appointment to the squadron command of Lieutenant Wilkes, promoted over the heads of senior officers, the status of the scientific corps on the expedition took a severe blow, despite his reputation for attainments in mathematics and astronomy.[23] Wilkes favored navy over civilian personnel for all scientific posts. In a memorandum circulated among President Van Buren and the secretaries of war and the navy, Wilkes outlined his idea of the proper relation of civilian science to navy and navigational needs, assigning to naval officers "all the duties appertaining to Astronomy, Surveying, Hydrography, Geography, Geodesy, Magnetism, Meteorology, and Physics."[24] Medicine having been for decades the profession most closely allied to natural history, and natural history itself having lacked a defined occupational status, Wilkes proposed that members of the navy medical corps would fulfill the natural history duties. Only when qualified amateur naturalists among the navy physicians could not be found did Wilkes resign himself to seeking civilian specialists. With the concurrence of Titian Peale, Wilkes reduced the number of scientific personnel.[25] For the task of pictorial documentation, Wilkes sought generalists and reduced the total number needed for the entire squadron to "two draughtsmen well qualified in all the Different departments of drawing, which number of persons are the most that can be accommodated on board vessels

without great inconvenience to their officers and crews."[26] Since Peale believed that naturalists should do their own drawing, his recommendations may have influenced Wilkes in the reduction of the number, and hence the status and importance, of painters and draftsmen on the voyage.

The secretary of war endorsed the new proposals, noting, however, that "it would appear injudicious to dismiss entirely the whole of the Scientific Corps," for it could not be done "without creating much clamour"; still, "every judicious effort ought to be made to reduce the minimum as low as possible."[27] The secretary, nevertheless, considered the voyage's mission to be primarily scientific, as he stressed in his instructions to William Hudson, commander of one of the squadron's vessels: "The objects . . . being altogether scientific and useful, intended for the benefit equally of the United States, and all commercial nations of the world, it is considered to be entirely divested of all military character."[28]

After the reshuffling and resignations among the scientific corps had finished, the staff included Dana as mineralogist; Horatio Hale (1817–1896), a Harvard graduate, as philologist and ethnologist; Joseph Pitty Couthouy (1808–1864), a Boston sea captain, and member of the Boston Society of Natural History as conchologist; Titian Peale as collector of birds and mammals; Charles Pickering (1805–1878), also a graduate of Harvard and a curator at the Academy of Natural Sciences, Philadelphia, as general naturalist; and William Rich and William Dunlop Brackenridge as botanists. Joseph Drayton (d. 1856), a Philadelphia engraver and lithographer with an extensive publication record in scientific illustration, and Alfred Agate (1812–1846), a young miniature painter from New York with a European tour to his credit, were appointed to the positions of artist.[29] Drayton's illustrations, both engraved and lithographed, had appeared regularly in the *Transactions* of the American Philosophical Society from 1830 until after his departure, in mid-

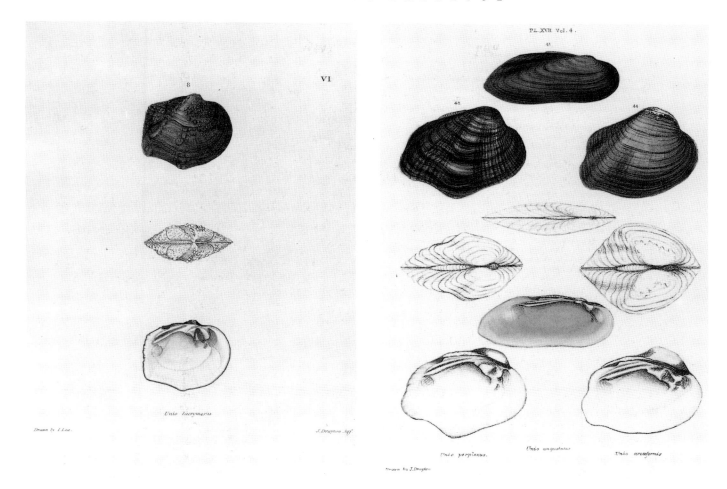

4.8–9 Two of Drayton's plates of shells, one an engraving (*left*), the other a lithograph (*right*), illustrating Isaac Lea's descriptions, from the *Transactions of the American Philosophical Society* 3 and 4 (1830 and 1834).

FIGURES
4.8–9

career, with the squadron. Drayton could engrave on copper and draw on stone. Only when the shell plates Drayton had prepared for the Philadelphia conchologist, Isaac Lea, finally ran out, did he have to resort to the services of a commercial lithographer.[30]

Even before the squadron's departure in 1838, disagreements between Wilkes and his scientific staff became the gossip of American scientific circles. On the voyage, the scientists found their work impeded by the ill will of the officers and Wilkes's arbitrary rulings. Early on, Wilkes's relations with Peale and Couthouy became openly hostile; they were to end in courts

martial. Couthouy's health failed, and he left the squadron to return home; Wilkes considered him dismissed. The commander's insistence on access to the scientists' journals and their relinquishment of all their collections, notes, and drawings to the government caused considerable discontent. Nevertheless, the scientific corps managed to make the most of opportunities to observe and collect in and between Madeira, Rio de Janeiro, Valparaiso, Samoa, Tahiti, Sidney, Tonga, "Feegee," and Hawaii, then called the Sandwich Islands, and the Oregon coast.[31]

Wilkes wished to command the publication

FIGURES
4.10–11

Plate 27. Rousette de Tonga. Color engraving from the zoology atlas of the *Astrolabe* expedition. Edited by J. Tastu, printed by V. Drouart, 1830–1835

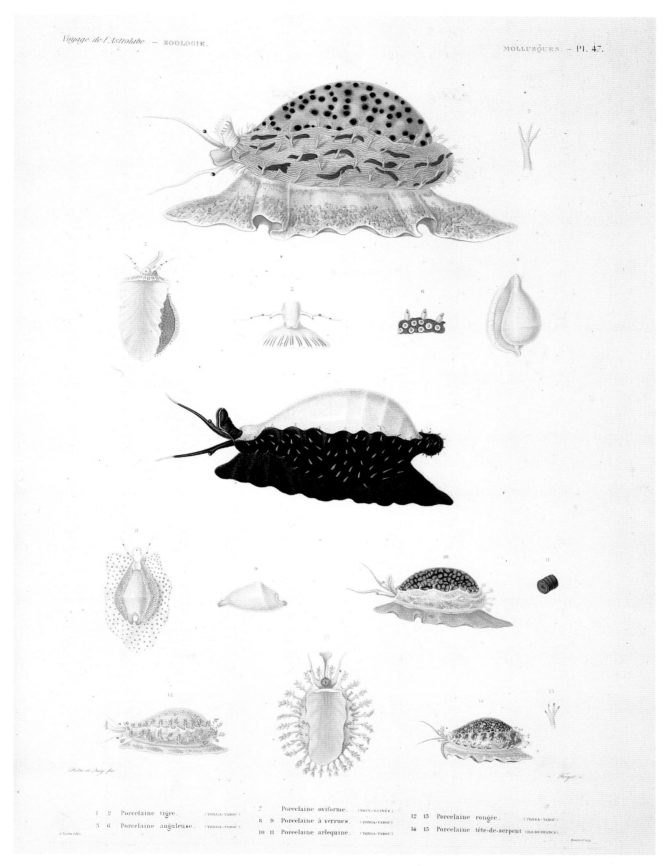

Plate 28. Porcelaine species. Color engraving from the zoology atlas of the *Astrolabe* expedition. Edited by J. Tastu, printed by V. Drouart, 1830–1835.

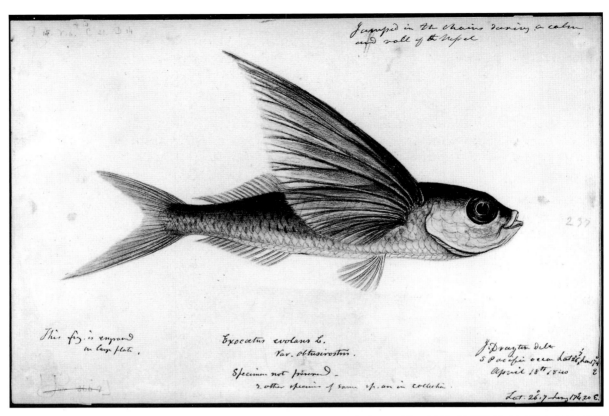

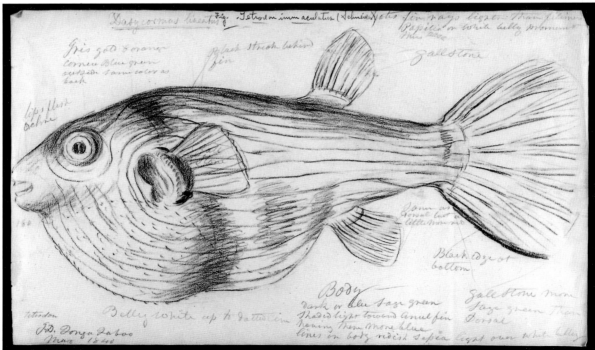

4.10–11 Two of Drayton's fish drawings made on the voyage of the U.S. Exploring Expedition.

of the expedition's scientific work as firmly as he had ruled his squadron at sea. The enmities that had arisen during the voyage entered the public record in the courts martial of Wilkes and three junior officers, and persisted both in print, especially in Wilkes's own account, and in the management of the publication program.[32] In 1842 Wilkes wrote to the secretary of the navy:

> I truly regret that anything should have occurred to dampen the ardor of those who are attached to the Expedition and absolutely necessary to the bringing out the results. . . . The reputation of our country is at stake and if what had been attempted and succeeded in, is not now finished from any notion of economy, or derangement of the organization all will be ruined and we will become the laughing stock of Europe and all the praise that has been lavished on our Government for its noble undertaking prove but satire in disguise.[33]

The naturalists resented interference in what they viewed as scientific decisions. The rigid and often ad hoc policies formulated by Wilkes and by the overseeing publication body, the congressional Library Committee initially headed by Ohio Senator Benjamin Tappan, created obstacles to the efficient study of the collections. The collections must never leave Washington; only new species were to be published; all monographs must conform to the same format regardless of subject; each report was to be printed in editions limited to one hundred copies.[34] Nevertheless, distinguished members of the scientific community, eager for access to the Pacific collections and a chance to influence the direction of the publication program, vied to take on the authorship of reports as some of the original members dropped out.[35]

The federal natural history collections became something of a national embarrassment. The first shipments had been sent to the Peale museum. But subsequently, the enormous volume of the expedition's scientific accumulations sent to Washington had literally over-whelmed the capital's storage capacities long before the squadron returned in 1842. The Great Hall of the Patent Office, alloted for storage and exhibition, had filled with barrels, crates, cans, and bottles faster than anyone could sort, prepare, or arrange the specimens.[36] Emergency decisions about how to cope with the collections, or what Peale called "a general scramble for curiosities," frequently sacrificed both their safety and their eventual use to science.[37] Public exhibition took higher priority than research, and mounting specimens for display usually spoiled their research value. Peale counted one hundred and eighty bird specimens, including new species, lost.[38] He despaired: "my two birds (male & female) made into one,—the legs of one put on another body.—hundreds of fine insects put in 'families' without localities, although they came from all parts of the world . . . all for the great end,— promotion of science."[39] Some of Couthouy's specimens of rare shells had been distributed among naturalists; his labels with locality information were removed from others.[40] Dana found that many of his specimens of crustacea had been dried for exhibition, separated from their labels, or broken: "Some were rendered wholly unfit for description, especially those of small size, which, without regard to their delicacy of structure, were taken from the bottles containing them and dried, and sometimes transfixed with pins, to the obliteration of many of their characters. Moreover, the larger species were rendered by this process unfit for dissection."[41] Conditions for preparation of the reports caused Couthouy and Pickering to resign from their positions. Their replacements, and those who remained, broadcast their frustrations to sympathetic colleagues. Dana complained to Asa Gray: "It is perfectly absurd that I should be able to prepare my reports in a city where there are no books!"[42]

*On pages 132–133, 136–137:*
Plates 29–32. Hand-colored engravings from Gould, *Atlas: Mollusca and Shells* (1856).

131

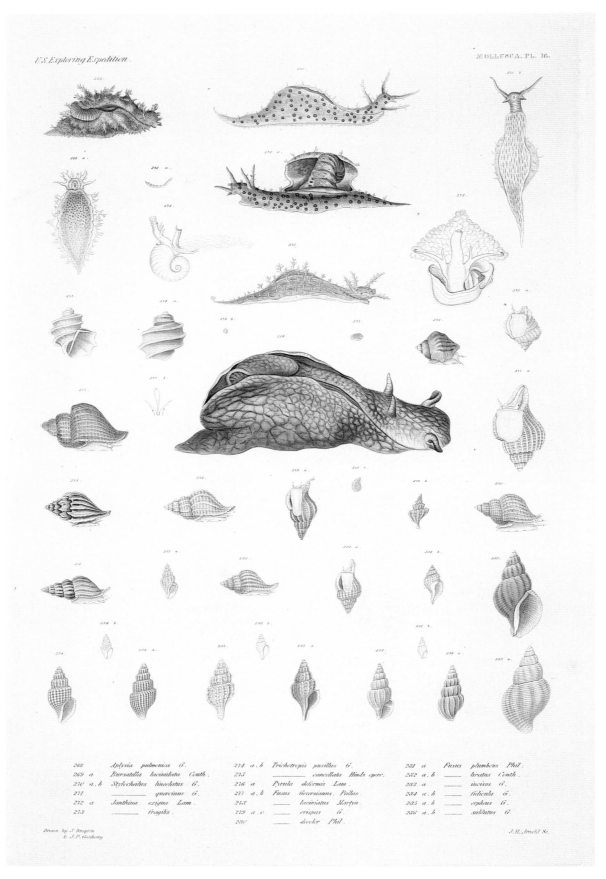

Plate 29.  Aplysia, Bursatella, and Stylocheilus species, drawn by J. Drayton and J. P. Couthouy, engraved by J. H. Arnold.

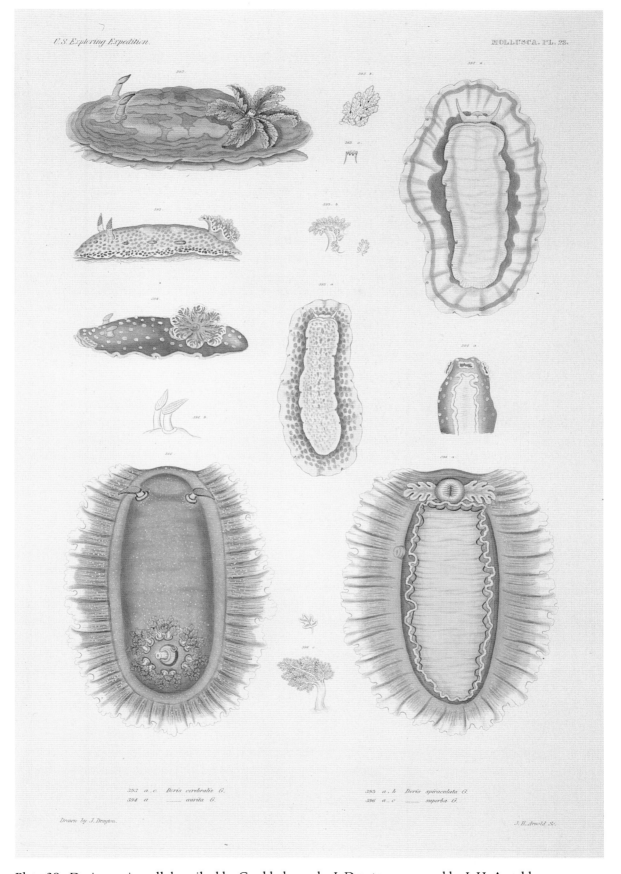

Drawn by J. Drayton.

J. H. Arnold Sc.

393 a, c  *Doris cerebralis* G.
394 a  ——— *aurita* G.

395 a, b  *Doris spiraculata* G.
396 a, c  ——— *superba* G.

Plate 30.  Doris species, all described by Gould, drawn by J. Drayton, engraved by J. H. Arnold.

To Joseph Drayton, who had sailed with the expedition and whom Wilkes considered "one of the most experienced artists in the U.S.," the commander assigned the major part of the burden and responsibilities for the publication program.[43] Drayton's official duties included preparing the drawings for the fish and shell reports as well as superintending the illustrations for all the other works. He chose and inspected paper stock for the letterpress, maps, and illustrations. He had charge of inspecting all the engraved plates before printing, approving the proofs and checking the complete press run for consistency. He also bore responsibility for supervising the hand coloring of all the engraving and for guiding the volumes through printing and binding. Together Wilkes and Drayton assigned printing contracts and screened and employed additional draftsmen and engravers for the work "under the strictest terms as respects cost, time, finish, style, &c."[44]

The 1842 bill that provided for the publication of the expedition's discoveries stated specifically the style and format the reports should take, "which account shall be prepared with illustrations and published in the form similar to the voyage of the Astrolabe, lately published by the Government of France."[45] The reports of the corvette *Astrolabe* under the command of Dumont d'Urville established the international standard for the American expedition's publications, both in style and substance. The engraved plates of the Astrolabe zoology atlas displayed French engraving at its most refined, portraying specimens in a combination of line and roulette techniques, printed in colored inks and finished with hand coloring. The illustrators and possibly also the engravers of the Exploring Expedition report atlases were clearly made familiar with the Astrolabe atlas of zoology, for many of the American plates strongly resemble the French illustrations. The Astrolabe volumes had made a strong impression on American administrators even before their own expedition sailed. In the late fall of 1837,

PLATES
27–28

Mahlon Dickerson, then secretary of the navy and an advocate of minimal government expenditure, had attempted to persuade President Jackson that the successful results of Dumont d'Urville's first expedition of a single vessel validated use of a small corps. So impressive were the volumes, however, that Jackson reached the opposite conclusion; if a small group of French investigators had produced material to fill seven volumes, then a larger American corps would achieve still more.[46] In 1836, while the United States Exploring Expedition remained mired in delays and disputes, Dumont d'Urville had departed on another expedition, with instructions to explore the same areas of the South Pacific as the American voyage.[47] The French party would challenge and dispute the priority of the Exploring Expedition's greatest achievement, the discovery of Antarctica.[48] The decision to emulate the appearance of the Astrolabe reports was in itself a competitive and nationalistic statement.

An 1845 resolution by Congress specified further that copies of the reports be distributed to the governments of France, Great Britain, Russia and twenty-five other countries.[49] These provisions indicated that all involved acknowledged the diplomatic purpose of the volumes. The elements of the acknowledged, if unofficial, international style of government science which the United States emulated with this project included the folio size of the atlases of engraved illustrations, embossed bindings, and heavy paper with gilt edges. The materials would be, of course, entirely of American manufacture, as would be the science; the authors were at first forbidden to use European material or aid in their reports. Yet the scientists were also to display their traditional erudition by writing their descriptions in Latin. The impressive presentation of the reports—large, heavy, multivolumed, time consuming and expensive to produce—together with the plan of distribution, established a clear meaning for scientific achievement in the federal context, advertising the government's authority and resources. The publications of the Wilkes expedition were intended to inform foreign recipients that the United States had achieved parity in navigation, in breadth of global commercial interests, in science, and in the industrial arts. Wilkes wrote in 1848: "We shall now produce a work that every American will be proud of, and which will show those across the Atlantic that we can compete with them in many more ways than they have as yet given us credit for and that too under every disadvantage."[50] The mandated style, however, proved to be perhaps the greatest disadvantage to the reports' scientific authors. Torrey, engaged to describe a portion of the botanical collections, wrote to Brackenridge: "I told him [Wilkes] that all the acts of the Committee in relation to the style of the scientific publications of the Expl Expedn showed how incompetent they were to judge such matters."[51]

The official stamp and intended audience of foreign governments helps explain in part the decision to engrave rather than lithograph the majority of the atlases. Only the plates, largely of fossils, illustrating Dana's geology atlas were printed in lithography.[52] Private publications such as Samuel George Morton's *Crania Americana* of 1839 and John Edwards Holbrook's *North American Herpetology*, published from 1838 to 1842, showed that American lithographic illustration could attain considerable refinement. Drayton, moreover, had personal experience with the medium's potential for scientific subjects. In 1842, however, lithography conveyed none of the imperial connotations of engraving. French state-published science was consistently illustrated with engravings, although lithographic plates illustrated the principal British competition, the scientific results of the voyage of the *Beagle*. In this instance, experimentation with the faster and potentially less expensive medium failed to appeal to the American legislators who, on other occasions, promoted technological innovations and were always cautious about federal expenditure. Moreover, the choice by Congress of the more formal medium suggests that France was the primary competition.

Drayton and Agate, Wilkes, Peale, Dana, Couthouy, and others on the voyage had prepared hundreds of drawings of specimens, views, land forms, and native portraits, customs, and implements.[53] The commander's narrative of the voyage, the first volumes published, incorporated firsthand accounts from the diaries of expedition members and, following the conventions for illustrating octavo editions, contained wood and copper engravings after many of their landscape and portrait sketches.[54]

Engraving the atlas plates from drawings of the collections made during the expedition and the scores of new drawings of preserved specimens would enlist over seventy American engravers in the slow process of translation onto metal. During the same period, the use of copper engraving in zoological publications virtually disappeared. Most journals and survey reports, including those of the federal government, had converted almost completely to lithography, except for the use of steel engraving, recently improved and capable of large print runs without wear, and, of course, wood engraving for text figures.[55] The volumes of the Exploring Expedition, then, constituted one of the last chapters in American zoological copper engraving. By the late 1850s the number of firms involved in engraving and printing the expedition reports had declined, and the engraver William H. Dougal (1822–1895) handled the bulk of the work. Even Dougal temporarily abandoned the profession. Tempted by the commercial opportunities of the Gold Rush, he sailed to San Francisco where he ran a grocery store and livery stable and made some money off the forty-niners before returning in 1851 to labor for science.[56] By the late 1850s Dougal received between $80 and $125 for engraving a single plate.[57]

FIGURES
4.12–14

135

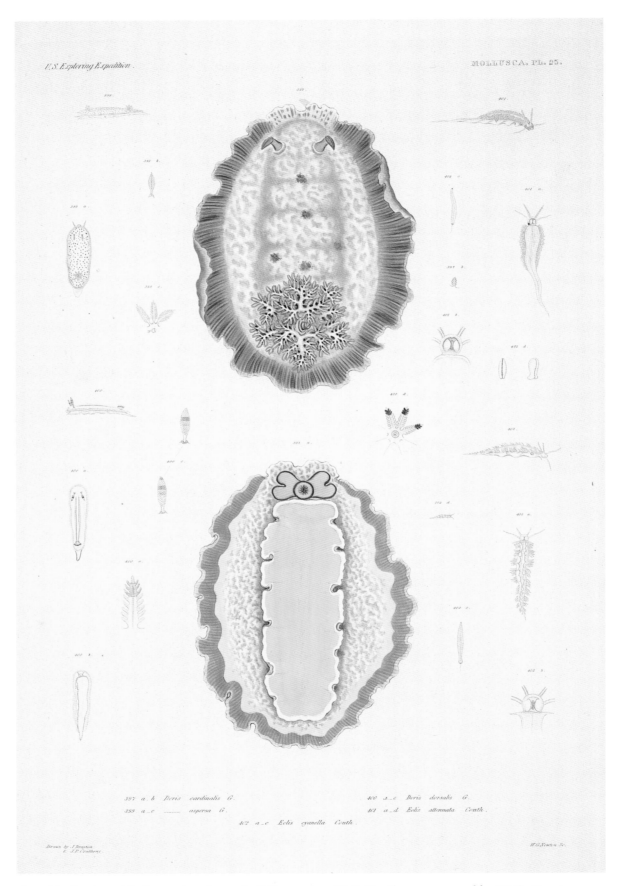

397 a, b   Doris cardinalis  G.                400 a_c   Doris dorsalis  G.
398 a_c  ———  aspersa  G.                      401 a_d  Eolis attenuata  Couth.

402 a_c  Eolis cyanella  Couth.

Drawn by J. Drayton.
& J. P. Couthouy.                                                    W. G. Newton Sc.

Plate 31.  Doris and Eolis species, drawn by J. Drayton and J. P. Couthouy, engraved by W. G. Newton.

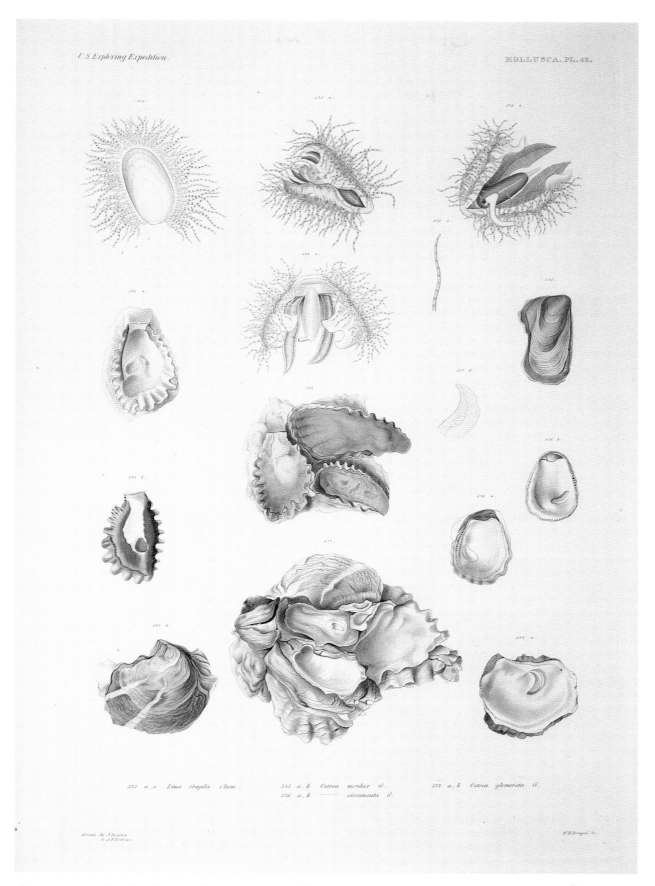

Drawn by J. Drayton
& A. F. Bellows.

W. H. Dougal Sc.

573 a,c   Lima fragilis   Chem.     575 a,b   Ostrea mordax   G.
                                576 a,b   ——— circumsuta   G.     577 a,b   Ostrea glomerata   G.

Plate 32. *Lima fragilis* Chem. and Ostrea species, described by Gould, drawn by J. Drayton and A. F. Bellows, engraved by W. H. Dougal.

4.12–14 Copper and wood engravings from Wilkes's *Narrative*, from drawings by Wilkes, Agate, and Drayton. From Charles Wilkes, *Narrative of the United States Exploring Expedition*, vols. 2, 3, and 5 (1845).

4.12 View of Antarctica.

Drayton's superintendence of the visual aspects of production often brought him into conflict with the authors. According to protocol and to salary, Drayton's status on the voyage ranked below that of the scientists; he worked in the service of the scientific staff.[58] When the squadron returned, Wilkes elevated Drayton to the position of his closest assistant, negotiator, and representative in unpopular decisions. He made an obvious target for the naturalists' dissatisfaction. It must have offended authors such as Dana, recognized as a brilliant and rising scientist, as well as Peale, a naturalist of the old school who considered himself as accomplished in natural history as in illustration, to find themselves with less power than an illustrator-engraver in decisions about their own publications. Furthermore, Drayton received a regular salary while the authors encountered difficulties obtaining reimbursements for their expenses.[59] Drayton's position was unusual in the history of scientific publishing. It was also clearly a thoroughly uncomfortable one. After eight years as shuttlecock between the irritable Wilkes and the offended authors, Drayton wrote in frustration to Wilkes: "It has been a long settled fact, that these d—d proffessors have all kinds of sence, but common sence."[60]

fathoms. I believe we were enabled to locate all the shoals in it, and I think it a safe passage. With the sun in the east, and steering towards the west, the dangers are distinctly visible. After passing through this channel, we kept the great reef in sight, sailing for Buia Point. When about half way to that point, we passed along a reef a mile in length, lying four miles off the large island. The water is so smooth within these reefs that it is necessary to keep a good look-out from aloft, as the smaller ones seldom have any break on them.

Beyond Buia Point the passage becomes still more intricate, and opposite Rabe-rabe Island it is quite narrow, though there is ample water for any vessel. We, however, went briskly on, having a fine breeze from the eastward. After getting sight of the Lecumba Point Reef, there is but a narrow channel into the bay, which we reached at half-past 3 p. m. The Peacock had just arrived from the north side of Vanua-levu, and anchored.

Mbua or Sandalwood Bay, though much filled with large reefs, offers ample space for anchorage. The holding-ground is excellent, and the water not too deep. The bay is of the figure of a large segment of a circle, six miles in diameter, and is formed by Lecumba Point on the east and that of Dimba-dimba on the west. The land immediately surrounding it is low, but a few miles back it rises in high and picturesque peaks. That of Corobato is distinguished from the Vitilevu shore, and has an altitude of two thousand feet. The shores of the bay are lined with mangroves, and have, generally, extensive mud-flats. There are few facilities here for obtaining either wood or water, as the anchorage is a long distance from the shore. Several small streams enter the bay in its upper part, flowing from some distance in the interior. This was the principal place where the sandalwood was formerly obtained, but it has for some years past been exhausted. I shall defer speaking of this district until I have given an account of the operations of the Peacock.

FEEJEE BASKETS, ETC.

4.13 "Feegee" baskets.

One of Wilkes's errors of judgment both damaged Drayton's repute with the scientific corps and demonstrated the changing standards of American science. In 1843 Couthouy had resigned, charging that the shell collections had been so badly handled that he could not sort or organize them for publication. A distinguished Boston naturalist, Augustus Addison Gould (1805–1866), author of the report on invertebrates for the Massachusetts zoological survey, applied for the privilege of preparing the expedition report in Couthouy's stead.[61] Wilkes expressed his faith in the ability of his assistant, and insisted that Drayton should do it. Wilkes argued that the artist had already completed many of the drawings for the conchology atlas and had the advantage over Gould of having been on the voyage and having visited the collecting localities. In other words, by the criteria of the naturalist-illustrator working primarily in the field, Drayton had the qualifications. By 1843, however, naturalists had reached a consensus on the separation of the role of the zoologist from that of the draftsman. The scientific community rallied to denounce Wilkes's decision. Amos Binney, one of Gould's Boston colleagues, wrote to Benjamin Tappan, chairman of the congressional committee overseeing publication: "In matters of Natural History, Capt Wilkes has not sufficient information to enable him to judge, what the arrangements in those departments should be, what the present condition of the Science requires, or what is due to the scientific character of the country."[62]

The most persuasive argument against Drayton's authorship concerned one of the objects of the publication program closest to Wilkes's heart: priority in description of new species. Wilkes naively believed that priority in collection established priority of discovery. Gould instructed Tappan: "I feel that no time is to be lost, inasmuch as every month will take something away from the novelties which we may hope to find in the collections. Every month brings in from abroad descriptions of new shells from the very regions visited by our Squadron; and if much more delay is made, there would be little inducement for any naturalist to undertake the task from the hope that he might contribute something new to the stock of knowledge."[63] Eventually the protests of Gould and his colleagues prevailed upon Wilkes to change his mind.

Before Wilkes yielded to pressure, however, he offered a compromise. Gould should write the descriptions, but only after Wilkes himself, with Drayton's help, had arranged the notes, prepared all the drawings, and selected which

4.14  Pine Forest, Oregon.

specimens should be engraved. This plan would have preserved Wilkes's authority in deciding which species deserved discussion or were new, leaving Gould to do no more than fill in the blanks. Wilkes's proposal implicitly inverted the conventional relationship of words to pictures in science. Gould's descriptions would be subordinate, after-the-fact verbal illustrations of the pictorial documentation. Gould dissuaded Wilkes from the plan.[64] Having achieved his goal, Gould found Couthouy's field notes obscure, and he complained of the difficulties of working with material he had not seen alive. But Couthouy proved to have been an astute observer, collecting mollusca without shells and making notes on the living animals, in contrast to most of his contemporaries, who focused exclusively on the shell.[65]

In another incident, an alliance between Wilkes and the judgments of the community of science turned against Titian Peale. Peale had been born into a family and culture that endorsed an image of science uniting the collector, describer, and illustrator—an image personified by Peale's father, and by Wilson, Lesueur, Audubon, and himself. In his official position as collector of birds and mammals, Peale accomplished good work, keeping up with the younger scientists under the rigors of expedition life. But the rigors of Washington life proved greater. Peale had to support a family on a small and irregularly paid salary; he complained that Drayton received twice as much as "the Naturalist and artist."[66] Books for reference and comparison were hard to obtain or unavailable. Complaining of restrictions that prohibited him from removing specimens from the collections for study, Peale concentrated instead on his paintings in oil and drawings of new species, completing thirty-

140

four of a projected fifty by the end of 1844.[67] He expressed frustration that he was not permitted to oversee the engraving of his work; indeed, he did not even know to which engravers Drayton had sent the pictures.[68] In writing up the collections, however, Peale not only could not produce the required Latin descriptions, he failed to meet current taxonomic standards as Wilkes understood them. Wilkes sought a way to intervene, to "preserve both the Country and Expedition from being disgraced."[69]

Peale's volume diverged from the Library Committee's injunctions on format. Specifically, he failed to include a catalog of species. He entitled the work "Zoology" instead of "Ornithology and Mammalia" because he had planned to include lepidoptera as well as birds and mammals. Although it broke the rule for the expedition reports, Peale had precedents for naming his volume "zoology"; Charles Darwin had similarly entitled his report on the collections from the voyage of the *Beagle*, published in 1840. Peale also caused offense by writing a preface in which he explained the obstacles under which he had worked.

> When we returned home from the voyage the "Scientific Corps" of the expedition was discharged from the service of the United States before we could collate the results of our respective labors. — Our notes, drawings, and collections were taken from us, and given to others; which occasioned much confusion, some loss, and greatly increased the labor. . . . I have had access to but few of the needfull books of reference: Those constituting my private library were lost in the wreck of the U S ship Peacock on the bar of the Columbia river in July 1841, since that time it had not been in my power to replace them.[70]

Wilkes found that "the tone of this in my opinion is very objectionable & not borne out by the facts to my knowledge."[71]

Above all, the illustrations that Peale designed proved too elaborate for the already-strained budget of the report program. In 1847 Wilkes reported on Peale's progress: "Fifty plates in Mr. Peales department of Mammalia and Ornithology were contracted to be engraved, but the engravers have abandonned their Contract, in consequence of the price being inadequate to the quantity of work."[72] Peale's extravagance lay not in his detail of rendering the animals, but in the elaborate landscape backgrounds. Wilkes pointed out that "these backgrounds have but a remote reference to the objects, and indeed are inappropriate to the plates; by omitting some part of them the cost is reduced to a reasonable sum." Wilkes wished, however, to "adhere with the most scrupulous exactness" to Peale's drawings of the animals; that aspect of the work he never questioned.[73] Peale objected to any alteration of his work. His conviction of the necessity of backgrounds, of their integral importance to the meaning of his illustration, dated from his childhood, from the landscape backgrounds painted behind the mounted specimens at his family's museum. The Peale museum exhibit style had entered zoological illustration through Wilson's plates, after which Audubon elaborated the relationship of landscape backgrounds to animal subjects. The ornithological plates Peale prepared bore a strong resemblance to the lithographic plates of the birds collected by Charles Darwin on the voyage of the *Beagle* by the British ornithologist and illustrator, John Gould. The similarities stemmed from the influence of Audubon on Gould's style. Peale envisioned an illustrated report "drawn from nature" in the tradition of the naturalist-illustrator. He would show nature's unity—the animal as if alive in its environment—and the unity of observation and picturing. He would neither abstract and isolate the animal from its natural context, nor relinquish his view of his own role as observer.

The scale of scientific operations had altered in the course of Peale's career. Although Peale and Wilkes agreed that new species added to national prestige, Wilkes wanted new species in bulk, while Peale wanted to dignify each one. The individual attention that Wilson and Audubon had lavished on each new American species had accorded to each the status of an

PLATE

33

individual, almost a personality. In their hands, especially Audubon's, illustration had approached the level of portraiture, paralleling the portraits of patriots and naturalists that Charles Willson Peale had painted and displayed above the natural history exhibition cases in his museum. The sheer bulk of later natural history collections such as those accumulated by the Exploring Expedition forced a change in styles of exhibition and publication. It had become impractical to mount each specimen for display. Misguided attempts to prepare the expedition collections for public exhibit before the completion of the scientific studies had ruined hundreds of specimens for dissection, anatomical study, and drawing. Public exhibition, a way of providing direct accountability for the use of public funds for science, had become incompatible with specialist taxonomic practice.

Similarly, it was no longer the most useful method of book illustration to isolate a single species on a page, even apart from considerations of economy. The growing recognition of the importance of geographic distribution of related species, and the search for a clearer understanding of the relationship between fossil and living species and for a clearer definition of the morphological boundaries between species and varieties made comparison all-important. In a bound volume, when a single species occupied a whole page, the pictures were separated visually because engravings and lithographs could be printed on one side of a sheet only. The viewer could compare the illustrations only in sequence, never in direct juxtaposition, which posed a severe limitation. When many related animals, usually species of the same genus, were pictured together on a single sheet, however, the visual information came already juxtaposed for comparison. Composite plates reflected the new comparative working methods of zoology.

Peale's text describing the mammals and birds of the expedition, illustrated with woodcuts after his drawings, went to press in 1848; Wilkes suppressed the work as it was printed.[74]

Shortly thereafter, Peale resigned from his position, leaving the illustrations for the atlas unfinished. Wilkes dismissed Peale, and then offered him twenty dollars each to complete the remaining drawings, an offer that reduced Peale's services to those of a mere draftsman. Peale responded: "I am personally interested in the proper completion of the drawings, considering them as I do identified with my scientific reputation," but he declined, claiming that each drawing took him two weeks to complete, and that he could not live on the work.[75] Wilkes's insulting offer, added to the loss of the family museum not long after Peale's homecoming, and the disappointment of not receiving a curatorial position with the federal collections, moved Peale to abandon attempts to continue to support himself as a naturalist. Instead, he accepted a position with the U.S. Patent Office. After retirement, Peale experimented with photography and translated his expedition sketches into oil paintings. He never published his most cherished work, a study of lepidoptera, illustrated in the glowing colors of his youthful triumphs such as those in Say's *Entomology*.[76]

Without a supervising ornithologist, the report could not be completed for publication. When Drayton approached John Cassin (1813–1869), an ornithologist at the Philadelphia Academy of Natural Sciences, about consulting on the atlas, Cassin informed him that birds Peale had marked as new were in fact well-known species. Wilkes decided that the volume required a complete revision and offered Cassin the job. When Cassin's new text appeared in 1858 it retained Peale's illustrations of skulls, bearing Cassin's attributions of genus and species. Peale had depicted the skulls faithfully; it was the change of caption that updated his work and rendered it acceptable.[77]

Cassin's revised atlas contained fifty-three plates, thirty-two of them Peale's. Wilkes had doubted that new drawings could be made from

FIGURES
4.15–17

4.15–17 Hand-colored engravings from Cassin, *Atlas: Mammalogy and Ornithology*, 1858.

1. Lupus gigas. Townsend.

2. Lupus occidentalis. Richardson.

T. R. Peale del.

Rawdon, Wright, Hatch & Edson Sc.

4.15 *Lupus gigas*. Townsend (lower figure), *Lupus occidentalis*. Richardson (upper figure). Drawn by T. R. Peale, engraved by Rawdon, Wright, Hatch, and Edson. Note the resemblance between this plate and Lesueur's illustration of wolves in Godman's *American Natural History* published in Peale's youth.

Pandion solitarius. (Peale)

T R Peale  del                                                        Dougal  sc

4.16 *Pandion solitarius* (Peale). Drawn by T. R. Peale, engraved by W. Dougal. Cassin had the bird's prey changed to a fish.

Todiramphus vitiensis. (Peale)

1. adult. 2. young.

T.R.Peale del.                    W.H.Dougal Sc.

Plate 33. *Todiramphus vitiensis* (Peale). Fig. 1, adult; fig. 2, young. Hand-colored engraving drawn by Titian R. Peale, engraved by W. Dougal, for Cassin, *Atlas: Mammalogy and Ornithology* (1858).

Procellaria nivea. Gmelin.

T.R Peale del. Rawdon Wright Hatch & Edson sc.

4.17 *Procellaria nivea* (Gmelin). Drawn by T. R. Peale, engraved by Rawdon, Wright, Hatch, and Edson. Peale's original background for this plate was hammered out and replaced with clouds and water.

the preserved specimens: "Had his [Peale's] descriptions been full enough in the text and the description corresponded with the specimen to be drawn an artist would have found less difficulty in giving a true representation, but neither is the description full enough, nor has the specimen been so prepared as will admit of this being done."[78] Indeed, Drayton had a low opinion of the plates that Cassin had had prepared: "Now so far from desireing Mr. Cassin employed, I wish the d—d book with his d—d bad plates would be annihilated."[79] In some of Peale's illustrations, the landscape backgrounds, already engraved but costly to print, had been laboriously hammered out and simplified. The savings, however, were doubtful; Dougal charged $70 to change one plate.[80]

In a couple of instances, Peale described birds of separate species as juveniles of other species. Yet Cassin also made taxonomic gaffes: he gave new names to species already described, and lumped with similar-looking relatives other species that had already been demonstrated to be distinct.[81] The errors of both authors arose from similar causes; missed references in the literature, probably through lack of access to books, the necessity of working from poorly preserved specimens, and misinterpretation of field observations. Cassin, for example, had the prey grasped in the talons of a Hawaiian hawk reengraved from a honeycreeper to a fish, based on his interpretation of the species' diet. But the constraints under which the two men worked affected their careers in opposite ways. For Peale, the work drove him from the occupation of natural history, while for Cassin it provided a stepping-stone to an active career at the center of his scientific generation.

Other publications of the same period suffered from transitional conflicts similar to those that plagued the publications of the Exploring Expedition. Like the expedition authors, John Edwards Holbrook, author of *North American Herpetology*, emulated French models, those of Cuvier, A.M.C. Duméril, Gabriel Bibron, and Achille Valenciennes at the Jardin des Plantes in Paris, where he had studied in the early 1820s. Finding contradictions between the established French taxa and the anatomical evidence of his new world specimens, Holbrook subjected his work to constant revision. Publishing at his own expense, he issued three volumes of the work between 1836 and 1838, but canceled the first edition to issue a replacement. The steady accumulation of previously undescribed specimens required the establishment of new genera, something the responsible Holbrook was reluctant to do without direct access to European collections and publications. In attempting his definitive revision of North American reptiles and amphibians, Holbrook had to contend with the careless work of others, a vast synonymy, and descriptions and illustrations based on preserved specimens.[82]

The backgrounds and working methods of Holbrook's illustrators again marked current shifts in approaches to zoological illustration. The work was originally grounded on the assumption of amateur participation in the American natural history project. A wide array of personnel drew for Holbrook, among them amateur naturalists plus draftsmen employed in the printing business. Maria Martin, a member of the family of John Bachman, Audubon's coauthor, had drawn for the *Birds of America* and contributed "some accurate and very spirited Carolina Reptiles" to Holbrook's volumes.[83] One of Holbrook's principal illustrators, J. Sera, an Italian who settled in Charleston, ton, South Carolina, Holbrook's home, advertised himself in the local newspaper as a musician as well as an artist, and he was known for his landscape painting and theatrical scenery, a typical mixed bag of skills for an illustrator.[84] For Holbrook, Sera produced watercolor drawings almost like miniature portraits. A Charleston naturalist remembered that Sera prided himself on investing his natural history subjects with physiognomic expressiveness; the "'peculiarly hard, cruel expression of the

PLATES

34–35

147

Anolius Carolinensis.

Jᵒ Queen del.

P. S. Duval, Lith, Philᵃ

8.

Plate 34. *Anolius Carolinensis*, hand-colored lithograph drawn by James Queen, lithographed by P. S. Duval, for Holbrook, *North American Herpetology* (1842).

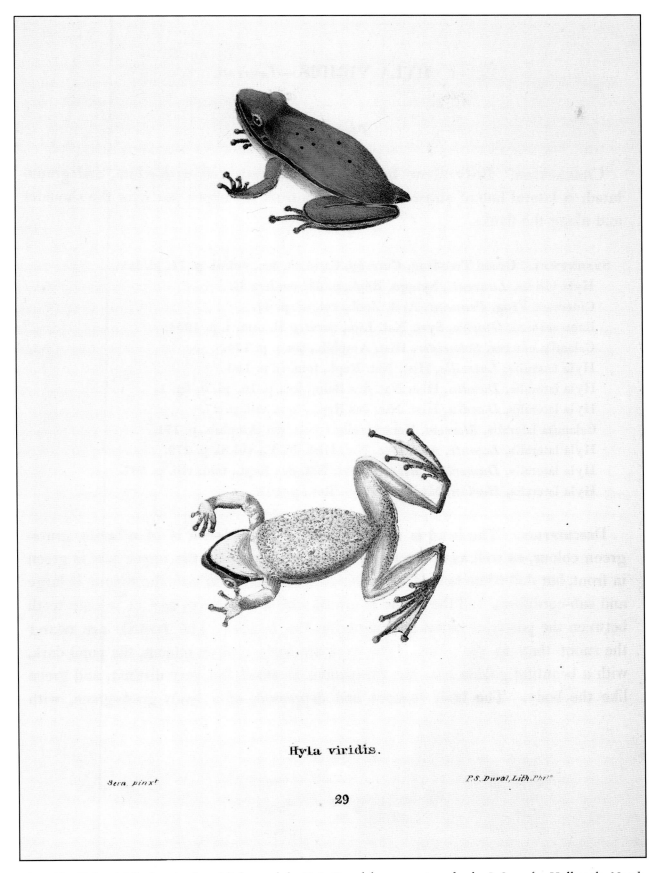

Hyla viridis.

P.S. Duval, Lith.Phila

29

Plate 35. *Hyla viridis*, hand-colored lithograph by P. S. Duval from a watercolor by J. Sera, for Holbrook, *North American Herpetology* (1842).

alligator's eye,' or the 'bright, deceitful look of the eye of the black snake.'"[85]

When Sera died in 1837, halfway through publication, John H. Richard replaced him, perhaps on the recommendation of the Paris herpetologists.[86] Richard maintained more detachment in his work than Sera. He concentrated on the patterned elegance of the reptiles rather than on their personalities. He also took a hand in printing the plates, working with Peter S. Duval, the Philadelphia lithographer of the project, and experimenting to produce color-printed lithographs by the lithotint method.[87] While Richard experimented with the lithotint, Duval assigned two of his best draftsmen to Holbrook's work, a practice that would increase with the use of lithography for zoological illustration. Albert Newsam, known for his portraits, drew turtles, and the young James Queen (d. 1886), a native Philadelphian apprenticed to Duval in 1835 at the age of fifteen, drew a substantial number of the plates.[88]

FIGURES
4.18–20

Although he experimented with printing techniques, Holbrook's illustrations continued many of the premises and conventions of the naturalist-illustrators.[89] Holbrook intended his illustrations to serve as primary documents— to record the live animal, especially its coloration, as the basis for the taxonomic attribution. Like Thomas Say and Titian Peale, Holbrook coped with taxonomic difficulties by emphasizing illustration. He studied the internal anatomy of his subjects, but limited his illustrations to external morphology. The plates depicted one species each, the smaller species represented life-size. Most were drawn from life. In the tradition of drawing for engraving, Sera, Richard and several others prepared drawings on paper; later illustrators would omit that step and draw from specimens directly onto the prepared lithographic stone.[90] Exceptionally, Richard signed his plate of *Alligator Mississippiensis* with bravado: "From Life On Stone." Holbrook informed the reader when an illustration had been made from a preserved specimen. There were no landscapes in Holbrook's plates. Separate figures displayed the dorsal and ven-

tral views of frogs, turtles, and salamanders, while the snakes twisted upon themselves to display scale patterns and to fit onto the rectangular page. Overall, Holbrook's work shared more with that of Thomas Say's generation than with that of James Dwight Dana's.

The transition in taxonomic practices and illustration conventions reflected in Holbrook's work lay behind the seeming contradictions in Wilkes's dicta. Wilkes, intolerant of Peale's outmoded approach to illustration, was equally intolerant of Dana's innovative treatment of the collection of corals, or zoophytes. In the matter of publishing only new species, the committee had emulated the rules of the Academy of Natural Sciences for its *Journal*.[91] Dana, however, saw his report as an opportunity to prepare a taxonomic revision of the corals, and insisted on the necessity of illustrating all his specimens: "Corals are so peculiar in their forms & so little known that descriptions, unless extended to an unwarrantable length, convey but little idea of the species: and figuring one species in a genus will not answer the purpose it does in Conchology."[92] Wilkes adhered stubbornly to an ideal of reporting and taxonomic description that neither digressed nor interpreted, and in this he remained closer to the approach to natural history of the early American naturalists. When Dana wrote a theoretical preface on the corals, Wilkes refused to publish it, although a few copies of the report with the preface were printed before he could enforce his edict. Asa Gray attempted to explain to Tappan the value of Dana's work, as an exemplar of current scientific standards:

His discoveries in the voyage have given him not only a vast number of new facts, but *new* principles also, to be applied to the old as well as to the new species. There is the highest propriety in his making the application of these new principles *himself* in this work, and thus connecting fully and forever, the name and history of this expedi-

PLATE
36

4.18–20 Plates from Holbrook, *North American Herpetology*, 1842.

150

From Life on Stone by J. H. Richard                    Alligator Mississippiensis.                    P. S. Duval Lith Phil.

7.

4.18  *Alligator Mississippiensis*, "From Life on Stone," by J. H. Richard, printed by P. S. Duval.

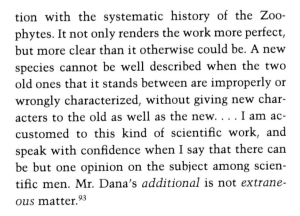

4.19 *Coronella sayi*, Schlegel, drawn by J. H. Richard, lithographed by P. S. Duval.

4.20 *Crotalus adamanteus*, lithograph by Lehman and Duval from a watercolor by J. Sera.

tion with the systematic history of the Zoophytes. It not only renders the work more perfect, but more clear than it otherwise could be. A new species cannot be well described when the two old ones that it stands between are improperly or wrongly characterized, without giving new characters to the old as well as the new. . . . I am accustomed to this kind of scientific work, and speak with confidence when I say that there can be but one opinion on the subject among scientific men. Mr. Dana's *additional* is not *extraneous* matter.[93]

Dana's plans for a similar taxonomic revision based on the expedition crustacea met with disaster. After the plates of the crustacea atlas had been printed, but before they had been colored, fire destroyed many of Dana's original drawings, and with them, the color information they contained. The preserved specimens had long since faded. The documentation contained in the drawings was irretrievable, and the plates had to be published largely uncolored. Dana had intended his plates to correct previous publications that attributed colors to

152

1. 1. a. b. c. b. c. Actinia flagellifera.          3.     Actinia veratra.
2. 2. a. b.          A.     pustulata.          4. 5. A.     clematis.

Plate 36.  Species of the genus *Actinia*. Drawn by Joseph Drayton, engraved by W. A. Wilmer. Hand-colored engraving from Dana, *Atlas: Zoophytes* (1849).

4.21–22  Hand-colored engravings from Dana, *Atlas: Zoophytes*, 1849.

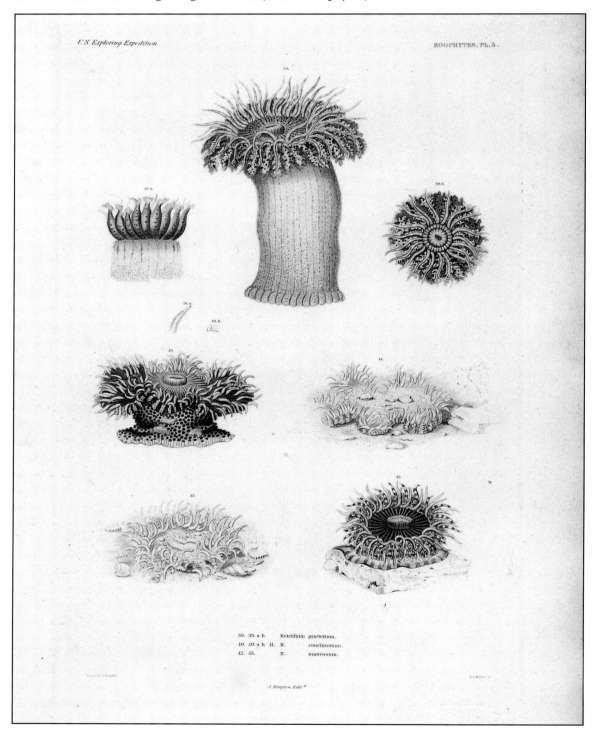

4.21  *Actinia* species. Drawn by Joseph Drayton, engraved by W. A. Wilmer.

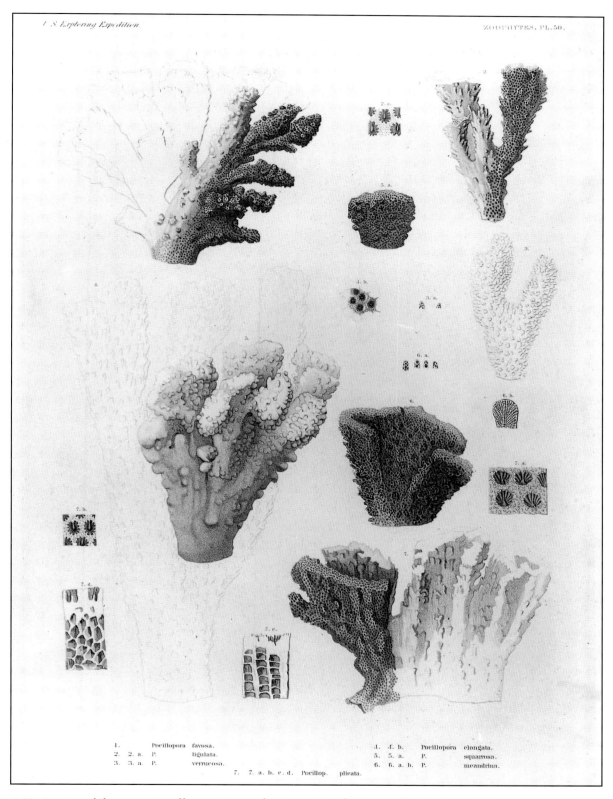

1.         Pocillopora favosa.
2.   2. a.   P.       ligulata.
3.   3. a.   P.       verrucosa.
7.   7. a. b. c. d.   Pocillop.   plicata.

4.   4. b.   Pocillopora elongata.
5.   5. a.   P.       squarrosa.
6.   6. a. b.   P.       meandrina.

4.22 Species of the genus *Pocillopora*. Drawn by James Dwight Dana, edited by Joseph Drayton, engraved by A. B. Walter.

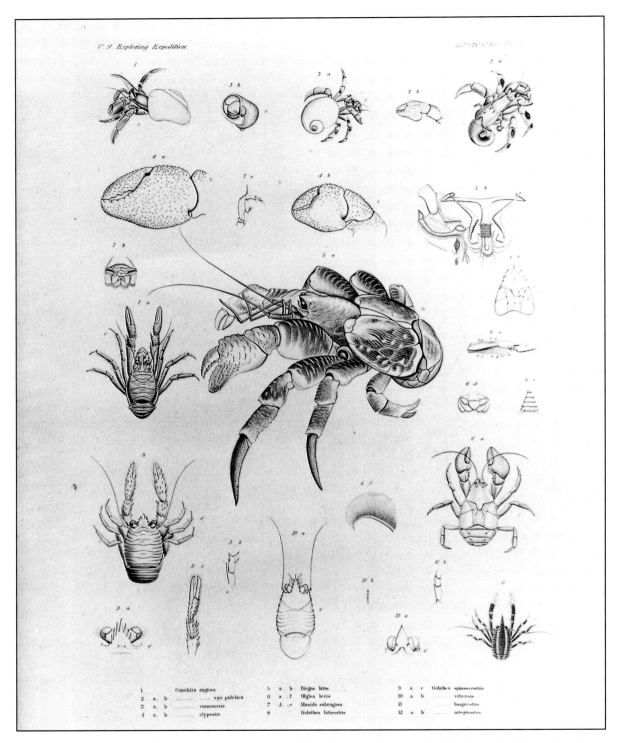

4.23 Hermit Crabs. Hand-colored engraving, unattributed, from drawings by J. D. Dana for his *Atlas: Crustacea*, 1855.

species based on faded specimens. He noted flatly: "As the coloured originals are gone, these figures have lost the principal part of their interest."[94]

Dana gave special credit to Joseph Drayton for his role as superintendent of publication, yet he also took care to note that the illustrations were his own work: "I take this opportunity to observe that, in the preparation of my several Reports, and the Atlases, with which they are illustrated, I have had, half a dozen plates excepted, neither the assistance of an amanuensis nor a draftsman. The plates, however, owe much to the artistic skill and taste of Mr. Joseph Drayton, Artist of the Expedition, who has superintended the engraving and printing, and contributed in many ways to the beauty of the work."[95] Dana, as his own draftsman, had seen to the science; indeed, his statement implies that he considered drawing a necessary instrument of science. His praise of Drayton points to the role he considered appropriate to the artist, the superintendence of style, taste, and beauty.

The policies and conflicts that attended the publications of the Exploring Expedition document the transformations in American science during the 1840s and '50s. Many assumptions and practices about the production of science prevalent at the time the squadron had sailed had changed by the time the collections entered the literature. The plates in the report atlases spanned two eras of zoological illustration. Peale's vertebrates harked back to the naturalist-illustrator ideal of animal portraiture. Gould's and Dana's invertebrates reflected the increasingly comparative approach to taxonomy and attention to geographic distribution.[96] In Dana's atlas on corals the expense of engraving and printing had made it expedient to abbreviate individual illustrations of complex structures; many figures had the details filled in only partially, with a mere outline to suggest the remaining form. The use of color in all but Cassin's atlas was restricted. Such strategies spoke to the exigencies of the budget; the scientific requirements for pictorial description took precedence over the prestige value of extravagant plates.

Later amendments to the initial limit of one hundred copies expanded the editions. Some of the authors published unofficial editions of their reports, making their contents available to a wider audience. Yet the protracted history of disagreements and delays rendered the publications of the Exploring Expedition largely anticlimactic. The reports' congressional audience became heartily sick of the project long before it was brought to a close.[97] And by the time the reports reached their scientific audience, priority for describing scores of species new to science when collected had been preempted by other authors. Dana himself placed his work in its historical context. He wrote of his report on the crustacea, published in 1853, the atlas in 1855: "The drawings for this Atlas, issued at this late date, were to a large extent made during the years 1838 to 1842, in the course of the cruise of the Expedition: and in the history of the Science, they would properly have their place in that period."[98]

*5*

# "A Better Style of Art": The Consolidation of Convention in Midcentury Illustration

CONGRESSIONAL patience for scientific expenditures wore thin during the wearisome debates over requests for additional appropriations for the reports of the Exploring Expedition. One senator compared the publication program to Charles Dickens's fictional law suit, Jarndyce vs. Jarndyce. Another urged: "Throw it into the Potomac; that is the best thing."[1] The scientists, moreover, seemed always to need more money for more projects. In 1861, in the midst of the Civil War, the Senate discussed an appropriation for distributing zoological specimens to educational and scientific institutions around the country. Senator Simon Cameron of Pennsylvania, a long-time opponent of federal support for science, expressed his opinion:

Here is an appropriation of $6,000 for a most worthless purpose: and what right have we to appropriate it? . . . We are here to appropriate $6,000 or $10,000 to preserve a parcel of what you call scientific specimens. A Senator over the way said they were toads and snakes, and I have no doubt that they are that sort of thing. They are no use to anybody now; they have served their day.

I am tired of all this thing called science here. It was only the other day we made another appropriation in regard to the expedition which Captain Wilkes took out to the Pacific Ocean. We have paid $1,000 a volume for a book which he published. Who has ever seen that book outside of this Senate, and how many copies are there of it in this country?[2]

The senator's expostulations make clear that although by 1860 the federal government had become an organizer and producer of science in its own right, that activity by no means enjoyed universal political acceptance.

Federal zoological publications, however, already exerted a strong influence in the process of consolidating pictorial conventions in the discipline. Beginning by emulating foreign and private science, the federal government was, by 1860, beginning to set the pace of what constituted the American midcentury institutional style. But it was less a pace of innovation than one of consolidation. State-sponsored illustration played an essentially conservative role in consolidating zoological pictorial convention for several reasons: the legislators who voted the money could be persuaded of the worth of the projects by their resemblance to existing work; the publication of survey reports took place within the linked and growing bureaucracies of government and of government-supported scientific institutions; the men who directed the survey publication programs had impressive longevity—their tastes were formed early in the century and their influence lasted late into the century. This interaction of individuals and institutions shaped the pictorial project with which Congress, like it or not, had become involved.

Continuing debate about the relationship of federal government to science bore directly on the initial scope and credibility of the Smithsonian Institution under Joseph Henry (1797–1878), its first secretary. While the Exploring Expedition had been at sea, Congress had continued to disagree about the disposition of the bequest of James Smithson, the English chemist who left his fortune to the United States for the foundation of an institution dedicated to "the increase and diffusion of knowledge among men."[3] During the ten years between the nation's acceptance of the money in 1836 and the passage of the Smithsonian Act, introduced by Robert Dale Owen, a variety of proposals came before Congress for its use. They included Owen's proposal for spending the money on popular education and publications, the idea of founding a national university, an observatory—the pet project of John Quincy Adams—an agricultural experiment station, and a meteorological bureau, these last reflecting the widely held view that priority should be given to practical applications over speculative science. Strong opponents of the extension of federal authority, like John C. Calhoun, opposed them all.[4]

A number of earlier institutions influenced the eventual form of the Smithsonian. The organizers of the National Institute, established in 1840 to raise the tenor of intellectual life in the capital, hoped to receive the bequest. The Patent Office exhibition of the collections of the Exploring Expedition set a precedent of public exhibition of government collections. Its displays, including ethnographic collections, David Dale Owen's geological specimens, modern copies of European masterpieces, and American landscapes, became immensely popular.[5] Public museums retained, however, connotations of charlatanism; miscellaneous assemblages of curiosities and fakes attracted crowds and their money throughout the country. Joseph Henry wanted to avoid the expense of curating and exhibiting bulky collections.[6] His conception of the Smithsonian emphasized research and publication for a specialist audience.

The role of integrating publication for a specialist audience and curating a public collection fell to Spencer Fullerton Baird (1823–1887), appointed assistant secretary of the Smithsonian in 1850. The mismanagement of the Wilkes expedition collections and the history of the publication program had given government science in general a reputation for inefficiency. Baird's tact, efficiency, and honesty would win back from Congress a respect for scientific publishing. Under his supervision, the regular production of handsome volumes with fine illustrations of zoology, botany, geology, and ethnography of the western territories and national boundaries played an important role in convincing Congress that their appropriations were worthwhile.

Baird also faced the challenge of presenting public science as a worthy national enterprise, above controversy and regional divisions. Support for federal expeditionary science and its publications was entangled in the issue of state versus federal powers. Science and scientific book production were both activities heavily concentrated in the northeast. Proposals to continue surveying the west, on the other hand, implied westward expansion; and expansion, during the 1850s, entailed the divisive issue of whether territories would enter the union as slave or free states.[7] (An amendment to the rules for distributing the Exploring Expedition reports ensured that new states would receive copies of them.)[8] Later, explorations for railroad routes—certain to bring prosperity to adjacent regions—were also subject to congressional negotiations. The routes of scientific surveys depended on projected railroad routes, and on events such as the Treaty of Guadalupe Hidalgo, which determined the border after Polk's war against Mexico. Thus, political maneuvering determined the type localities—the

159

place where a type specimen was collected—as surely as it set railway routes and national boundaries.[9]

Baird had to pick his way through this maze, promoting his programs for the collection and publication of natural history specimens, and hoping that congressionally supported volumes would advance his cause. To a remarkable degree he was successful. By 1860, in a Senate debate on distributing zoological specimens, although Senator John P. Hale of New Hampshire doubted the value of "pictures of bugs, snakes, and reptiles," his colleague Senator Fessenden of Maine trusted the stewards of the Institution: "As to the Smithsonian Institution itself, what it has done for science, and what it is doing for science. I have no doubt that it is doing much; how much, I do not know," but, he confessed, "It is my own fault."[10] Although his colleague, Senator Rice, firmly believed that appropriations for the Smithsonian were unconstitutional, he added: "I know that the Smithsonian Institution has done great good for the country. I am applied to daily for books published by it, and I know that they are valuable."[11] Indeed, the volumes that Baird worked to produce formed the core of midcentury American science.

Baird's scientific judgment, visual tastes, and business acumen exerted a strong influence on almost forty years of American scientific illustration. Acting as editor in chief for a vast program of scientific publications, and carrying out the Institution's mandate to "diffuse knowledge among men," Baird assigned the authorship of monographs, chose illustrators, and disciplined a booming printing industry to the needs of science. He accomplished these things in addition to organizing the scientific work of the continental surveys, overseeing the accumulation of collections in Washington, convincing Joseph Henry, the Institution's secretary, of the value of the natural history programs, and persuading Congress to vote continued support for his plans. Throughout, he continued to produce his own zoologi-

cal monographs. Understanding the importance of illustration in zoology, Baird used his position at the center of federal science to foster and encourage young men to become competent illustrators. From the 1850s through the 1880s Baird's attentions resulted in scores of volumes describing the zoology, botany, geology, paleontology, and archeology of the United States as well as publications in the physical and exact sciences.

Baird's controlling hand insured a consistent, conservative style for government science during a period of dramatic social and technological change. While Baird directed the natural history programs for the Smithsonian, not only biology but also social theory completed their conversion to the evolutionary model.[12] American science achieved an identity as a profession as the graduates of new degree programs took positions at centralizing research institutions.[13] In parallel with the federal programs and with the increasing availability of academic training and employment, the state surveys offered training and the credentials of experience to aspiring scientists. The potential for earning a living in the pursuit of science opened the field to a wider range of aspirants. In the same decades, printing became an industry. The periodical press, addressing a mass market, rushed toward the goal of complete mechanization. Illustrated scientific publication was affected by new technical possibilities at every stage of image production. Above all, the spread of photography and of photographic reproduction methods recast the roles of the draftsman and the printer, and profoundly reshaped the taxonomy of representation.

Institutional science, a new theoretical basis for taxonomic zoology, new forms of pictorial representation, and an industry with new capacities for speed and quantity changed the standards and appearance of zoological illustration. Baird, whose scientific decisions were required to represent the best interests of the nation, and whose available funds were at the discretion of Congress, would incorporate those changes into his publishing program; but his tastes had first been formed by the work of his friend Audubon, and he never relinquished some of his earlier ideas about how science should look. Similarly, James Hall (1811–1898), director of the paleontological survey of New York, would guide his publications through five decades of change and growth from a virtual cottage industry to an institution, carrying into the late century pictorial conventions dating from the 1830s and '40s.[14] The publications of the state surveys worked with the federal publications to establish standards of style as well as substance. While government-sponsored science formed the core of American scientific production, refinements of printing techniques and stylistic innovations tended to enter the literature through privately published monographs. Baird, Hall, and their counterparts responsible for other publicly funded institutions and surveys worked to adapt those refinements and innovations to the requirements of their publications. In so doing, they effected what hindsight can identify as a midcentury consolidation of conventions in zoological illustration. Those conventions established a continuity of illustration practices so closely wedded to the institutionally based taxonomic monograph that they transcended theoretical controversy and resisted technological change.

The representational elements that combined to define the midcentury conventions had developed in the changing practices traced so far in this study. The early formation of Baird's pictorial taste was firmly rooted in the naturalist tradition. Like the independent naturalists of a previous generation, Baird had learned his favorite science on his own. A dedicated bird-watcher and collector from boyhood, Baird and his brother formed a comprehensive collection of birds of the region of their home, Carlisle, Pennsylvania. At the age of seventeen, Baird initiated a correspondence with Audubon concerning the identification of a flycatcher, which proved to be a new species. Audubon

was one of many naturalists who befriended and encouraged the young Baird, and from whose work he would form the taste in pictures that would guide him in his adult editorial policies.[15]

During the winter of 1841–1842, which he spent in New York to study medicine, Baird in fact devoted most of his time to exploring the city's natural history resources. He attended meetings at the New York Lyceum of Natural History, met the personnel of the New York State scientific survey, visited the city's taxidermists, and became acquainted with his relatives the LeContes, a family of naturalists. He also studied drawing with Audubon, who had settled in New York and was preparing the illustrations of North American mammals for his collaboration with the Reverend John Bachman. Baird, who spent all his spare money on natural history books and specimens, noted with enthusiasm that Audubon's mammal plates would be printed in lithography and be available at half the price of the bird engravings.[16] In a letter home to his brother, Baird described his work with Audubon: "The old man continues to be as clever as ever; he even offered the other day to teach me to paint & draw after his own peculiar manner, on the condition of telling no one, and I have already commenced with him."[17] Baird practiced by copying from Audubon's original drawings for the *Birds of America*. The student described his teacher as "now drawing *Vespertilio noctavigans* [the silver haired bat], and just finished a rabbit; they are the most exquisite things in the world."[18] The quality of movement with which Audubon invested his bird illustrations also animated his mammal series, and made a profound impression on Baird. His letter continued: "Being so much at Mr. Audubon's I have an opportunity of seeing a great many North American Quadrupeds. He has made a most beautiful drawing of our squirrel. . . . It is in the attitude of leaping from one bough of a hickory to another and you expect every minute to see it in the air."[19]

PLATE 37

While he learned the secrets of Audubon's style, Baird also studied the delicate insect drawings owned by his relative, Major John Lawrence LeConte. The major had made many of the drawings while stationed in the southwest, but others he had bought from John Abbot, the Georgia naturalist who had shared specimens and observations with Alexander Wilson. LeConte's collection of drawings amounted to about nine thousand sheets, which Baird thought were "most beautifully executed."[20]

Later that winter, on a visit to Philadelphia, Baird took tea with Isaac Lea and viewed his collections of mollusks. He also followed with interest the latest news about Holbrook's progress on his *North American Herpetology*, one of the first American zoological monographs illustrated with lithographs.[21] Baird's early exposure to the style of privately produced science would influence the standards he set for federal science, although experience would teach him that controlling style and standards at the federal scale required some compromises and proved more complex than in the private sphere.

In the summer of 1842, after abandoning his medical studies, Baird visited Washington at the urging of his brother. With intense interest, he made repeated visits to the Exploring Expedition collections exhibited at the Patent Office to study the birds and the methods of taxidermy and preservation. Neither his desire to become a curator of the collections, nor Audubon's invitation to Missouri to collect mammals for the *Quadrupeds* materialized; but the two possibilities reveal Baird torn between institutional work and field natural history.[22]

At the time of Baird's first visit to Washington, the Patent Office still housed much of the expedition material in the basement. The public exhibition improved and became more elaborate, and on subsequent trips Baird, among the stream of visitors, would have seen there a collection of oil copies and probable forgeries of

5.1 Figures of Coleoptera such as Baird would have seen at the LeContes'. Lithograph by G. and W. Endicott, New York, from drawings by John Lawrence LeConte.

European masters such as Andrea del Sarto, Rubens, and Titian in addition to the natural history collections.[23] Such paintings, along with landscapes and seascapes by American painters, had become the principal art fare of

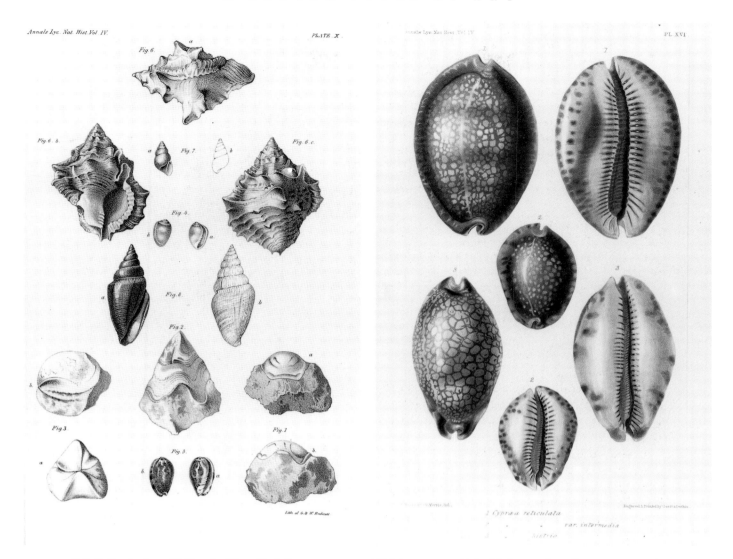

Annals Lyc. Nat. Hist.Vol IV.

5.2–3  In the late 1840s, the New York Lyceum was publishing accomplished lithographic illustration, such as that by G. and W. Endicott, illustrating shell descriptions by John H. Redfield, *Annals of the Lyceum of Natural History of New-York* 4 (1848). But the society also continued to rely on engraving, as in this plate of shells, engraved by Gavit and Duthie from drawings by C. Brevoort and M. Morris, *Annals of the Lyceum of Natural History of New-York* 4 (1847).

the American public during the 1840s. The mix of natural history specimens with European religious painting and American landscapes speaks to the values by which American critics and their reading public responded to works of art. Categories of nature, art, and spirituality overlapped in critical statements about truth, beauty and the sublime. For picture-viewing Americans, the idea that painting should reflect nature meant that nature should display qualities associated with painting.[24]

By the time of his appointment to the Smith-

5.4 Cat Squirrel. Drawing by J. J. Audubon, watercolor and graphite, 9 December 1841, for Audubon and Bachman, *The Viviparous Quadrupeds of North America*, vol. 1 (1845).

sonian, Baird enjoyed friendly and professional relations with almost every naturalist in the country. He was sympathetic to both the older and younger generations of naturalists. A close friend of John Cassin's, Baird nevertheless urged restraint in negotiating for the revision of Peale's mammals and birds, both to protect Peale's feelings and to avoid playing into the hands of Wilkes. He established his place in the growing network of institutional natural history from his position as professor of natural history at Dickinson College. Elected to the more specialized societies established since 1830, such as the Boston Society of Natural History, the Entomological Society of Pennsylvania, the American Association of Geologists and Naturalists, and the National Institute, Baird had prepared himself for the Smithsonian position he coveted by corresponding and traveling widely to visit colleagues and study collections and by keeping abreast of scientific news. Moreover, he had already undertaken his first major publishing project, translating and editing an American edition of the *Iconographic Encyclopedia*, a work that replaced extended descriptions of architecture, natural history, and technology with a large number of plates.[25]

Baird arrived in Washington to take up his Smithsonian duties in 1850, in the middle of the Exploring Expedition publication program. The mismanagement of the collections and of the publication program served as a perfect example of how not to administer a museum, negotiate for funding, supervise scientists, or edit a scientific publication program. Baird seemed to model his policies with Wilkes's pitfalls in mind. He would play a role similar to that of Wilkes's and Drayton's combined, but he would play it with tact and humor toward the authors. Baird knew that he bore responsibility for all aspects of government scientific publications, including the national prestige they could bring; new species, furthermore, seemed to provide arguments in favor of continued congressional support for the scientific work of the

continental surveys. To ensure the prestige of priority in description for government collections, Baird permitted his authors to publish their work as articles in journals of local scientific societies. Meanwhile, he gave warm encouragement to his illustrators and ruled his printers with a firm hand.

As the de facto editor in chief of federal science, Baird became an expert on paper grades, engraving tools, and new mechanized printing technologies, skills at midcentury as necessary to a scientific administrator as to the production manager of a publishing house. Writing to George P. Marsh, a regent of the Smithsonian and family friend, Baird described his new responsibilities: "In addition to my Natural History operations I have entire superintendence of the publishing department, revising memoirs, fighting printers and engravers, correcting proofs, distributing copies."[26] A few months later he added: "I have to visit all the printers, binders, lithographers, etc., in the city almost every day.... I sometimes feel as if I were wasting my time attending to these details; but then ... if I do not attend to them no one will, and I flatter myself that the publications of the Smithsonian Institution could not go on without me."[27]

Baird exhibited a remarkable ability to remain above controversy.[28] Early in his Smithsonian position, he was confronted with the issues raised by Samuel George Morton's theory of the separate creation of the human races. Morton's theory touched on two central American social institutions, religion and slavery. Separate creation, since it clearly contradicted Genesis, implied the independence of science from religion. At the same time, it played into the hands of proponents of slavery, providing grounds for rationalizing the supposed inferiority of Africans to Europeans.[29] To propose, as Morton did, that human races constituted separate species contradicted the biological definition of species established by Linnaeus, that of the infertility of hybrid offspring. Yet the wide recognition that Morton's work received in

Europe enhanced the status of American science abroad. Baird responded to the debate with an enthusiastic call to his far-flung correspondents to collect canids—dogs, foxes, and wolves. Separate species of canids, like humans, produce fertile offspring, and in calling for their collection, Baird defined his role and that of the Smithsonian not as arbiters but as the curators of the physical evidence on which the debate could finally be decided. Similarly, Baird would maintain a rigorous professional discretion in later controversies over religion and evolutionary materialism. This neutrality informed the meaning of the illustrations he sponsored; they constituted a public record of national resources, and a pictorial catalog of the national collections.

As soon as he took up his duties Baird needed to recruit illustrators. Already familiar with the work of most of those currently working for zoology, Baird needed to engage several at once

5.5–7 Lithographs by John Collins for Morton, *Crania Americana*, 1839.

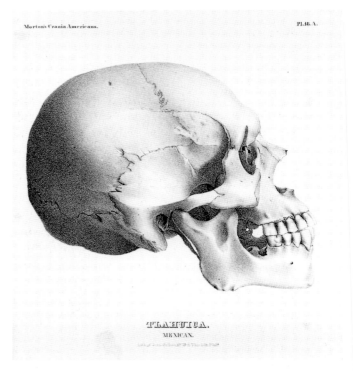

5.5 Tlahuica. Mexican.

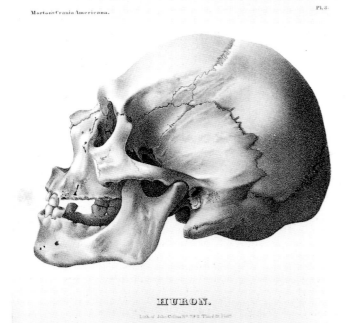

5.6 Huron.

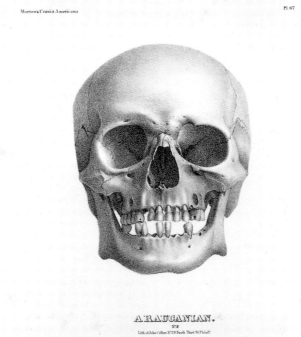

5.7 Aruacanian.

to work on simultaneous publications. Unlike Wilkes, who had insisted that authors and illustrators reside in Washington while working on the collections, Baird enlarged his field of choice by allowing collections to be sent away for description and illustration. For this reason, Baird needed illustrators who could work independently and care for valuable and fragile specimens and for living animals.

In addition to his Smithsonian duties, Baird contracted to write the report on the Exploring Expedition's herpetological collections, although he would turn the work over to Charles Girard. Because all reports had to conform to a stipulated format and style, the illustrations had to be copper engravings. For other Smithsonian reports, however, Baird planned to use either steel engraving or lithography, because with these methods he could stretch his publication budget to buy more illustrations at lower prices and publish larger editions in less time. Baird would need illustrators who could work in the fine line style appropriate for translation into engraving and who could either prepare drawings in continuous tones or draw directly onto lithographic stone.

FIGURE
5.8

John H. Richard, who met all of Baird's requirements, was one of the first illustrators he engaged. Baird was familiar with Richard's work and knew that it met the well-known perfectionist standards of his former employer. Holbrook had insisted that his illustrators draw from live specimens whenever possible. Baird could send Richard shipments of living reptiles both for drawing and for exchange with Europe. He assured Richard: "There are no poisonous snakes in the lot sent. The keg is in the big box. You ought to open this and look after the turtles," he instructed, confident that the artist could handle them.[30] Richard was engaged to draw the fish for the Exploring Expedition report atlas, and he spent hundreds of hours preparing the drawings. He preferred, however, to vary his subjects. He wrote to Baird in his heavily accented phonetic spelling: "Der ser . . . if you haf Reptillien to to I woult laig to half a Chaens. . . . I get sig to nothing bout fisch."[31]

FIGURE
5.9

Baird also needed engravers. Whereas over twenty engravers had contributed to Dana's *Zoophytes* atlas published in 1849, only Dougal engraved the official issue of the herpetology atlas, published nine years later. To engrave some of the additional plates for the unofficial issue of the herpetology atlas, Baird engaged R. Metzeroth, who as a young and less experienced engraver no doubt charged less than Dougal. Baird would take advantage of Richard's skill in drawing for engraving and Metzeroth's growing experience with zoological subjects to employ them together on the steel engravings for further federal publications.

PLATES
38–40

FIGURES
5.10–24

PLATE
41

At the same time that Baird oversaw this engraved work, he was commissioning lithographic illustrations for survey reports and the *Smithsonian Contributions to Knowledge*. His generation of American scientists proved as eager to try color printing as the earlier naturalists had been to use lithography. Lithographic firms like Duval's and James Ackerman's in New York were beginning to offer commercial color lithography.[32] Thomas Sinclair had won industrial awards for his color printing as early as the 1840s. The color process, however, required even greater quality control than monochrome printing, including the specialized skills of color separations, done by eye, and the careful justification of the separate plates. The price of the plates rose with the number of colors required for printing. The expense of color lithography insured that few mainstream zoological monographs appeared with full-color plates; of those that did, most limited the full-color treatment to the first or the first few plates. One ambitious government-sponsored ethnographic monograph illustrated with lithographs printed in colors by Duval and Ackerman, Henry Rowe Schoolcraft's report on American Indians, failed to meet the standards of both Schoolcraft and the illustrator Seth Eastman, who returned to engraving to complete the work.[33] Baird commissioned prints with backgrounds of a single color for the landscapes of the Stansbury Great Salt Lake report in 1852. Meanwhile, Baird was trying to attract

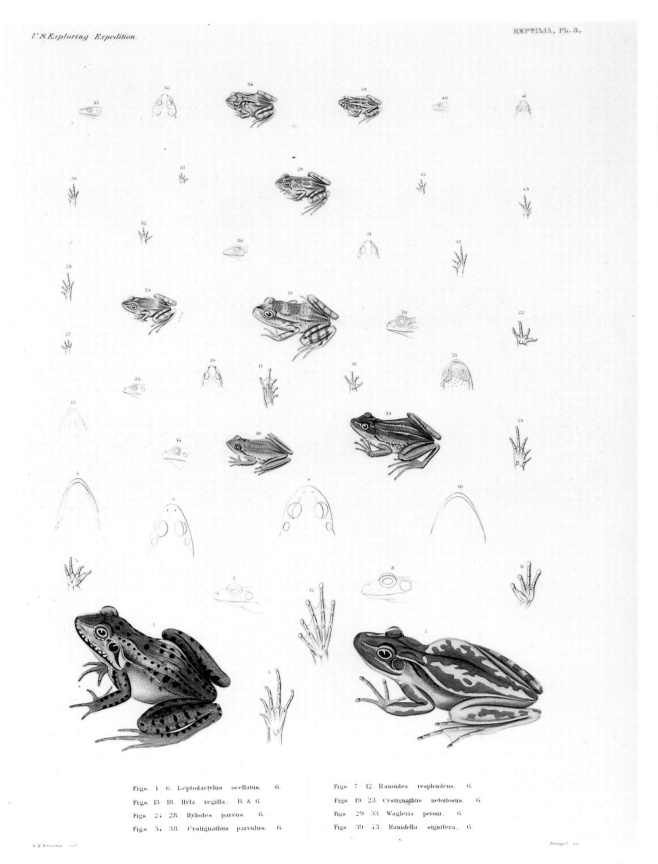

Figs. 1-6. Leptodactylus ocellatus. G.
Figs. 13-18. Hyla regilla. B. & G.
Figs. 24-28. Hylodes parvus. G.
Figs. 34-38. Cystignathus parvulus. G.

Figs. 7-12. Ranoidea resplendens. G
Figs. 19-23. Cystignathus nebulosus. G.
Figs. 29-33. Wagleria peroni. G.
Figs. 39-43. Ranidella signifera. G

5.8  New species of frogs. Hand-colored engraving, drawn by W. W. Winship, engraved by W. Dougal, for the unofficial issue of Baird and Girard's Exploring Expedition *Herpetology* atlas (1858).

5.9 John H. Richard posing with his work in his Cambridge, Massachusetts, studio.

a master lithographer to the Smithsonian. He had in mind Auguste Sonrel (d. 1879), an associate of the Swiss, Louis Agassiz.[34]

Agassiz (1807–1873), protégé of Cuvier and Alexander von Humboldt, was a specialist in fossil and living fish and author of the theory of glacial advance and retreat. He had arrived in Boston from Switzerland in the fall of 1846 to deliver a public lecture series at the Lowell Institute. Agassiz stayed to take both the American public and scientific community by storm with his broad knowledge, his infectious enthusiasm and gift for popularizing science, and the authority of opinion he wielded as an eminent European. Within a few months of his arrival, Agassiz had visited most of the leading American zoologists, studied their collections,

and embarked on several collaborative projects.[35] Baird had written to Agassiz from Carlisle and offered to cooperate with him in any scientific capacity the visitor cared to name.[36] Although Baird and Agassiz would never publish together, they became key figures in the development of museum-based zoology.

Agassiz also became a force in shaping American zoological publishing at midcentury. In Europe he had been accustomed to working at a scale unheard of in nongovernment science

5.10–21 Steel engravings by R. Metzeroth and W. H. Dougal, from drawings by J. H. Richard, for the mammal and reptile reports by Spencer F. Baird, in Emory, *Report on the United States and Mexican Boundary Survey* (1858).

Figure 5.10.

J. H. Richard del.                                                              R. Metzeroth sc.

*Sciurus castanonotus ?*

Figure 5.11.

Figure 5.13.

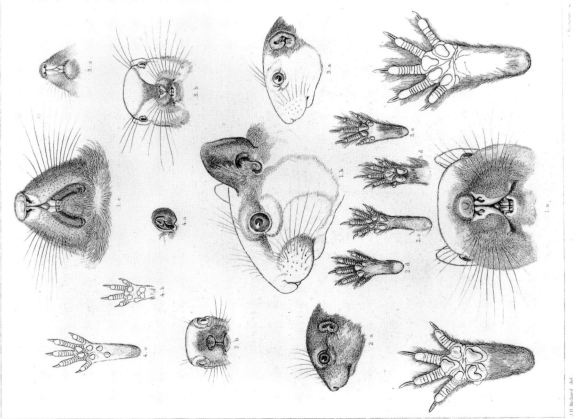

Figure 5.12.

Figure 5.15.

Figure 5.14.

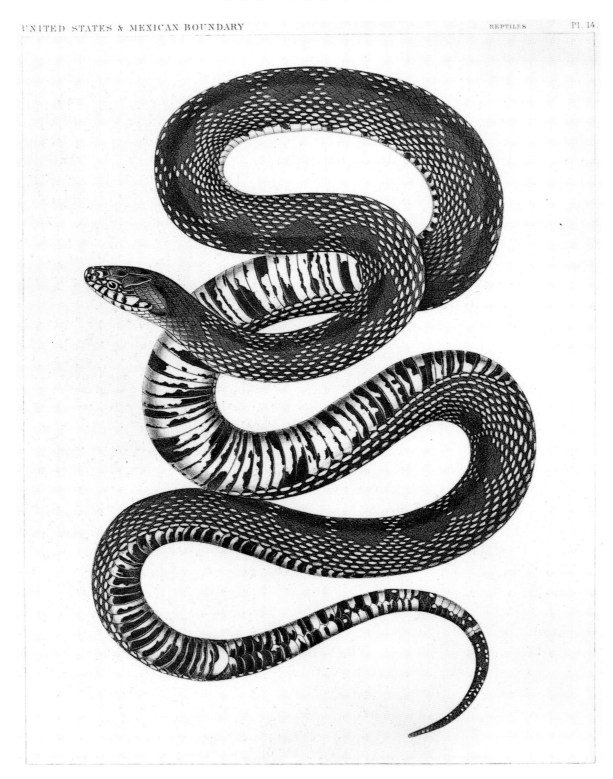

Figure 5.16.

UNITED STATES & MEXICAN BOUNDARY

REPTILES PL. 17.

Figure 5.17.

J.H.Richard del. W.H.Dougal sc.

Figure 5.18.

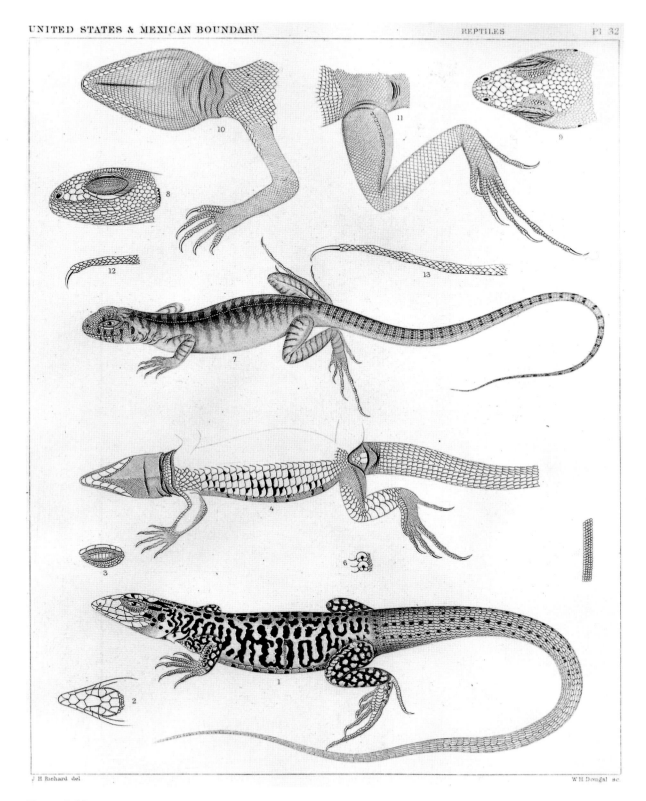

J H Richard del         W H Dougal sc.

Figure 5.19.

UNITED STATES & MEXICAN BOUNDARY
REPTILES
Pl. 36
J. H. Richard del.
R. Metzeroth sc.

Figure 5.20.

UNITED STATES & MEXICAN BOUNDARY
REPTILES
Pl. 40
J. H. Richard del.
R. Metzeroth sc.

Figure 5.21.

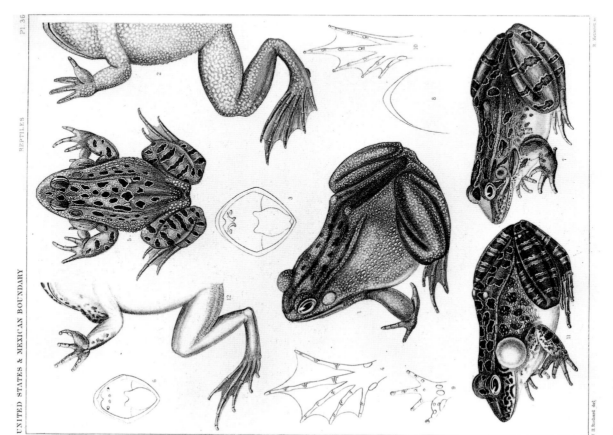

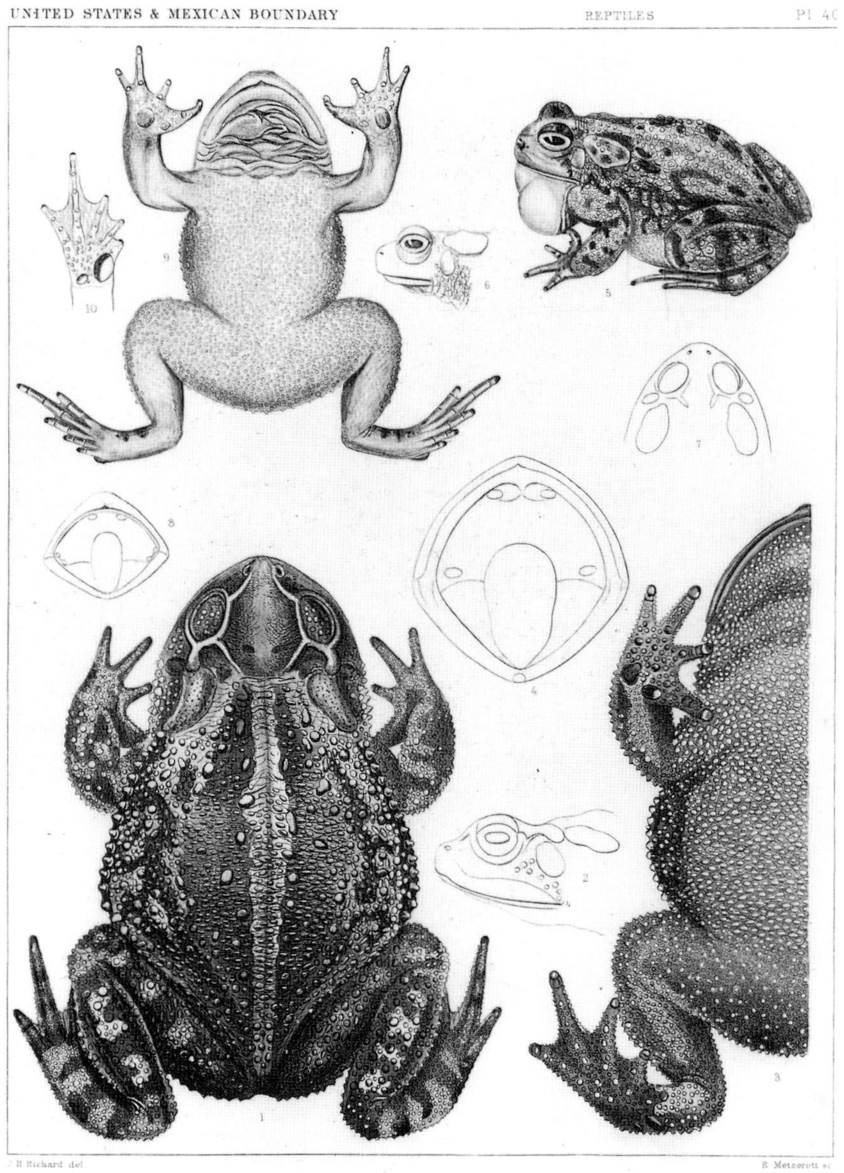

5.22–24 Birds of the U.S. and Mexican boundary. Note the use of landscape and color in the bird illustrations, versus the uncolored, and in some cases abbreviated, treatment of the mammals and reptiles in the same publication. Hand-colored lithographs by Bowen and Co., for the bird report by Spencer F. Baird, in Emory, *Report on the United States and Mexican Boundary Survey* (1858).

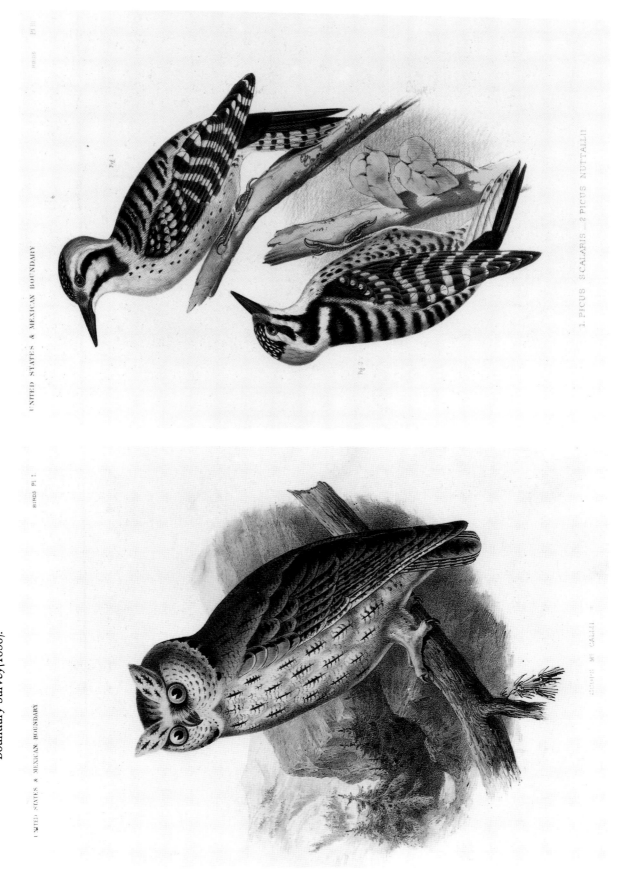

UNITED STATES & MEXICAN BOUNDARY

BIRDS PL. I

UNITED STATES & MEXICAN BOUNDARY

BIRDS PL. III

Fig 1

Fig 2

1. PICUS SCALARIS _ 2 PICUS NUTTALLII

Figure 5.22.

Figure 5.23.

1 LOPHOPHANES WOLLWEBERI   2 AEGITHALUS FLAVICEPS.   3 PSALTRIPARUS MELANOTIS

Figure 5.24.

in the United States. In Neuchâtel, Agassiz had organized his large staff of assistants to produce numerous simultaneous scientific publications on echinoderms, fossil fish, and glaciers. Having so much to publish, he became involved in ownership of a lithography business. Sonrel had been among the many lithographers, printers, illustrators, collectors, and assistants who formed Agassiz's bustling scientific atelier. Once in business, Agassiz discovered that he must continue to feed the presses. He hired more assistants, undertook more publications, issued bibliographies of scientific literature, and pirated translations of important works. The operations of a scientific publishing house cost more than it earned, however, and the concern went bankrupt. Agassiz hoped that his American lecture tour would help extricate him from debt.[37]

Agassiz's scientific workshop reformed in Boston. Sonrel, originally from Nancy, arrived there in the fall of 1848, one of the many European lithographers who immigrated to the United States that year, some of them displaced revolutionaries. Julius Bien (1826–1910), for example, the lithographer who produced one of the lithographic editions of Audubon's *Birds of America*, was a forty-eighter; so was Louis Prang, whose Boston firm became a leader in color lithography.[38] The arrival of Sonrel along with other accomplished lithographers at that time brought to American lithography a new technical polish, and American commercial lithography began to acquire a more cosmopolitan look. Regional idiosyncrasies, subjects, and personal styles endured and flourished in popular genre pictures, but commercial lithography of the eastern seaboard began to rival European work.[39] For over forty years the same medium and industrial structure both made available to a broad public the most popular form of picture in America and produced the bulk of American scientific illustration for a specialist audience.

Having met most American naturalists and discussed their publication plans, Agassiz knew

that a lithographic illustrator of Sonrel's caliber would receive more commissions than he could accommodate, and even before Sonrel's arrival Agassiz began to advertise his accomplishments. In 1847 James Hall, the leader of the New York state survey, had just published the survey's first volume of paleontological studies. Some of the drawings, prepared in part by Hall's wife and sister-in-law, had been redrawn on stone by Frederick Swinton and printed by the Endicott company of New York. Some had been engraved. Hall, overly sensitive to criticism, interpreted Benjamin Silliman's review of the volume to mean that the illustrations were "utterly bad."[40] Silliman had merely written: "The plates are in the main very beautifully executed. They exhibit an historical fact connected with the work—that lithography which was first adopted for the figures was afterward rejected, from the unsatisfactory manner in which the plates were finished. A large part of the plates were reengraved on steel, in which style the later engraving was done."[41] When Hall turned to Agassiz for advice, Agassiz promised the services of Sonrel. Sonrel may have come on the strength of Hall's potential patronage; or perhaps Agassiz, viewing Hall's plates, decided that he himself would need better lithographic illustration than he saw published in the United States, and insisted that Sonrel resume his place in the naturalist's transplanted scientific ménage. Sonrel never did draw or print for the New York survey. Hall had the next volume illustrated with steel engraving.[42]

In Boston, Sonrel rapidly established himself as one of the leading scientific lithographers in the country, thanks partly to Agassiz's reputation. Acting as a sort of agent whose own publications served as testimonials, Agassiz recommended Sonrel to other naturalists, and after 1848 the lithographer's work appeared in the major journals of American zoology. Baird first saw Sonrel's work late in 1848, and met the lithographer the following summer.[43] Master of both the medium and of the peculiar require-

PLATE 42

ments of scientific drawing, Sonrel brought to his work an elegance that informed Baird's standard.

Baird and Agassiz had agreed to write a collaborative study of American fish. Although Baird proceeded with his portion of the agreement, Agassiz was characteristically too busy initiating more scientific schemes to complete those already begun—such as the description of the fish collections of the Exploring Expedition. Realizing that he had lost his coauthor, Baird was loath to waste any of the work, particularly the plates which Sonrel might already have made. He inquired directly of Sonrel whether, if any of the fish illustrations had been drawn on stone, he could purchase them for use in another context.[44]

A year later, in October 1851, Baird engaged Sonrel to draw plates for Captain Stansbury's reports on the government exploration of the Great Salt Lake. Sonrel drew vertebrates, fossils, insects, plants, all from specimens that Baird forwarded to Boston. Baird was only in his first year at the Smithsonian and eager for completed volumes of survey reports as evidence of the effectiveness of his program coordinating exploration, collection, description, and publication. Waiting impatiently for Sonrel's illustrations of reptiles for the Stansbury report, Baird wrote: "I trust these plates of yours to make such an impression on Congress as will induce them to continue such explorations; and publish the results in creditable style." Sonrel's plates would "also materially affect the question of the style of publishing the zoological results of Major Emory's expedition" along the newly negotiated United States and Mexican boundary. Baird needed the best for his initial volumes, and he wished to engage Sonrel for the next six months to illustrate collections already at hand.[45] He would repeatedly invite the lithographer to move to Washington to facilitate their arrangements. Sonrel resisted the move. He had found employment with the Boston firm of Eben Tappan and Lodowick H. Bradford, for whom he drew commercial work

and views.[46] But he also maintained an independent business in scientific illustration, working largely from his home in Woburn, ten miles north of Boston.

Printing and proofing lithographic plates was a complicated technical process made more complicated by the structure of institutional publishing. Delays, misunderstandings, and extra expense were almost inevitable. When the printers sent the proofs back to the Smithsonian for corrections and the insertion of figure numbers to correspond with the captions, the author saw his illustrations in final form for the first time and often requested alterations. Changing lithographic drawing was difficult and expensive; substantive changes required another round of proofs. Even after the author or editor had approved the proofs and the printer had run the full edition, each print had to be checked for consistency. Atmospheric fluctuations affected the chemical processes of lithography. During printing the stones were reinked repeatedly, and changes in temperature, humidity, the presence of dust, and variations in the paper stock all affected the even application of ink to paper. Baird sometimes rejected an entire edition of plates.[47]

Overall, however, Baird was delighted with the quality of Sonrel's work and eager to complete publication. Problems arose not over proofs but over charges. Sonrel negotiated his own contracts based on his own estimates of time and costs, and then subcontracted with Tappan and Bradford for printing from the stones on which he drew. Sonrel's position as an independent lithographer raised his prices, but Baird accepted that. Baird found, however, that after negotiating the estimates directly with Sonrel, he had been overcharged by Tappan and Bradford. A spokesman for the firm tried to justify the bill. He explained that Sonrel had subcontracted the printing to them because the cost of printing the plates himself consumed his profits from the drawing. Tappan and Bradford, not having been party to the original contract, had charged Sonrel's estimated amount for printing, but had taken a loss. They claimed that Sonrel's fine drawing required custom printing at extra expense to them. Sonrel's detailed work also required a better paper grade than did ordinary work. To compensate for their loss in printing, Tappan and Bradford had added a surcharge for the finer paper and for the waste trimmed from the paper stock, which did not come in the correct size for the edition.[48] Tappan and Bradford had hoped that their surcharge would go unnoticed, but Baird requested an itemized bill. The congressional appropriations had specific designations—one amount for drawing, another for paper, another for printing, and so forth; Baird had to account to Joseph Henry for all his expenses.[49] Tappan and Bradford found that meeting Smithsonian standards for lithographic illustration cost more trouble than the contract was worth:

> The price for the printing we consider ruinous so far as profit is concerned. As the work is of such character as to render it difficult in the extreme to obtain the requisite number of first quality proofs, the work always being drawn so fine, the difficulty is so great that we consider our only profit in the whole business, what we gain in reputation, trusting that we shall hereafter get simpler work from other sources, being well aware that but little work comes from your vicinity north that is not of a similar character and void of profit.[50]

The attitude of Tappan and Bradford toward Smithsonian work complicated further arrangements with Sonrel. To retain him as a draftsman, Baird needed to solve the logistics of transporting Sonrel's drawings to a printer for whom the long-term value of a government contract would justify a very narrow margin of profit. If the artist continued his practice of drawing directly on stone, then the slabs would have to be moved at high cost and risk of damaging their surfaces. If Sonrel drew on paper instead, then Baird could have the drawings applied by the transfer method to stones located anywhere.

For transfer lithography, the draftsman drew in lithographic crayon on thin paper prepared with a solution of gum, starch, and alum. The moistened paper was then pressed against the stone, which had been heated so that the drawing and the paper's coating adhered to the surface, permitting the paper to be peeled away. After the coating had been washed off, the design remained on the printing surface. For maps and other drawings such as scientific illustration that must not stretch, the stone was moistened instead of the paper. The drawing at first appeared weaker than a design made directly on stone, but after treatment with an acid etch, it took the lithographic ink and was printed in the standard manner.[51]

Baird believed that the use of transfers would permit larger editions; he needed a technique that could withstand editions of up to seven thousand copies, for which, he estimated, Sonrel would have to make six or seven successive lithographic drawings.[52] Tappan and Bradford discouraged the use of transfers. They advised that engraving Sonrel's drawings on steel, although initially more expensive, would prove more economical in the long run than using transfer lithography. The printers doubted that work as fine as Sonrel's could be transferred without loss of detail, "the transfer being weaker than the original drawing."[53] The printers also warned Baird that pulling the same number of prints from a stone that could be printed from steel would exceed the cost of steel engraving. Furthermore, they doubted that Baird would be satisfied with the results of a process that lost the details so essential to scientific illustration. Baird decided nevertheless to go ahead with the transfer method. He made arrangements with James Ackerman in New York to complete the plates for Captain Stansbury's Great Salt Lake reports, using the transfers prepared by Sonrel. Baird seemed satisfied with the results.

FIGURES
5.25–27

While most of the time Baird and Sonrel understood each other's requirements, the editor and illustrator constituted only two sides of a

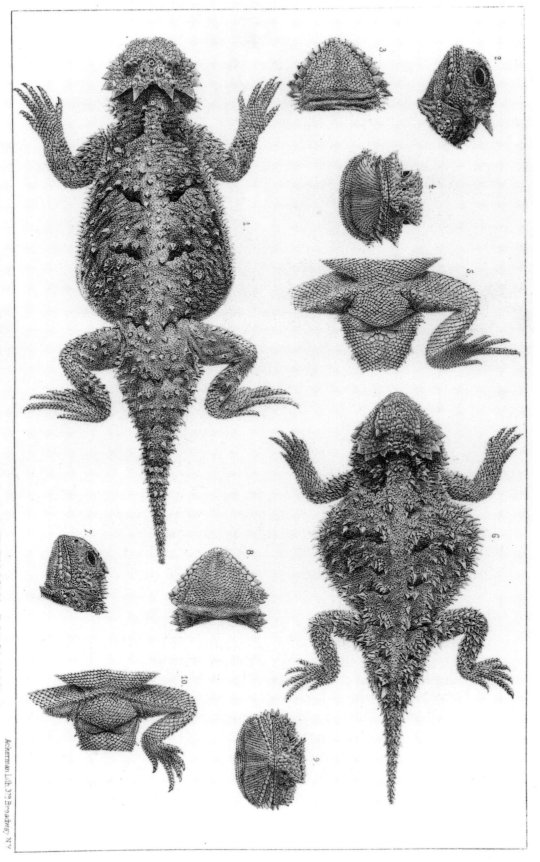

Fig. 1-5. PHRYNOSOMA PLATYRHINOS, Girard.

Fig. 6-10. PHRYNOSOMA DOUGLASSII, Gray.

Ackerman Lith 3ᵈ Broadway N.Y

5.25 (*left*),
5.26 (*opposite, top*),
5.27 (*opposite, below*)
Reptiles and insects. Lithographs drawn by A. Sonrel and printed by the firm of J. Ackerman, for Baird and Girard's report on the reptiles and S. S. Haldemann's report on the insects in Stansbury, *Exploration and Survey of the Valley of the Great Salt Lake of Utah* (1852).

1 3. Labidus saji. — 4-6. L. harrisii. — 7-9. L. melshajmeri. — 10. Euphoria cernii. — 11. Cotalpa granicollis.
12-14. Henous techanus. — 15. Megaderus corallifer. — 13. Cicada striatipes. — 17. C. ref.

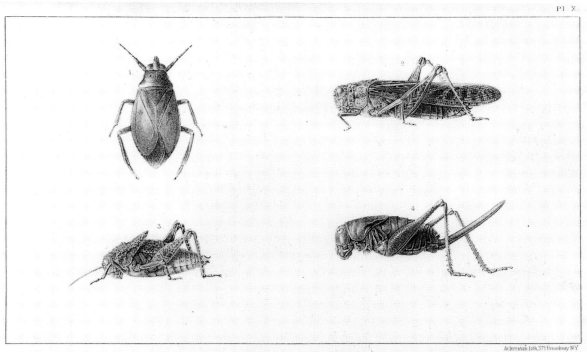

1. Zaitha bifoveata. — 2. Oedipoda corallipes. — 3. Ephippiger trivavensis. — 4. Anabrus simplex.

three-way process. In 1852, to Baird's surprise, Joseph Leidy (1823–1891), a paleontologist writing on the fossils of the Nebraska Territory, took exception to Sonrel's work. Leidy, professor of anatomy at the University of Pennsylvania and a member of the Academy of Natural Sciences, published ably on a wide range of topics, from vertebrate paleontology to freshwater protozoa. Leidy did not maintain separate specifications for illustrating living forms and fossils. His own studies encompassed both fields, with an emphasis on anatomy. The shared criteria and pictorial conventions for the illustration of fossil and living vertebrates dated from Cuvier's work, which had made anatomical structure the basis of classification. It had become standard practice, moreover, to include vertebrate skulls in taxonomic illustration of living forms. A fastidious worker, Leidy made his own study drawings, and had seen them published in the Philadelphia scientific journals as lithographs printed by the city firm of Thomas Sinclair.

Baird and Leidy seem to have given contradictory instructions to Sonrel for drawing the fossils. Baird, wishing to economize, requested that Sonrel simplify his technique, "namely to confine the high finish in the teeth," and work the bone and stone matrix in outline to reduce his hours and simplify printing. Leidy complained to Sonrel about the results. Sonrel, who preferred to take his instructions directly from Baird, forwarded Leidy's complaints to Baird, and asked whether he should finish the drawings according to the paleontologist's wishes. Sonrel reminded Baird that the price would increase with the amount of work.[54]

Leidy complained directly to Baird that one figure "could not, with the exception of the teeth, be done much worse. How much did this plate cost?"[55] On hearing the price, he wrote back: "The price of Sonrel for the last plate I think very exhorbitant considering its execution, and I am satisfied that they can be done almost as well by an artist of Sinclair's for less than half Sonrel's price."[56] Baird, who consid-

ered Sonrel the best lithographer of scientific subjects in the country, must have been surprised by Leidy's preference for a staff lithographer at Sinclair's establishment. The Smithsonian had already done business with the Philadelphia firm. Although only a few years before Sinclair's draftsmen had drawn plates of variable quality, Baird knew that Sinclair could now produce fine illustration.[57] But he also knew that Sinclair was stubborn and could conveniently ignore instructions in order to preserve his profit margin.

Sonrel's "exhorbitant" charge of $45 per plate had to cover the time and trouble of transporting the stones to and fro between Boston and his home in Woburn for drawing and proofing. Sinclair could charge half Sonrel's price because he paid his draftsmen half price for low-profit medical and scientific work.[58] Sinclair's draftsmen had to deliver high-quality drawing or lose their positions. For Sonrel, an independent craftsman, no charge could compensate adequately for the personal inconvenience of late payments for services rendered—almost inevitable with government work. A large firm, on the other hand, could ride out congressional recesses or the Smithsonian's temporary shortness of funds. Throughout the episode of Leidy's plates, the Smithsonian owed Sonrel money for past work, money that Sonrel reminded Baird he needed badly.[59]

Leidy seemed to be overreacting; only one figure in three plates failed to meet his standards. The paleontologist may have manufactured objections because, having established a comfortable working relationship with Sinclair, he resented the inconvenience of dealing with a different lithographer and of sending the fossils to Boston for illustration. Leidy was prepared to accept drawings "almost" as good as Sonrel's for the convenience of Sinclair's accustomed service. In other instances, however, Leidy set standards so high that only he himself could meet them. In a later report on freshwater protozoa, he inserted a note to explain that the lithographs by Sinclair's firm were not

FIGURES
5.28–31

as faithful to nature as his original drawings, and that the plates were colored too deeply throughout.[60]

Sonrel, accustomed to finishing his drawings to his own high standards, may have bridled at the suggestion that he confine his style to mere outline. Until Leidy took exception, Sonrel's work had justified the high price. As an independent businessman, Sonrel could not reduce his prices, and as an artist he took offense that when he simplified his work on request he should be blamed for the results. His business reputation was at stake. "I will endeavor to redraw the objectionable figure and make it more finished," he conceded. But, he reminded Baird, the Smithsonian was still behind in paying his account.[61]

From Baird's perspective, the reputation of the Smithsonian depended on both the author and the illustrator. Leidy was one of the Smithsonian's most prolific contributors of articles. Equally, the value of Sonrel's understanding of the requirements of scientific illustration outweighed the extra cost or inconvenience. Baird did not consider Sonrel's high prices "above the beauty of the work."[62] Sonrel knew that his work set Baird's expectations for the performance of the commercial companies. Baird called his drawing "exquisite," "prodigiously fine." "Please do up in your usual beautiful style," Baird instructed, and praised other work as "splendid, the handsomest I ever saw."[63]

At the same time, Baird made constant efforts to economize, and eventually those efforts threatened Sonrel's control over the plates printed from his drawings on stone. When Tappan and Bradford had refused to lower their prices for printing Sonrel's stones, Baird decided he must use another printer. Sonrel responded: "After this, as the Institution may, in regard to printing of plates to come, adopt measures that might not suit me, if so I should respectfully decline undertaking any further work for it, for I wish not to resign the superintendence of the printing of my plates."[64] Sonrel considered himself a lithographer in the sense

of both draftsman and printer; careful printing technique was as important to his finished illustrations as the original drawing. At Tappan and Bradford's he could oversee the printing of his work. Baird continued to send work to Sonrel, but he requested drawings on paper instead of on stone, to be engraved or lithographed elsewhere. By the time that Baird had found a way to afford the artist's work, however, Agassiz had embarked on a publication that would virtually monopolize Sonrel's time.

With Sonrel occupied by drawing for Agassiz, Baird sent more work to Thomas Sinclair, who had immigrated from Scotland before 1833 and established his own business in 1838.[65] From his first days in office, Baird had sent work to Sinclair, and had promised more if the printer put on a good show. Sinclair, in spite of his large commercial operation, at first enumerated the same objections to Smithsonian standards in illustration as had Tappan and Bradford: scientific work for the government required custom work and turned a marginal profit. Sinclair was, perhaps, counting on Baird's inexperience; but he soon learned that he could not manipulate the assistant secretary, who managed to bring the lithographer into line with his requirements.

On one occasion, Sinclair refused to cut the printed plates. Depending on the size of the volume—folio, quarto, or octavo—an uncut sheet of paper could accommodate two or more

5.28–29 *Anchitherium Bairdii*, Leidy (5.28); *Rhinoceras nebrascensis*, Leidy (5.29). The plates about which Leidy complained. Lithographs, drawn by A. Sonrel and printed by the firm of Tappan and Bradford, for Joseph Leidy, "The Ancient Fauna of Nebraska," *Smithsonian Contributions to Knowledge* 6 (1854).

5.30–31 Oreodon species. Plates almost as good for half the price. Lithographs drawn by A. J. Ibbotson and A. Frey, printed by the firm of Thomas Sinclair, for Joseph Leidy, "The Ancient Fauna of Nebraska," *Smithsonian Contributions to Knowledge* 6 (1854).

Plate XI.

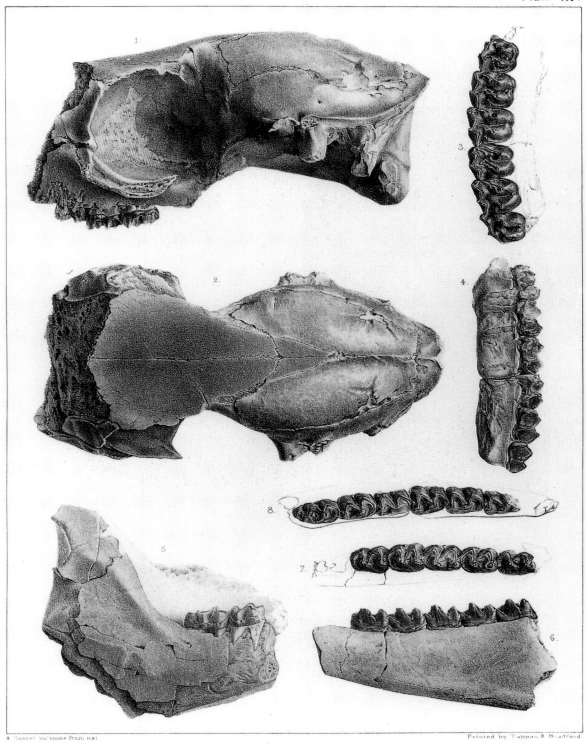

A Sonrel lith'stone from nat.

Printed by Tappan & Bradford

ANCHITHERIUM BAIRDII, Leidy

Figure 5.28.

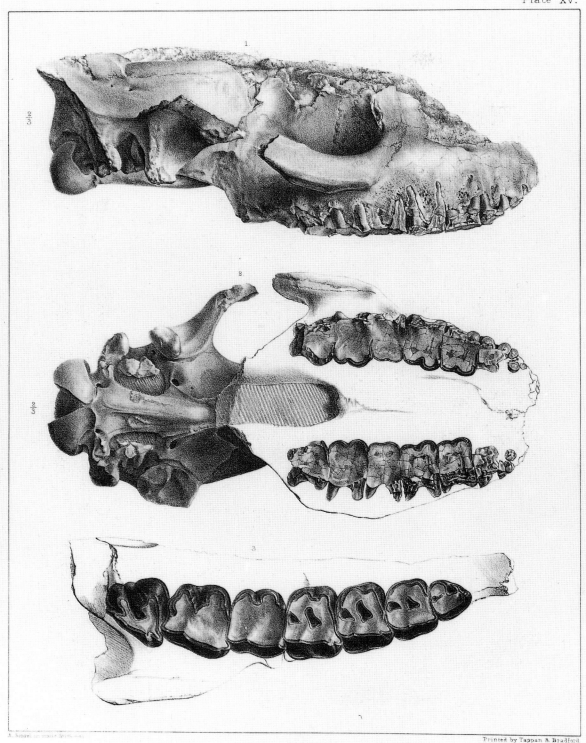

Plate XV.

RHINOCEROS NEBRASCENSIS, Leidy.

Figure 5.29.

Plate IV.

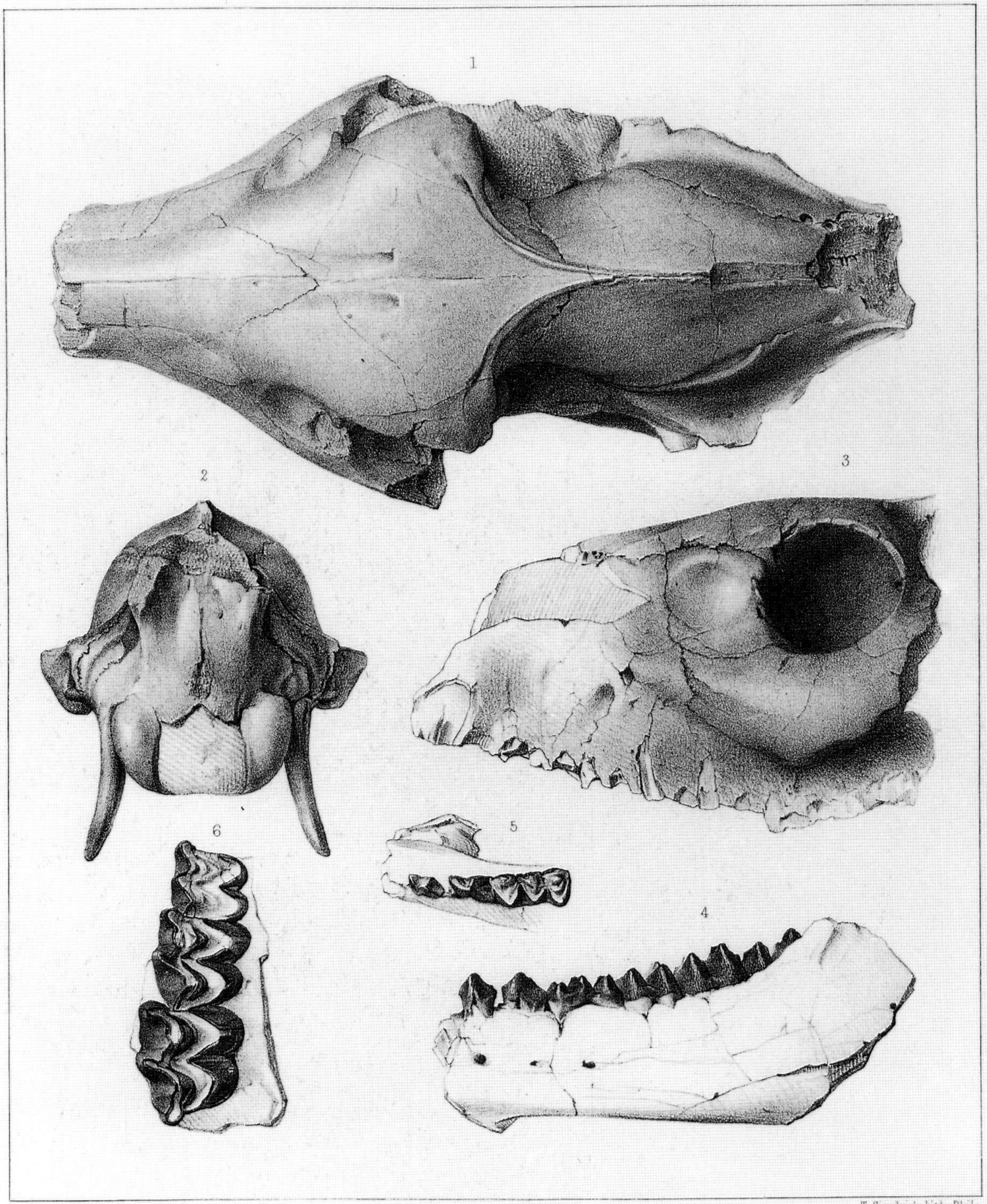

A.Frey Del.

T. Sinclair's Lith. Phila.

1-5. OREODON CULBERTSONII, Leidy.

6. OREODON MAJOR, Leidy.

Figure 5.30.

Plate V

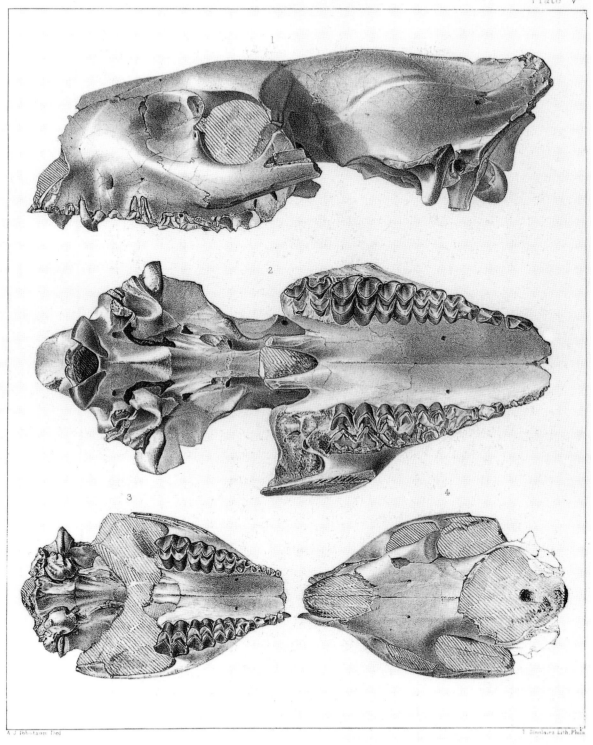

1,2, OREODON CULBERTSONII, Leidy.
3,4 OREODON GRACILIS, Leidy.

Figure 5.31.

Plate 37. *Canis (Vulpes) virginianus*, Gmel. Grey Fox. 5/7 natural size. Drawn by J. J. Audubon. Lithograph printed and colored by J. T. Bowen, Philadelphia, for Audubon and Bachman, *The Viviparous Quadrupeds of North America*, vol. 1 (1845).

prints, the increasing mechanization of the paper industry having extended the available sizes of paper.[66] Plates for the whole sheet were drawn on a single stone, and the printed sheet was later cut for binding. The printer insisted that cutting was the binder's job, but Baird replied that the binder would insist that cutting was the printer's job, and that Sinclair had better do it.[67] In another confrontation, Sinclair, boasting twenty-five years' experience in lithography, complained that Baird expected more from the medium than it could deliver.

Baird's reply summed up all that the assistant secretary had learned about printing and about managing printers during his first year on the job:

You will please allow me to differ with you in regard to what you term unnecessary minutiae. Nothing is unnecessary which tends to [sec]ure symmetry and neatness to a plate. So far is it from there being but little hope of success in your work that I have to congratulate you on the excellence of your work and its general accuracy, far superior

Figs. 1—8. Gehyra vorax. G.           Figs. 9—16. Gekko indicus. G.

Figs. 17—26. Naultinus punctatus. Gray.

J.H.Richard del.                                             Dougal sc.

Plate 38. Geckos. Hand-colored engraving, drawn by J. H. Richard, engraved by W. Dougal, for the unofficial issue of Baird and Girard's Exploring Expedition *Herpetology* atlas (1858).

to that displayed by others. Your "experience of 25 years" has not been thrown away, and if we are a little more critical than is [usual] you should consider it as much for you own reputation as for ours. This Institution intends as far as it can to be surpassed by none in the excellence of its works, and in nothing is extraordinary care required more than in the lithographic plates.[68]

"Extraordinary care" could secure "symmetry and neatness." On the style appropriate for maps, Baird instructed: "The work is to be done well, and clear, but the lines should be as delicate as possible consistent with proper proportion."[69] These were the elements of style for which Baird strove and demanded that his printers achieve.

Baird's pictorial concerns were consistent with his taste and expectations of landscape painting as well as with strictly scientific requirements. The lithographic landscape illustrations in the federal expedition reports portrayed the type localities of the new species that figured in the zoological plates as vast expanses and rugged escarpments, which dwarfed the foreground figures of the surveyors and collectors. These landscape illustrations contributed significantly to the context in which the zoological plates appeared. They maintained a close connection between taxonomic description and representation and the larger social and cultural contexts of both westward expansion and landscape painting. They affirmed the link between art and science celebrated alike by American painters and critics.[70] The implications of these links were further reinforced by, for example, Baird's use of the New York lithographic firm of Edward Bierstadt, brother of Albert Bierstadt, preeminent painter of the sublime in the American west. Baird complained to his printers of a lithographic plate of a western view: "There is no perspective . . . it looks more like an upright wall than a gently sloping bank."[71] And, "the distant shrubs on the hill sides look too much like stones." Both Baird and the author wanted them "made more

shrubby."[72] He criticized drawings of western topography as too tame, plates of Indian antiquities as "stiff" and having "no grace."[73]

Lack of perspective, drawings too stiff or too tame, all these criticisms were grounds for rejecting proofs of illustrations to scientific reports. Sinclair grumbled, but he complied. Baird anticipated Sinclair's objections with the explicit warning: "Put it as low as you can afford as you are threatened with competition."[74] With a shipment of four gallons of crabs, to be illustrated for the journal of the Boston Society of Natural History, Baird instructed: "I want them done in the best manner to show what you can do in this line, as there may be much in it. Also to show the Bostonians what Philadelphia can do."[75]

Sinclair, among other Philadelphia lithographers, would continue to print a large share of American zoological and paleontological illustration.[76] Eventually he ran the fourth largest lithography business in the city. The largest, that of the French emigré Peter S. Duval, had printed the lithographs for Holbrook and continued to take scientific work. As long as Philadelphia remained a center of science, the city printers made a specialty of scientific jobs; the skills and equipment were in place for handling the large editions of mid-century government publications. And as long as many of the zoologists and paleontologists whom Baird engaged were Philadelphians, it was in his interest to cultivate and discipline the city's printers to his standards.

In some respects, the replacement of engraving by lithography had not changed the look of Philadelphia science. Especially in bird illustration, and to some extent with mammals, the large quarto serials of the Philadelphia institutions still used plates representing single species with landscape backgrounds. By the 1850s such lavish illustrations connoted a conservative point of view and the money to indulge it. John Cassin became a partner in the Bowen lithographic company; as both partner and client, he could afford the expense of such plates.

Baird, faithful to his early training, perpetuated the style as much as he could in federal reports. In 1867 he suggested that Richard go to the Academy of Natural Sciences to study bird illustrations in preparation for some Smithsonian work. Richard was impressed. He wrote: "I wend to the Agatamy of N. sciance and sie thos book[.] thos ar wery neisly don[.] wath seiz you want them don[,] laig thos for the Bataent offis the sem seis laig thos of Auduband with baeg ground [or] on graund and staem of threis."[77]

The composite plate of comparative morphology and geographic relationship, however, predominated in most serial publications and in the federal, state, and private monographs of the period. In place of the background and foliage to convey the authority of life, the illustrators of the new style created hyper-real representations of each specimen that impressed by their detail rather than by the weight of environmental evidence.

Two staff draftsmen of the Philadelphia firms, A. J. Ibbotson and A. Frey, both known and requested by zoologists, drew in a manner similar to Sonrel's "high finish" and completed the monograph for which Sonrel's controversial plates were intended. Staff draftsmen like Ibbotson and Frey were not confined to scientific work; they also drew for the firm's more profitable lines of portraits, fashion plates, commercial views, and city scenes. Extraordinary detail characterized the scientific lithographs by the staff draftsmen, detail achieved by the closest scrutiny of the individual specimen. All the torques, knobs, and splinters of, for example, fossil bone or shells were rendered to convey both contour and texture. Using the black of the lithographic crayon and the white of the paper, their drawing style gave the figures a vivid, crisp presence. Their style, like Sonrel's, was also distinguished for its smoothness. Unlike Drayton's first lithographs, or the plates in Hall's first volumes, the work of Ibbotson and Frey and some of their colleagues scarcely betrayed the action of the draftsman's hand. The marks of the crayon could not be dis-

tinguished, as if to conceal that the images had been drawn. Competition with photography, hailed since its invention in the 1830s as the medium that promised complete objectivity, may have provoked this emphasis on seamless verisimilitude in scientific drawing.

The practice of sending zoological specimens directly to the lithographer meant that the draftsman worked in the commercial rather than the scientific context. Clearly, the staff draftsmen of the large companies had a different relationship to the specimens from those who drew in an institutional setting or who specialized in zoological illustration. Some drew scientific subjects against their will. Matthias S. Weaver, an aspiring artist among Sinclair's employees, confided to his diary: "I can do better things than I have been employed at—drawing shells for crabbed conchologists at 1/2 price and lettering all the stones for S[inclair], dancing when he whistles."[78] Lithographic draftsmen, to whom were assigned scientific jobs among other kinds of drawing, produced pictures that the specialist readership judged by the same criteria as it judged plates produced by their professional colleagues, or illustrators associated with institutions, such as Richard with the Smithsonian, or the illustrators working for Louis Agassiz at his newly founded Museum of Comparative Zoology in Cambridge, Massachusetts. From the comments of the fastidious Joseph Leidy one can see how commercial production of scientific pictures created new scapegoats for the allocation of blame for illustrations that failed to meet standards.

A move toward order and symmetry prevailed in and unified the collective national scientific oeuvre. By ordering the scientific specimens into neat and symmetrical arrangements preferred by their scientific clients, commer-

FIGURES

5.32–36

5.32–35 Unionidae. Lithographs by J. L. Magee and A. Frey, printed by the firm of Thomas Sinclair, for Isaac Lea's descriptions of the family Unionidae, in the *Journal of the Academy of Natural Sciences, Philadelphia*, n.s. 3 and 4 (1858 and 1859).

Pl. 33

33  *Triquetra  contorta*

Figure 5.33.

Pl 21

1  *Unio  Buccinatus*

Figure 5.32.

120 *Unio Cumingii*

Figure 5.35.

31 *Unio rubellinus*
32 *Unio coronatus*
33 *Unio umbrans*

Figure 5.34.

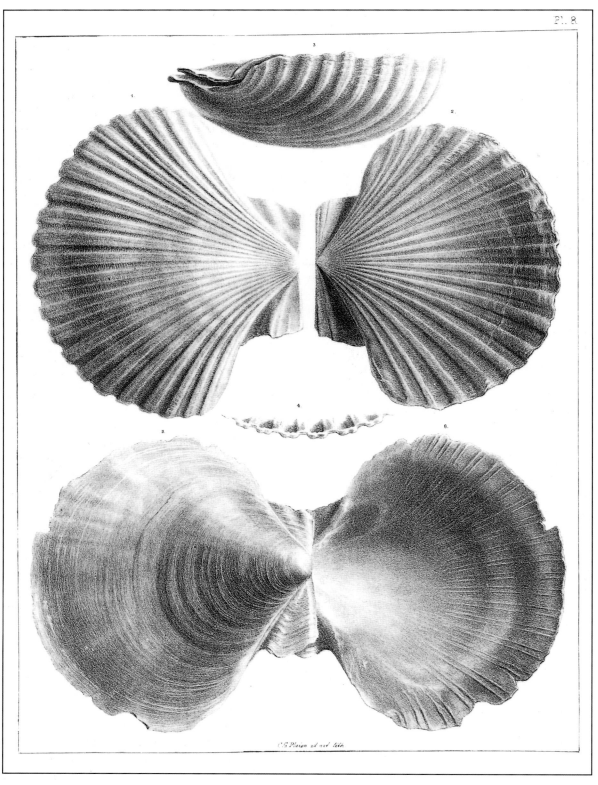

5.36 Fossil shells. Lithograph by C. G. Platen for Toumey and Holmes, *Pleiocene Fossils of South-Carolina* (1857).

cial lithographers made pictorial sense of unfamiliar material. Outside of the principal centers of science, science still recruited illustrators from the ranks of fine artists and amateurs. For the fine artist, the ordered and symmetrical style tended to cast the natural objects, especially bones and shells, into the role of ornament, as in the plates by C. G. Platen, an artist at the College of Charleston, for the South Carolina state geological survey.[79] The emphasis on symmetry gave the work classical overtones.

Zoologists and paleontologists gave no specific name to the detailed and symmetrical style that emerged during the 1850s. They invoked its conventions only obliquely. The supervisor of Platen's work thanked the liberality of the South Carolina legislature for its subscriptions, enabling "us to complete it in the best style of art."[80] James Hall echoed his southern colleague in his preface to volume three of the New York paleontological survey, in which he reiterated for his readers the history of delays in funding that had postponed publication of the next installment of his work. He explained: "Some of the plates of this volume were engraved before the completion of Vol. II in 1850; a considerable number, between that time and 1855, while the work was not in authorized progress, except by the existence of contracts between the State and the engravers. The remaining portion, much of which has been executed in *a better style of art*, has been done since 1855."[81]

Hall's separation of the volume's plates into three periods and his evaluation of the quality of illustration in each indicate how crucial the 1850s were to the formation of a standard of scientific representation. In 1843, while acknowledging his partiality, Hall had praised the illustrations produced for the fourth part of the *Geology of New York* by Mrs. Hall and her sister Mrs. Brooks for their "fidelity and precision." Hall had stated that "the figures would enable any person to identify the fossils of western New York."[82] Hall based his evalua-

tion of the quality of work for the first volume of the *Paleontology*, published in 1847, on similar criteria: "I can only content myself with having represented in the engravings, as accurately as possible, every object described in this volume; thus affording, to those who desire to do so, the means of comparing species, and of correcting any erroneous references."[83] Silliman's comments on the plates of that volume, however, had shaken Hall's confidence in a standard based on legibility alone.

The division that Hall noted in the production of the plates for the *Paleontology*'s third volume referred not only to style but also to personnel. Fielding Bradford Meek (1817–1876), a veteran of Owen's geological surveying team, had worked with Hall until 1855, when a dispute between them over the terms of their co-authorship of a report terminated their collaboration.[84] Another young geologist-illustrator, Robert Parr Whitfield (1828–1910), had taken over and finished the plates for the volume. Hall's reference to a before and an after in the volume's style contained, therefore, a tacit criticism of Meek and praise for Whitfield's collaboration with the project's principal lithographer, Frederick Swinton, a veteran of Hall's workshop.[85] Hall's statement, nevertheless, constituted a recognition of what had happened in American zoological and paleontological illustration. The plates produced from Whitfield's drawing and Swinton's lithographs, illustrating the third volume of 1859 and its second part of 1861, conformed to the standards of graphic fluency established by federal and private publications. FIGURES 5.37–42

Other conventions besides the style of drawing and degree of detail also consolidated during the same period. An implied light source from beyond the upper left-hand corner of the plate illuminated all the objects arranged rows in a shallow space. Arrangement of the figures not only worked with the symmetries of the rectangular page, but with the symmetries—bilateral or radial—of the subjects. Bilaterally symmetrical animals such as mammals and

Figs. 1–7. Teius teguixin. Schinz.    Figs. 8 & 9. Brachylophus fasciatus. Cuv.
Figs. 10–16. Hoplodactylus pomarii. G.

Plate 39.  Lizards. Hand-colored engraving, drawn by J. H. Richard, engraved by W. Dougal, for the unofficial issue of Baird and Girard's Exploring Expedition *Herpetology* atlas (1858).

Figs. 1–5. Dolophis flaviceps. G.    Figs. 6–12. Contia mitis, B. & G.

J.H.Richard del.    Dougal sc.

Plate 40. Snake species named by Baird and Girard. Hand-colored engraving, drawn by J. H. Richard, engraved by W. Dougal, for the unofficial issue of Baird and Girard's Exploring Expedition *Herpetology* atlas (1858).

5.37–38 Engravings by Gavit and Duthie, from drawings by Mrs. Hall and Mrs. Brooks, for James Hall, *Paleontology of New York*, vol. 1 (1847).

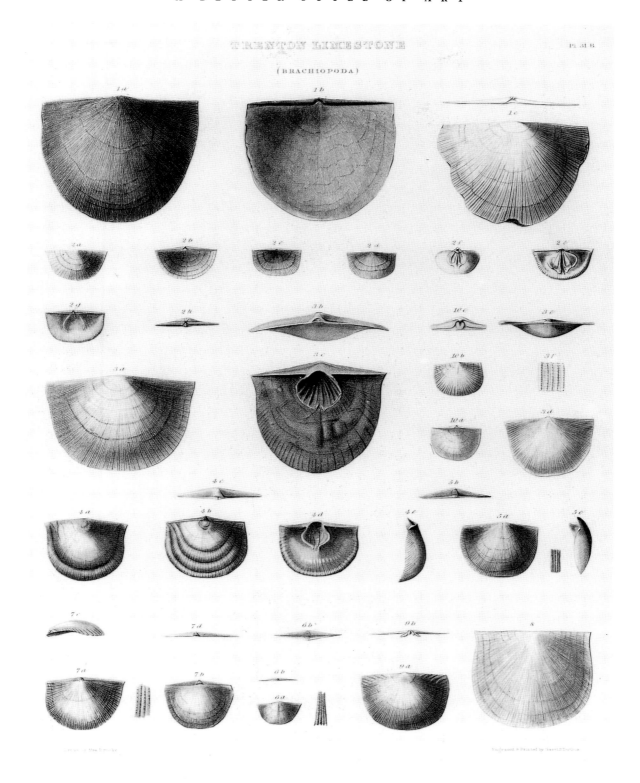

Figure 5.38.

VULPES (UROCYON) LITTORALIS, Baird

Plate 41. *Vulpes littoralis*, Coast Fox. Hand-colored steel engraving by R. Metzeroth, from a drawing by J. H. Richard, illustrating Baird's description, for his report on the mammals collected on the Pacific railroad surveys, 1857.

fish faced to the left, oriented toward the inner margin, to display the left side of the body. (Lithographic plates were usually printed on the recto side of the paper only.) The internal composition of the plate space referred to and defined the edge of the plate. That edge began to be marked by a border. Some illustrators used the edge to abbreviate the figure, its symmetrical other half implied beyond the inked border. Occasionally, a limb broke the boundary of the border, a playfulness that acknowl-

edged and reinforced the convention. No one mentioned the consolidation that was taking place, perhaps because many of the graphic and pictorial devices had been around for a long time. Not every plate adhered strictly to all the elements that defined the style; the conventions provided the outlines within which the field produced its working pictures. Throughout the discipline, the rectangular space of the plate increasingly became an internally organized frame of reference, establishing the en-

A. Sonrel on stone from nat.      P. S. Duval & Cº Steam lith press Philad.

Plate 42. *Pomotis* and *Bryllus* species. Hand-colored lithograph by A. Sonrel, printed on a steam press by the firm of P. S. Duval, for John E. Holbrook's descriptions in *Journal of the Academy of Natural Sciences of Philadelphia*, n.s. 3 (1855).

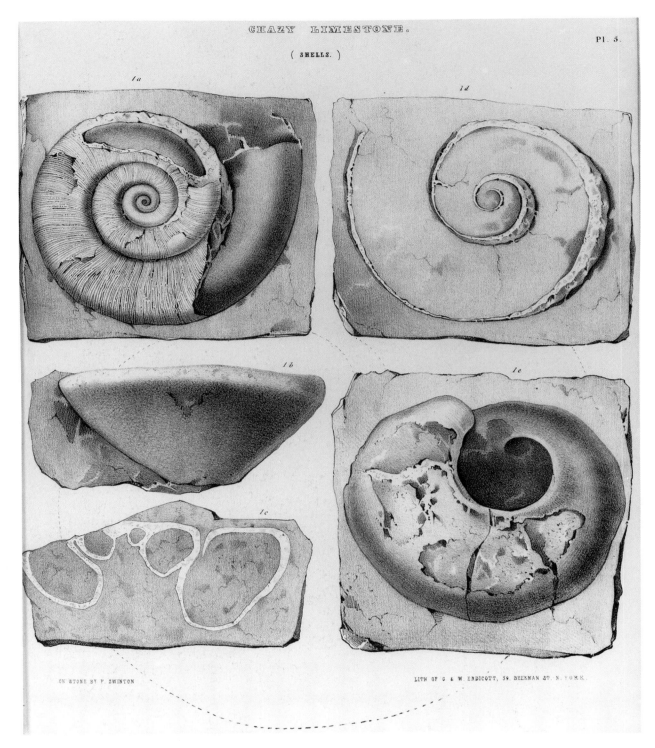

5.39–40 Lithographs by Frederick Swinton, printed by G. and W. Endicott, for Hall, *Paleontology of New York*, vols. 1 and 2 (1847 and 1852).

206

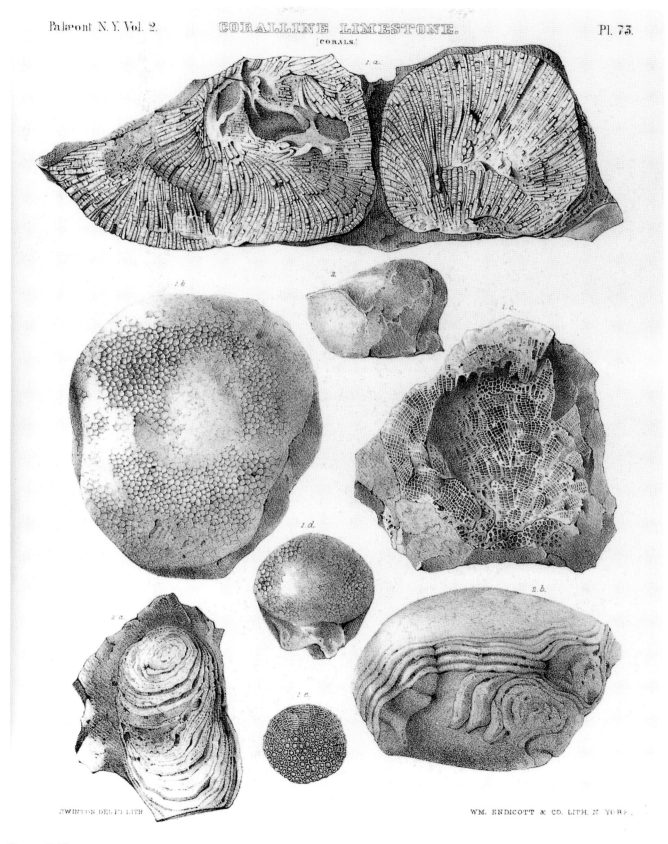

Figure 5.40.

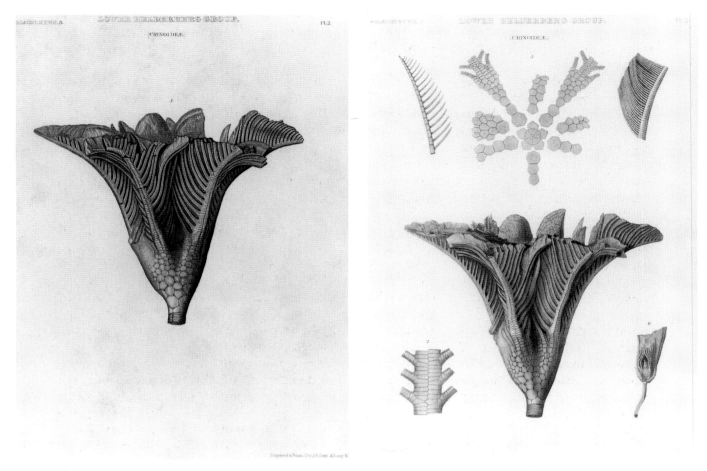

5.41–42  Steel engravings by J. E. Gavit and J. Duthie, from drawings of the same specimen of a fossil crinoid, by F. B. Meek and R. P. Whitfield, for Hall, *Paleontology of New York*, vol. 3 (1861).

vironment and the terms of scientific authenticity. Although each figure of each plate was intended to establish distinctions between species and among varieties, each plate conforming to these conventions also referred to all others of its kind, defining by accumulation the consensus of practices of the field.

The more science produced, the more it came to look the same, in the United States, England, and Europe. American zoologists and paleontologists, savoring newfound international recognition, also matched the stylistic conventions of their publications to those of European colleagues. Many factors had contributed to the movement away from the delib-

erate distinctiveness characterizing American illustration during the early decades of the century. The emulation of the *Astrolabe* reports for the Exploring Expedition atlases, the arrival of European lithographers in the United States, the influence of individuals such as Louis Agassiz, Europe's informal delegate to American science, among other events, all contributed to the mainstream participation of American science in the consolidation of graphic conventions in the descriptive disciplines. From mid-century onward, the international status and style of American zoology was cemented by study and travel abroad, networks of personal friendship, professional correspondence, American memberships in foreign professional societies, the private exchange of specimens, and the official exchange of state-sponsored scientific reports.

The mass of fossil specimens collected by Hall and his assistants, and by the other state surveys, like the broad geographic collections that Baird caused to accumulate in Washington, provided the raw materials for ongoing taxonomic descriptions of American fauna. The cataloguing of North American animals and plants had been the self-assigned mission of the first American naturalists. But even with the support of state legislatures and with the federal government's unprecedented resources—army and navy personnel, Baird's own corps of energetic young assistants, and the growing number of amateur and professional collectors—the task begun early in the century seemed only to expand.[86] As the pace of western settlement accelerated, Baird worried that he was missing collecting opportunities because he lacked funds to outfit more agents with preservatives, packing vessels, and money for railroad freight. His policies for collecting enlarged the task further. In 1852 he had written to a former student in Texas: "Don't suppose that you could ever get enough specimens of a species. Collect every confounded individual . . . for lizards and smaller serpents, get 500 each if you can." In short, he concluded, "Collect everything you can find, especially the very COMMONEST species."[87] With the collecting equipment he sent out to aid their work, Baird passed along to his assistants the instructions he had first received from Agassiz, who, on first arriving in the United States, had written to Baird: "I prefer to have a great number of specimens of *the most common species in all their ages* than to have a few specimens of many rare species."[88] In 1861 Baird wrote to Agassiz: "In accordance with . . . policy and my own special predilections, I am anxious to see collected here [as] complete a series of Vertebrata and alcoholic invertebrates of North America as possible, and such others from the rest of the world as are necessary to properly elucidate their study. . . . When these are all determined, labelled and reported on, I am willing to let them stand as types only caring to prevent their destruction."[89]

Baird's policies provided the grist for continuing description and illustration of species, the basic activity of midcentury zoology. In 1861, after ten years of supervising the collecting and publication programs, he wrote: "There are so many new and unfigured species of animals, especially fishes and shells, that I hope plenty of money can be had for illustrations."[90]

CHAPTER

6

# Illustrations of Theory, Illustrations of Practice

AFTER Louis Agassiz had decided to settle permanently in the United States, he inaugurated a ten-volume work of his own, *Contributions to the Natural History of the United States*.[1] Supported by $300,000 in twenty-five hundred advance subscriptions, the four completed volumes published between 1857 and 1862, with illustrations by Sonrel, Jacques Burkhardt, and the microscopist Henry James Clark, were perhaps the most elaborately produced American zoological publication of midcentury.[2] The appearance of Charles Darwin's *On the Origin of Species* in 1859 caused Agassiz, in midpublication, to give heightened polemical emphasis to his own morphological theories, which were incompatible with Darwinian modification by descent. The emphasis on symmetry and detail in Agassiz's plates can been seen as an important element of his antievolution argument; the illustrations offered vivid evidence of the organic forms which Agassiz believed were primordial and permanent. At the same time, their elegance can also be seen as delivering the value of the subscription price to the largely nonscientific audience supporting publication.

Agassiz's theoretical arguments in the *Contributions* had virtually no influence on subsequent zoology, which adopted an evolutionary model. Agassiz's illustrations, on the other hand, although they reflected his theory, shared the current graphic conventions of the discipline and influenced the style of subsequent publications; there is a traceable dynasty of plates illustrating jellyfish that begins with Sonrel's lithographs for Agassiz.[3] Agassiz's work in the *Contributions*, therefore, provides an interesting case for exploring the relative influences of theory, institutions, and production on zoological illustration.

The diverse studies that Agassiz had pursued to date culminated in his "Essay on Classification," published in the first volume of the *Contributions*. Agassiz intended the "Essay" to be the definitive statement on the existence of natural groups and a natural system of classification—a central issue in the philosophy of science that also bore on the day-to-day practice of taxonomy. He used his studies of the fossil and living fish of Europe and America, and of echinoderms and other invertebrate groups, as well as his geological researches and embryological investigations, to support his argument that the four divisions, or branches, of the animal kingdom defined by Cuvier—vertebrates, mollusks, arthropods, and radiates (radially symmetrical invertebrates)—represented the primordial divisions at creation and had no historical relationship with each other.[4] Most naturalists, with Agassiz, still assumed divine creation of original life. But Agassiz differed from his colleagues in his insistence on the frequency of intervention of the act of creation during the history of the earth, and in his emphasis on the

210

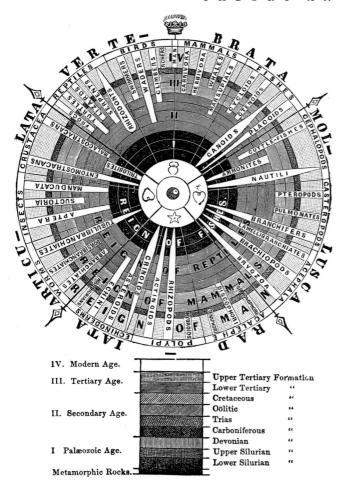

IV. Modern Age.

III. Tertiary Age.

II. Secondary Age.

I Palæozoic Age.

Metamorphic Rocks.

Upper Tertiary Formation
Lower Tertiary    "
Cretaceous    "
Oölitic    "
Trias    "
Carboniferous    "
Devonian    "
Upper Silurian    "
Lower Silurian    "

## CRUST OF THE EARTH AS RELATED TO ZOÖLOGY.

6.1 "The Crust of the Earth as Related to Zoology," a diagram of the four Cuverian branches and their fossil records, was the wood-engraved frontispiece in one of Agassiz's popular texts. The symbols at the center represent the fundamental symmetry of each group. Louis Agassiz and Augustus A. Gould, *Principles of Zoology*, 2d ed. (1851).

FIGURE

6.1

existence of a divine plan in the organization of nature. From the four major branches to the finest distinction between species, Agassiz saw plan, structure, design, pattern, and symmetry. The fundamental anatomical structures and the minutest ornamentation all represented, according to Agassiz, thoughts of God.[5]

In nature Agassiz saw evidence of the "Supreme Intelligence" creating increasingly com-plex organisms, as if God were composing variations on the four original themes. Beginning with the simplest organisms in each group, every advance elaborated on organization, detail, and ornamentation of the original symmetry. According to Agassiz, each of Cuvier's four branches was constructed on a general plan or type. For example, vertebrates expressed the idea of bilateral symmetry; the radiates, that is, echinoderms and coelenterates, expressed the idea of radial symmetry. These symmetries constituted the simplest level of structure. Within each branch, the subdivisions of classes, orders, families, genera, and species represented levels of refinement and elaboration of the basic structure. The classes were distinguished by the way the general plan was carried out—among vertebrates, by the differences among vertebrates between fish, reptiles, birds, and mammals. The orders within each class were organized by degree of structural complexity. Families, Agassiz said, could be recognized by their forms as determined by structure, genera by the detail of execution of the plan in certain organs. Species differed according to the proportions of their parts, their ornamentation, and by behavioral relations of members of species to one another, as in association and breeding. Field observations of behavior and mating were, therefore, necessary to determine species.[6] Agassiz did not support a single, linear hierarchy of organic complexity; indeed, the Cuvierian model of four branches had made an important break with linear hierarchy. Nevertheless, he believed that the relative rank of two groups—which one was "superior" to or "higher" than the other—should be determined by their structural complexity.

Agassiz planned that each volume of the *Contributions* would elaborate one aspect of his vision. He appended a classification of North American turtles to the "Essay" and devoted his second volume to a discussion of turtle embryology. Jacques Burkhardt (1808–1867) had made hundreds of colored drawings of turtles, mostly alive, sent from all over the country. The Swiss painter, after studying in Mu-

211

On stone by Sonrel after Jacques Burkhardt's colored drawings.                                L. H. Bradford & Co print.

PTYCHEMYS RUGOSA Ag.

Plate 43. Variations of Species, *Ptychemys rugosa* Ag. On stone by A. Sonrel, after Jacques Burkhardt's colored drawings, printed by L. H. Bradford & Co., for Agassiz, *Contributions to the Natural History of the United States*, vol. 2 (1857).

On stone by Sonrel after Jacques Burkhardt's colored drawings.

L. H. Bradford & Co. print.

PTYCHEMYS RUGOSA Ag.

Plate 44. See caption on p. 215.

nich and Rome, had drawn for Agassiz since the 1840s, and served in Boston as Agassiz's principal illustrator.[7] Some of Burkhardt's watercolor studies were reproduced in chromolithography, which required a separate stone for the application of each color, and printed by Lodowick H. Bradford, formerly in partnership with Eben Tappan. Henry James Clark, a former student of Agassiz's, studied the development of turtle embryos and drew his observations through the microscope "with untiring patience and unsurpassed accuracy."[8] His studies and their illustrations documented the symmetrical development of the embryos. Sonrel, too, studied and drew turtle development through the microscope. He also transposed Burkhardt's watercolor and Clark's pencil drawings into lithographs, and drew live turtles and turtle eggs directly onto stone, endowing the figures with an almost photographic intensity. In Agassiz's introduction, to which mention of illustrators and printers was customarily limited, he wrote of Sonrel: "The mastery he has attained in this department, and the elegance and accuracy of his lithographic representations, are unsurpassed, if they are anywhere equalled."[9]

Agassiz acknowledged no intermediate forms—varieties—within species. He therefore saw variations of form and color as marking species distinctions, although as an astute field zoologist he had a keen eye for recognizing individual variation within a species and behavior that distinguished closely related species.[10] Agassiz had set up a private network for collecting that rivaled Baird's. Working from the mass of material solicited from around the country, Agassiz believed that the extensive series of specimens allowed him to perceive and describe species with precision.

Agassiz declared that "species, genera, families, etc., exist as thoughts; individuals as facts."[11] He believed he had proven in the "Essay," "that individuals alone have a definite material existence, and that they are for the time being the bearers, not only of specific characters, but of all the natural features in

which animal life is displayed in all its diversity; individuality being, in fact, the great mystery of organic life."[12] Agassiz's emphasis on distinction and individuality translated into the illustrations. His illustrators drew primarily from living specimens, and the turtle figures convey a startling effect of life. Although many other authors, including Wilson and Holbrook, had drawn, or had their illustrators draw, from individual specimens to represent species, or, like Say, to represent entire genera, for Agassiz the practice had special theoretical significance. Sonrel and Burkhardt literally illustrated Agassiz's thinking in their portrayal of individual turtles.

Agassiz enlarged his scope in the third volume to take on the entire class of acalephs, called ctenophores or hydroids in modern terminology. It was while he was preparing this volume that Darwin's *On the Origin of Species* appeared. Agassiz felt compelled to use his study to oppose Darwin and to defend his own theory of classification. He believed that species existed as categories of Divine thought and as such were permanent. Darwin proposed that species were subject to change over time. Agassiz interpreted arguments for evolution, or what Darwin called modification by descent, to imply that natural groups did not exist in the animal kingdom. For Agassiz, for whom individuals were the only natural facts, the only conclusive evidence for modification by descent would be an individual that generated offspring different from the parent, or that altered its plan of organization during its lifetime. He argued: "Had Darwin or his followers furnished a single fact to show that individuals change, in the course of time, in such a manner as to produce, at last, species different from those known before, the state of the case might be different."[13] Agassiz protested that if organisms had evolved in the way Darwin proposed, through modification in relation to the environment, then all animals were related through descent and "differed from one another only in degree, all having originated from successive differentiation of a primordial organic form."[14] Agassiz wrote: "Far from agreeing with these views, I have, on the contrary, taken the ground that all the natural divisions in the animal kingdom are primarily distinct, founded upon different categories of characters, and that all exist in the same way, that is, as categories of thought embodied in individual living forms."[15]

The theories of both Darwin and Agassiz recognized and attempted to explain diversity, the subject of taxonomy, or what is now called systematics. Agassiz proposed that diversity was original and fundamental, Darwin that it was the product of divergent evolution. Darwin's view seemed to Agassiz to deny nature any structure and thus to suggest that science itself was a vain effort. "There are naturalists," Agassiz protested, "who seem to look upon the idea of creation . . . as a kind of bigotry; forgetting . . . that whenever they carry out a thought of their own, they do something akin to creating."[16] He had written in the "Essay": "All organized beings exhibit in themselves all those categories of structure and of existence upon which a natural system may be founded, in such a manner that, in tracing it, the human mind is only translating into human language the Divine thoughts expressed in nature in living realities."[17]

When Sonrel drew the acaleph illustrations directly onto stone from live specimens swimming in seawater he was "tracing" or "translating" "living realities" into pictorial vocabulary. He probably took his lithographic stones to Agassiz's summer home and laboratory on the coast at Nahant to draw specimens freshly netted from the sea. Sonrel organized the layered transparencies of tissue and the tangled tentacles of these creatures by emphasizing their geometry, symmetry, and ornamentation. In these plates, Agassiz and Sonrel reached their closest unity of descriptive and pictorial intention. The medusae, with their alternating generations of sexual and asexual reproduction and the drastic transformations in form and symmetry that occurred during the complete life cycle, posed a challenge to Agassiz's theories of classification based on a distinct and an unchanging structural plan: "Does this not seem at first as if we had before us a perfect exemplification of the manner in which different species of animals may orginate, one from the other, and increase the number of types existing at first? And yet, with all this apparent freedom of transformation, what do the facts really show? That all these transformations are the successive terms of a cycle."[18] Emphasizing the radial symmetry of the group, Sonrel's illustrations vividly support Agassiz's interpretations. It is impossible to determine whether Agassiz wrote his description of *Cyanea arctica* from Sonrel's lithograph illustration of the living animal, or whether Sonrel had read Agassiz's description, which scarcely conformed to standard taxonomic prose, before embarking on the plate:

From the lower surface of this disk hang, conspicuously, three kinds of appendages. Near the margin there are eight bunches of long tentacles, moving in every direction, sometimes extending to enormous length, sometimes shortened to a mere coil of entangled threads, constantly rising and falling. . . . These streamers may be compared to floating tresses of hair, encircling organs which are farther inward upon the lower surface of the disk. Of these organs, there are also eight bunches, which alternate with the . . . tentacles, but they are of two kinds; four are elegant sacks, adorned, as it were, with waving ruffles projecting in large clusters, which . . . might also be compared to bunches of grapes, by turns inflated and collapsed. These four bunches alternate with four

*On p. 213 and following pages:*
Plates 44–52. Color separations of the plate illustrating *Ptychemys rugosa*, for Agassiz, *Contributions*, vol. 2 (1857). The image on the left of each pair is a proof of one of the colors used; on the right is the same image in black as it was drawn on the stone. When all of these images were combined, the result was Plate 43.

215

Plate 45.

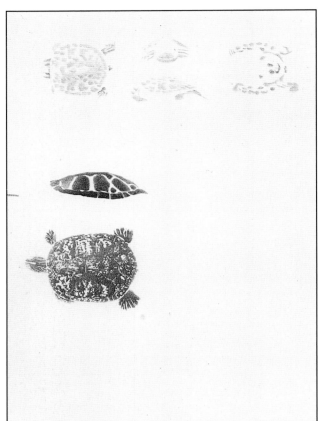

Plate 46.

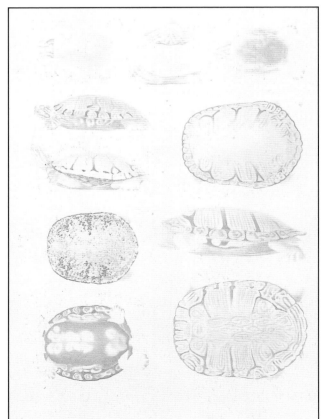

Plate 47.

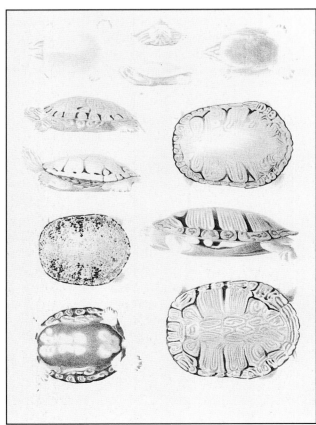

Plate 48.

Plate 49.

Plate 50.

Plate 51.

Plate 52.

Pl. I

A Sonrel on stone from nat.

L.H. Bradford & Co. print.

1-5. CHRYSEMYS PICTA Gr. — 6. CHR. MARGINATA Ag. — 7-9. NANEMYS GUTTATA Ag. — 10-12 MALACOCLEMMYS PALUSTRIS Ag
13. PTYCHEMYS CONCINNA Ag. — 14-16. DEIROCHELYS RETICULATA Ag

6.2–5 Young Turtles. Natural size. A. Sonrel on stone from nature, printed by L. H. Bradford & Co., for Agassiz, *Contributions to the Natural History of the United States*, vol. 2 (1857).

Pl. II.

A. Sonrel on stone from nat'

L.H. Bradford & Co. print.

1-3. DEIROCHELYS RETICULATA Ag — 4-6. PTYCHEMYS CONCINNA Ag. — 7-9. GRAPTEMYS GEOGRAPHICA Ag.
10-12. GRAPTEMYS LeSUEURII Ag — 13-15. TRACHEMYS SCABRA Ag

Figure 6.3.

Figure 6.5.

Figure 6.4.

6.6–7 Eggs of turtle species. Drawn by A. Sonrel and Vogel, and unsigned, printed by L. H. Bradford & Co., for Agassiz, *Contributions*, vol. 2 (1857).

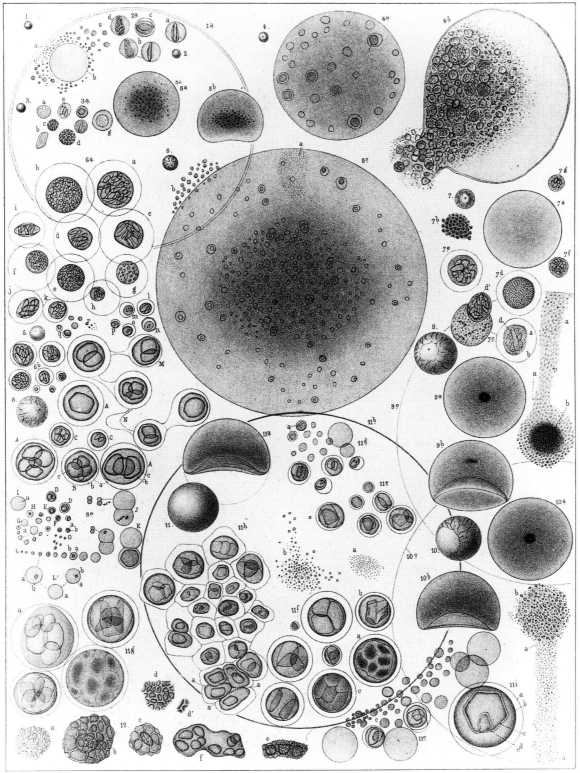

Clark from nat.    Ibbotson on stone.

L. H. Bradford & Co. print.

OVARIAN EGGS

6.8–13
Clark's and
Sonrel's
drawings of
microsopical
studies of
turtle
embryology.
On stone by
A. J. Ibbotson,
Sonrel and
Baner, printed
by L. H.
Bradford &
Co., for
Agassiz,
*Contributions*, vol. 2
(1857).

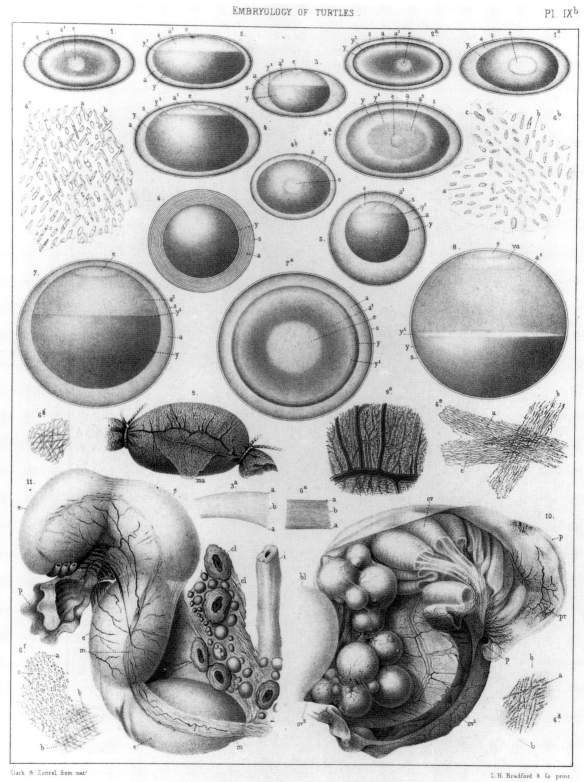

Clark & Sonrel from nat.

L.H. Bradford & Co print.

THE OVARY & OVIDUCT AND THE ABSORPTION OF THE ALBUMEN.

Figure 6.9.

Figure 6.11.

Figure 6.10.

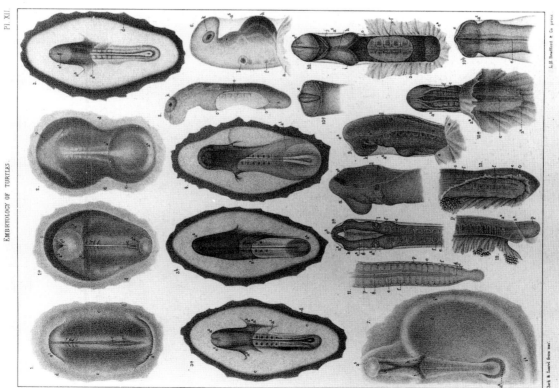

Figure 6.13.

Figure 6.12.

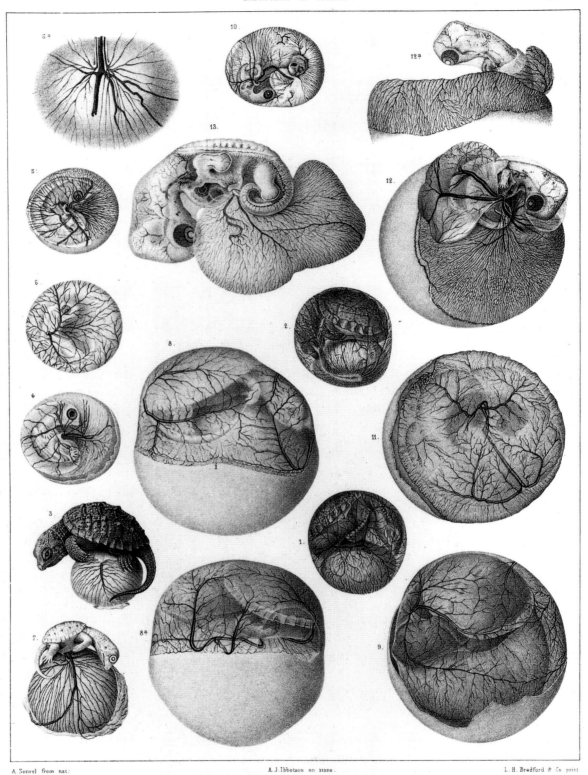

A. Sonrel from nat.

A. J. Ibbotson on stone.

L. H. Bradford & Co. print

VITELLINE AND ALLANTOIDIAN CIRCULATION.

6.14  Sonrel's studies of the developing circulatory system. On stone by A. J. Ibbotson, printed by L. H. Bradford & Co., for Agassiz, *Contributions*, vol. 2 (1857).

masses of folds, hanging like rich curtains, loosely waving to and fro . . . recalling . . . the play of streamers of an aurora borealis. All these parts have their fixed position.[19]

FIGURES
6.15–24

For much of the contemporary and later illustration influenced by Sonrel's work, symmetry merely regulated the conventions of plate composition without touching so centrally on the author's theoretical argument. In Sonrel's lithographs for Agassiz, symmetry represented a fundamental natural truth. In his preface, Agassiz (recalling with praise the illustrations of marine invertebrates by Lesueur) wrote of Sonrel: "I do not know of representations of Acalephs executed with greater accuracy, patience and skill." He continued, "I can truly say, that, without the aid of [Sonrel's] persevering zeal, I would not have accomplished what I aimed in this volume."[20]

For Agassiz, science, as the study of nature, reflected the plan of creation itself. Rearticulating the tradition of natural theology, he wrote: "There runs throughout nature unmistakable evidence of thought, corresponding to the mental operations of our own minds, and therefore intelligible to us as thinking beings, and unaccountable on any other basis than that they owe their existence to the working of intelligence." He insisted: "No theory that overlooks this element can be true to nature."[21] For Agassiz "all natural divisions in the animal kingdom are primarily distinct" because "they yet exist as categories of thought."[22] Such an argument elevated the volume's illustrations to representations of a sacred level of truth. In the context of the ideas that Agassiz presented in opposition to Darwinian theory, "A. Sonrel on stone from nature" provided more than the graphic equivalent of verbal description; with their depersonalized finish, his illustrations stood as tangible, authoritative evidence in support of a worldview. If nature expressed a divine plan, and if science, the human effort to understand nature, emulated the intelligence of creation itself, then scientific illustrators bore the responsibility for drawing pictures of divine thought expressed in individual organisms.

Publication of the *Contributions* was for Agassiz both professional crescendo and dead end. With his scientific colleagues turning away from the creationist assumptions on which he insisted, he withdrew from the central debates of his discipline, and until his death in 1873 concentrated his considerable money-raising skills on building his museum at Harvard and filling it with specimens. In the role of institution builder, public educator, and popularizer of science, however, Agassiz exerted a profound and lasting influence on American taste and feeling. In 1863 he collected a series of his articles that attacked Darwinian evolution, originally published in the *Atlantic Monthly*, in *Methods of Study in Natural History*, which saw nineteen editions between 1863 and 1884.[23] The appeal of his public lectures and popular writings lay in his emphasis on plan expressed in nature's structures. Probably more Americans read Agassiz's essays than, for example, a highbrow and specialized publication like the *Crayon*, the chief organ promoting the ideas of the English art critic, John Ruskin, in the United States. Ruskin's teaching that painters could perceive truth through close observation of nature resonated with Agassiz's similar ideas. The editors of the *Crayon* read and were influenced by Agassiz. William James Stillman, co-editor of the magazine, was a colleague of Agassiz's in the elite Saturday Club of Boston.[24]

FIGURE
6.25

The strongest pictorial expressions of ideas related to Agassiz's natural theology emerged in American landscape painting of the 1860s. Influenced by the holistic natural science of Alexander von Humboldt (1769–1859), who had been a mentor during Agassiz's youth, and by John Ruskin's emphasis on natural form, a school of American painters began to claim a close relationship between painting and natural science, their paintings, of course, in no way resembling printed scientific illustration. In 1804, after five years of exploration in Spanish

America, von Humboldt had been made welcome in Washington and Philadelphia before returning to Paris to publish his multivolume illustrated travel account. Frederick E. Church (1826–1900), who considered himself a disciple of von Humboldt and followed his example in traveling through South America, made close botanical studies of plants and foliage in preparation for his paintings. There was, however, stylistic and ideological latitude in this return of painting to nature. Church's enormous, light-filled exotic landscapes were widely acclaimed; but the Ruskinian critic, James Jackson Jarves, who wrote that painting "should exhibit a scientific correctness in every particular," criticized them as "astonishing rather than instructing."[25] They did not meet his precept that a painting should "as a unity, be expressive of the general principles at its center of being"—a way of thinking that was close to Agassiz's doctrine that organisms expressed the primordial design principles of their creation.[26]

Agassiz's emphasis on the meaning of form gave his illustrations in the *Contributions* an important role in his theoretical argument, but despite his differences from his colleagues the plates resembled more than they differed from other illustrations of the period. They shared with the plates of contemporary zoological publications all the current conventions: use of symmetrical composition to organize the page, a drawing style of detailed realism, a single light source, shallow space, an inked border. These conventions persisted despite the disciplinary shift to evolutionary theory; indeed, illustrations alone could not reveal whether an author adopted or resisted the theoretical change.

An evolutionary framework for taxonomy changed the *meaning* of generic and specific characters because it called attention to the process through which characters came to vary. Demarcation of lines between species had al-

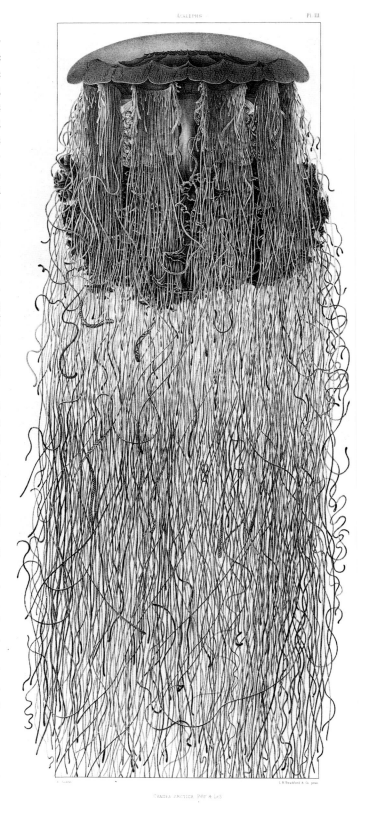

CYANEA ARCTICA PÉR & LES.

6.15 *Cyanea arctica* Péron and Lesueur. Lithograph drawn by A. Sonrel, printed by L. H. Bradford & Co., for Agassiz, *Contributions*, vol. 3 (1860).

A. Sonrel.                                                    L. H. Bradford & Co print.

CYANEA ARCTICA Pér. & LeS.

6.16 *Cyanea arctica*, from a dissected specimen. Lithograph drawn by A. Sonrel, printed by L. H. Bradford & Co., for Agassiz, *Contributions*, vol. 3 (1860).

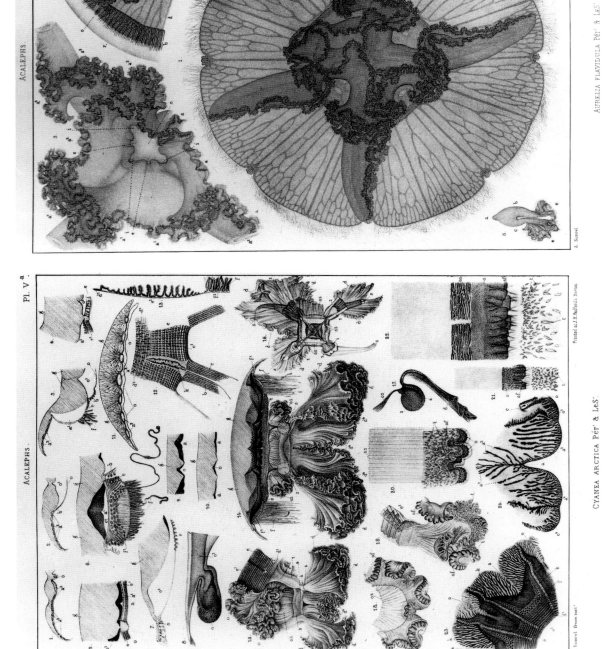

6.17 *Cyanea arctica*, anatomical studies. Lithograph drawn by A. Sonrel, printed at J. H. Bufford's, for Agassiz, *Contributions*, vol. 3 (1860).

6.18 *Aurelia flavidula* Péron and Lesueur. Lithograph drawn by A. Sonrel, printed by L. H. Bradford & Co., for Agassiz, *Contributions*, vol. 3 (1860).

6.19–22  Lithographs drawn by A. Sonrel, with J. Burkhardt, H. J. Clark, and E. Burrill, printed by J. Bufford and L. H. Bradford, for Agassiz, *Contributions*, vols. 3 and 4 (1860 and 1862).

Figure 6.20.

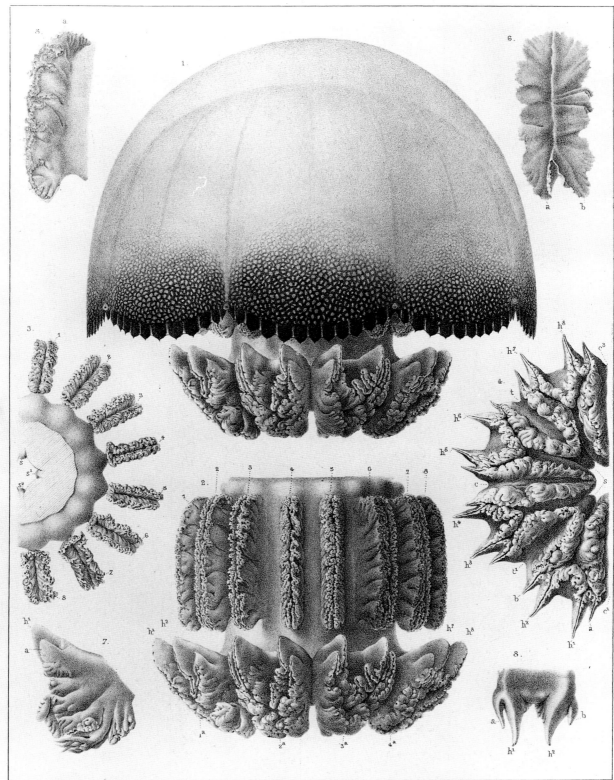

Sonrel & Burkhardt from nat'. Printed at J.H.Bufford's Boston.

STOMOLOPHUS MELEAGRIS Ag.

Figure 6.21.

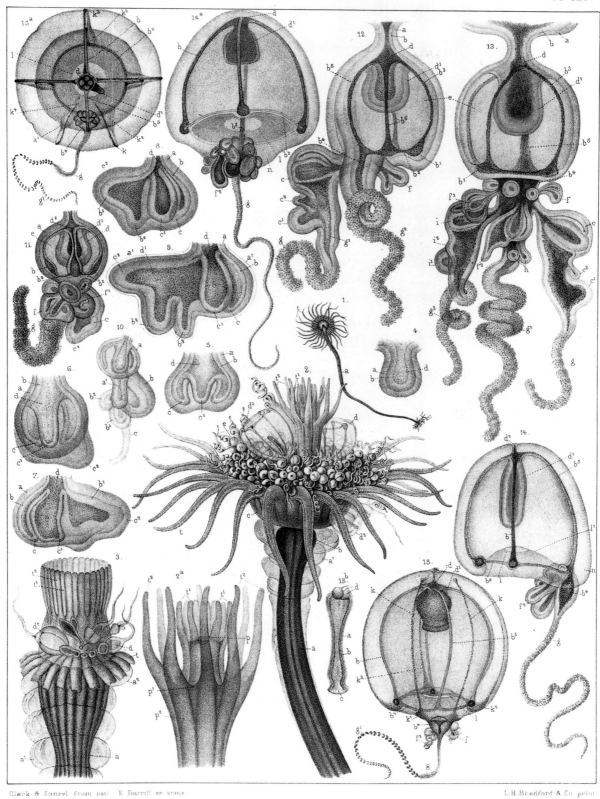

Clark & Sonrel from nat.   E. Burrill on stone                          L.H. Bradford & Co. print.

HYBOCODON PROLIFER Ag.

Figure 6.22.

ways been problematic, but now species no longer reflected an unchanging order. The question was whether a variety had yet become a distinct species or remained within its parent group. Evolutionary theory heightened the taxonomic problem of determining the boundary between species and variety, but it did not alter the methods of taxonomic study, which remained largely a matter of personal judgment. Darwin himself acknowledged the intuitive aspect of taxonomic practice in distinguishing and determining domestic and wild species: "[T]here are hardly any domestic races, either amongst animals or plants, which have not been ranked by some competent judges as mere varieties, and by other competent judges as the descendants of aboriginally distinct species. . . . naturalists differ most widely in determining which characters are of generic value; all such valuations being at present empirical."[27] Whether an author was a "lumper" who read varieties as geographic variants of a single species, or a "splitter" like Agassiz, for whom varieties were independent entities, had little effect on the graphic conventions that organized illustration. Splitting and lumping was more likely to affect the content of the captions or the number of illustrations needed. At the beginning of the New York paleontological survey, for example, Hall had anticipated that the publication would require 250 plates. Not only did the survey find more fossils than Hall anticipated, but he was a splitter. When the series was completed, the number of plates ran to 770.[28]

Agassiz's insistence on the permanence of species and his public opposition to Darwinian

6.23 *Physalia arethusa*, Til. A. Sonrel on stone from nature. Printed at J. H. Bufford's, for Agassiz, *Contributions*, vol. 4 (1862).

*On opposite page:*
6.24 Coral. Lithograph drawn by A. Sonrel, printed by A. Meisel, prepared for Louis Agassiz, and published posthumously in Agassiz, "Report on the Florida Coral Reefs," *Memoirs of the Museum of Comparative Zoology* 7 (1880).

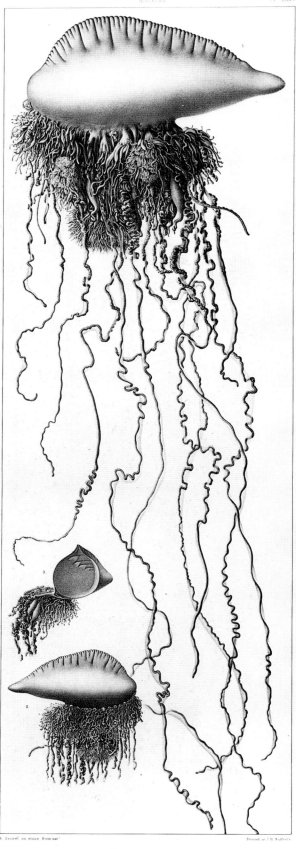

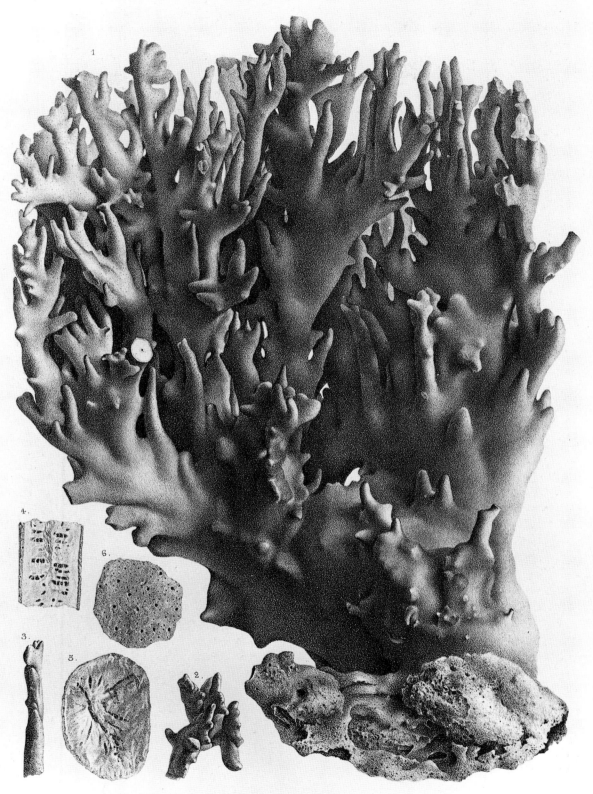

6.25 Louis Agassiz was known for his fluent and ambidextrous sketching as he lectured. Photograph by Watkins.

Scudder popularized Agassiz's method of observation in an account of his own initiation. Immediately on Scudder's arrival at Agassiz's museum, the professor left the new student alone to confront a single fish for days on end, until he had observed and analyzed every relevant aspect of its structure.[30] The early grounding of Scudder's work in Agassiz's taxonomic method may well have influenced his adherence to the illustration conventions associated with it. The plates in Scudder's major monograph on the butterflies of the northeastern United States, published in 1889 (discussed in the next chapter), resembled in many ways those of the *Contributions*.

Where the historian of ideas might expect to find a watershed between the pre- and post-Darwinian criteria and manner of zoological taxonomic illustration, one sees instead marked continuity.[31] Persistence in taxonomic practice provides a partial explanation of the continuity, as does the continued use of lithography, with which the smooth and crisp drawing style had developed. But the persistence of the conventions of the formal taxonomic description, with which illustration was historically associated, as well as the consolidation of institutional structure, constrained change in illustration conventions. The prose description and accompanying plates of the monograph characterized institutional zoology. The manner and elements of style in which Agassiz's illustrators worked, and which they shared with other museum illustrators and with commercial draftsmen producing zoological and paleontological illustration, were associated not with a particular theoretical framework, but with the very structure of disciplinary production in the United States, England, and Europe. If the shift to an evolutionary theoretical framework exerted any influence on illustration conventions, it was probably to discourage change. Although American zoologists themselves perceived in their adoption of evolutionary theory a watershed in their discipline, it was, perhaps, because of the initial controversy over the im-

evolution isolated him as many of his peers joined the younger generation in the Darwinian camp. Agassiz's students, who embraced evolutionary theory, nevertheless considered him an inspired teacher of taxonomic practice. One of Agassiz's students, Joel Asaph Allen, would become one of the leading students of geographic variation within the evolutionary framework. Allen's experience with the extensive geographic collections at Agassiz's museum in Cambridge brought him to formulate "Allen's Law," which stated that geographical varieties of species changed size and color according to latitude and longitude.[29] Samuel Hubbard

plications of Darwinian evolution, considered by many mechanistic and materialistic, that zoological and paleontological illustrations preserved for so long their classical, depersonalized appearance. That style connoted harmony; adherence to it connoted consensus.

In 1867, when John Strong Newberry addressed the American Association for the Advancement of Science (AAAS), he opened his speech with this clarion pronouncement: "Every day of our lives we hear that this is a new age of progress; and that it is so we find evidence at every turn."[32] The "railroad speed" of progress effected as much change in a single year as used to come about during an entire lifetime, he claimed.[33] According to Newberry, no idea characterized the materialist tendency of modern science better than Darwin's hypothesis of evolution through natural selection. "The wide question of the origin of species . . . has been shaking the moral and intellectual world as by an earthquake."[34] Like many of his generation, Newberry worried that the Darwinian hypothesis would undermine religious values, but he noted that "many of our best men of science look upon his theory as not incompatible with . . . religious faith."[35] Indeed, to those scientists who accepted Darwinian evolution "it is a proof" of God's omnipotence that "he should endow the vital principle with such potency that . . . all the economy of nature, in both the animal and vegetable worlds, should be so nicely self-adjusting that, like a perfect machine from the hands of a master maker, it requires no constant tinkering to preserve the constancy and regularity of its movements."[36] Newberry's machine analogy reflected the prevailing confidence in technological progress; for many Americans, science and industry were benign allies in the taming of nature. But Newberry resisted the materialist implications he saw in Darwinian evolution, and invoked beauty and the mysterious origins of life and of human consciousness as evidence for divine creation.

Nine years later, in 1876, the nation's cen-

tennial, Edward Sylvester Morse, a former student of Agassiz's, proclaimed to the same forum that American science had come of age. The criterion for its maturity was the universal acceptance of evolutionary principles, "that species vary, that peculiarities are transmitted or inherited, that a greater number of individuals perish than survive . . . that the physical features of the earth are now and have been constantly changing, and that precisely the same conditions never recur."[37] Although the centennial year might have tempted Morse to give a retrospective of one hundred years of American science, he dismissed natural history before 1859 as primitive, a mere inventory, and equated the motivations of the early naturalists to the childish collecting of postage stamps. With the adoption of Darwinian theory, however, American zoology had outgrown the inventory stage. "Never before has the study of animals been raised to so high a dignity as at present," Morse asserted.[38] No longer need zoology take an inferior place to other sciences: now it was governed by universal laws, like physics, and was useful, like chemistry or geology. While chemistry contributed to the arts, such as photography, and geology to the exploitation of mineral wealth, zoology had become "the pivot on which the doctrine of man's origin hinges."[39] No longer was nature the open book of God's creation, but rather the preliminary study of mankind. Now that zoologists had accepted the premise that "the varieties which survive are those . . . more in harmony with the environments of the time," or what would be popularized as the survival of the fittest, biology offered insights into the human present.[40]

Some of the most dramatic support from American science for the premises of evolution came from paleontology. "I am sure I need offer no argument for evolution; since to doubt evolution today is to doubt science, and science is only another name for truth," asserted Yale paleontologist Othneil Charles Marsh (1831–1899) to the 1877 gathering of the AAAS.[41] A

237

year earlier Morse had cited Marsh's fossil discoveries as the most exciting of all evidence offered by American science in support of evolution. Marsh's discoveries during the 1870s and 1880s provided Darwinian proponents with some of their most valuable corroboration. The fossil series illuminating the evolution of the horse became a textbook case and one of the favorite examples of Thomas Henry Huxley, Darwin's principal English advocate and popularizer.

Disagreements between Marsh and Philadelphia paleontologist Edward Drinker Cope over priority of description and interpretation of fossil finds became a well-publicized scientific rivalry, its episodes a great favorite in the popular press.[42] Within the discipline, the contest between Marsh and Cope heightened the importance of illustrations. Marsh, indeed, concentrated his resources on producing illustrations. After his death colleagues found hundreds of lithograhs, drawings, and wood engravings for projected monographs, but no manuscript text.[43] To protect his priority, Marsh guarded his discoveries in such strict secrecy that few of his professional colleagues had seen the fossils until they appeared as printed illustrations. Cope, too, relied on illustration to prove the correctness of his own interpretations. Over the identification of Cope's *Eobasileus*, which Marsh considered a new mammal order but which Cope was convinced belonged with the elephant-mastodon family, he wrote to a correspondent: "I suppose my paper on the fossil Proboscidians of Wyoming is now out. In it Prof. Marsh will find something to digest, though nothing indigestible. I hope to make matters so plain that he will have to swallow them whether he will or no. His charges are quite strong and I may have to prove them false by a special note, but I hope the internal evidence of the descriptions, plates, etc. will be sufficient."[44]

Competition for publication resources played a significant role in the contest between Marsh and Cope. Successful in the politics of government appointments, Marsh managed to maintain continued federal support from the United States Geological Survey for publishing his work.[45] Congressional appropriations to the U.S. Geological Survey paid for the fifty-six lithographs for the *Dinocerata*, printed in 1886 in an edition of thirty-five hundred and costing $11,000. Marsh also paid for a private edition with gilt-edged paper and wider margins.[46] For some of Marsh's reports for the U.S. Geological Survey, Sinclair printed the illustrations. But Marsh, whose early career had benefited from the patronage and wealth of his uncle, George Peabody, enjoyed the convenience of an in-house illustrator. Marsh hired Frederick Berger of the local New Haven lithographic firm of Emil Crisand. Berger, like Sonrel, drew to his employer's specifications and under his supervision. Marsh also employed a wood engraver, William F. Hopson, to execute figures for the text pages.

Cope, too, had private wealth, but after some financial setbacks he could not keep pace with his competitor. Publishing through the Academy of Natural Sciences and the American Philosophical Society, Cope relied on Sinclair's establishment for the lithographic illustration for his early work both for government and independent projects. In mid-career, Cope lost the support of the Geological Survey. His first volume of *Tertiary Vertebrata*, issued in 1884, "was found to far exceed in length, [and] number of plates" the anticipated volume, and to have cost so much to produce that the Geological Survey refused to publish its sequel, estimating that the second volume would cost six times as much as the first. Cope attempted in vain to obtain a personal appropriation from Congress for continued publication. With the appointment of Marsh to the position of vertebrate paleontologist for the the Survey, Cope was left to his own insufficient resources. The second volume was not published in his lifetime.[47]

FIGURES
6.26–33

6.26–28 Reptiles and amphibians of Costa Rica. Lithographs by the firm of Thomas Sinclair for E. D. Cope in *Journal of the Academy of Natural Sciences*, n.s. 8 (1876).

Figure 6.26.

Figure 6.27.

In contrast to the anniversary proclamations of decisive change in the discipline, the plates illustrating the work of Marsh and Cope were indistinguishable in their conventions of rendering, composition, and arrangement from Leidy's earlier plates. The special importance that Marsh placed on illustrations did not lead him to seek new techniques of picture making. On the contrary; to give authority to his pictorial documentation, Marsh relied on lithographs firmly rooted in the midcentury style.

The lithographic plates of the monumental *Dinocerata* and of *Odontornithes*—toothed bird-like creatures considered the "missing link" between reptiles and birds—of 1880, depicted minutely drawn, symmetrically arranged forms, printed on a tinted background.[48] The massive forms and surfaces of the fossils were drawn in scrupulous detail and portrayed at one-half and one-quarter size on huge sheets of paper that unfolded like maps from the royal quarto volumes. Of the reconstructions of the skeletons,

Pl. 25.

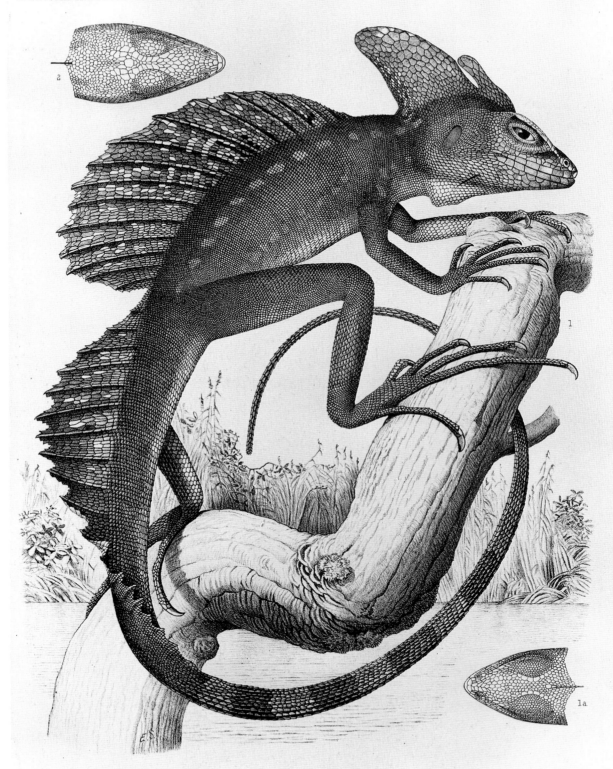

T. Sinclair & Son. lith. Phila

1. Basiliscus plumifrons. 2. B. mitratus.

Figure 6.28.

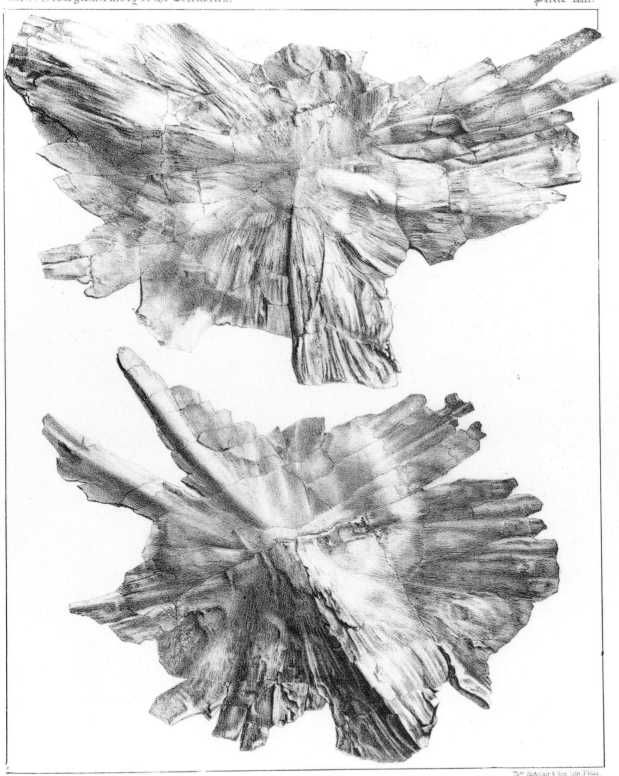

PROTOSTEGA GIGAS.

6.29  Dermal bones of *Protostega gigas*, one-third natural size. Lithograph by the firm of Thomas Sinclair and Son, for Cope, *Vertebrata of the Cretaceous Formations of the West* (1875).

U.S. GEOLOGICAL SURVEY OF THE TERRITORIES. PLATE

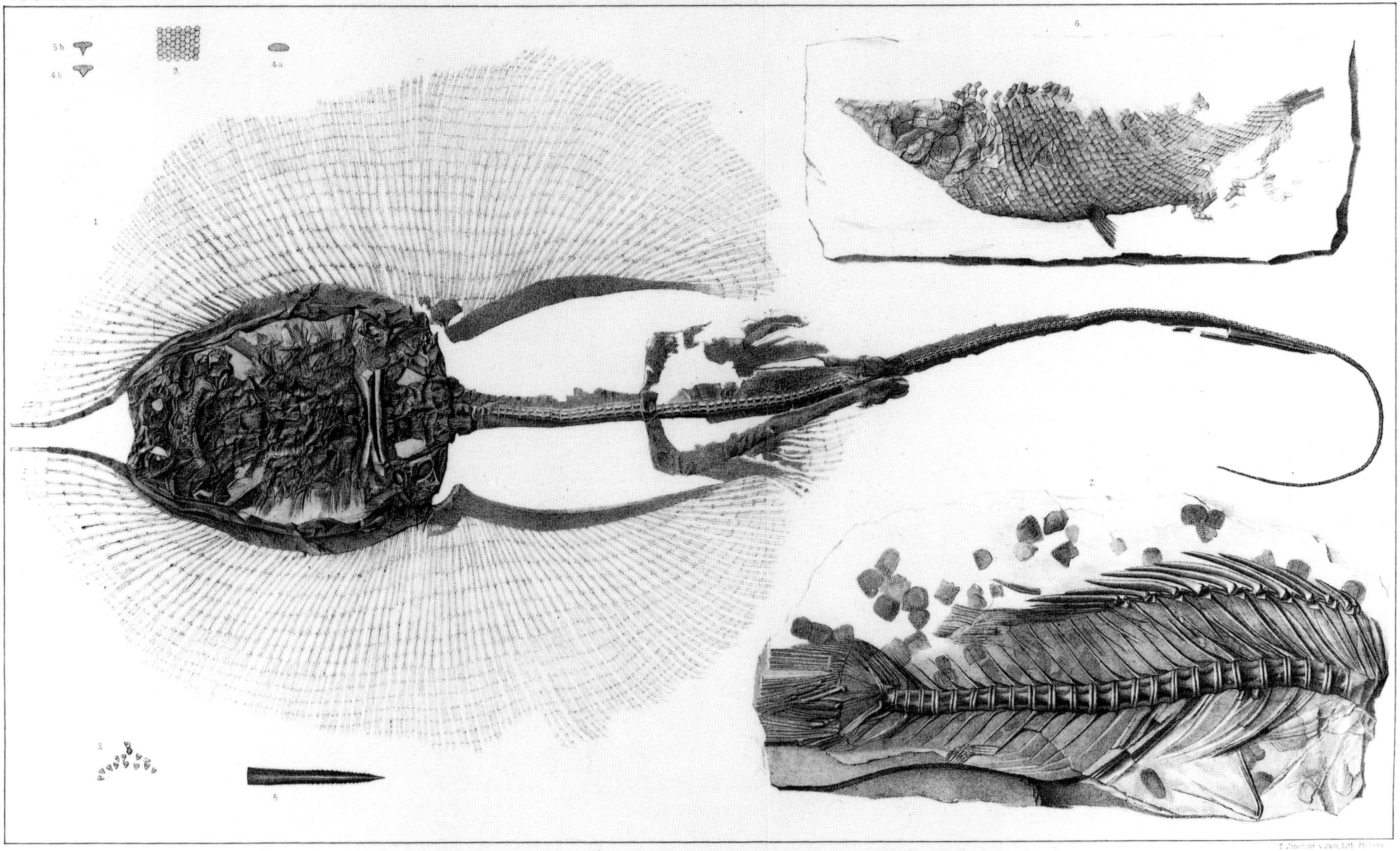


T. Sinclair & Son, Lith. Phila.

1-5 XIPHOTRYGON ACUTIDENS 1/1. 6 CLASTES CUNEATUS 2/3. 7 PRISCACARA TESTUDINARIA 2/3.

Figure 6.31.

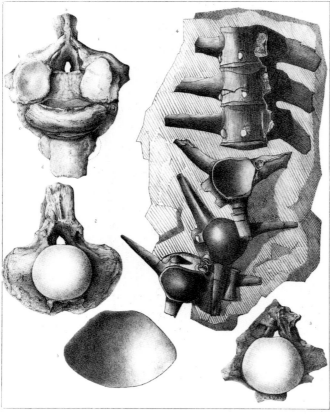

LIODON DYSPELOR.

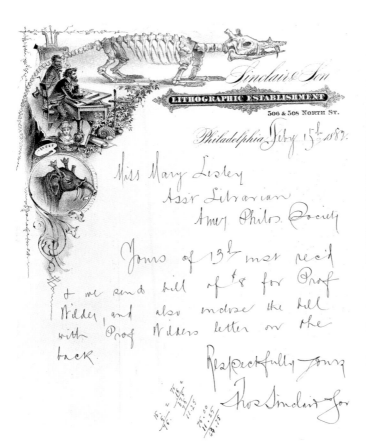

*On opposite page:*
6.31 Fossil fishes. Lithograph by the firm of Thomas Sinclair and Son, for Cope, *Vertebrata of the Tertiary Formations of the West* (1884).

*Top left:*
6.30 Vertebrae of *Liodon dyspelor*. Lithograph by the firm of Thomas Sinclair and Son, for Cope, *Vertebrata of the Cretaceous Formations of the West* (1875).

*Bottom left:*
6.32 Letterhead of Thomas Sinclair, 1882. Note the gentleman scientist observing the staff draftsman as he draws from the specimen directly onto the lithographic stone.

*Bottom right:*
6.33 The romanticized reconstruction of the fossil on Sinclair's letterhead resembles Cope's 1873 sketch of *Eobasileus cornutus*, "The Dawn-Emperor" of Wyoming, subject of one of Marsh and Cope's many disputes.

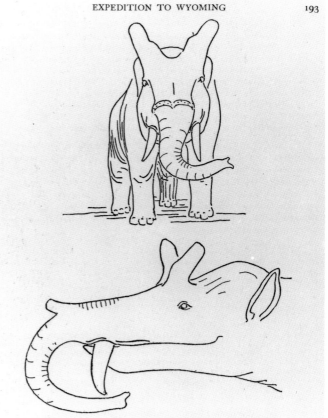

EXPEDITION TO WYOMING                                                    193

*Fig.* 16. Cope's First Sketches of *Eobasileus*, the "Dawn-Emperor" of Wyoming. Sketches from a letter of January 12, 1873, as reproduced by Persifor Frazer. Of Figure *a* (above) Cope wrote: "They had a proboscis I am now quite sure, and walked with the knee far below the body as elephants do. A form different enough from elephants generally, but reminding one more of the hog." Of Figure *b* he added: "The horns and head are rather too large in this drawing." At the time Cope thought *Eobasileus* was a Proboscidean.

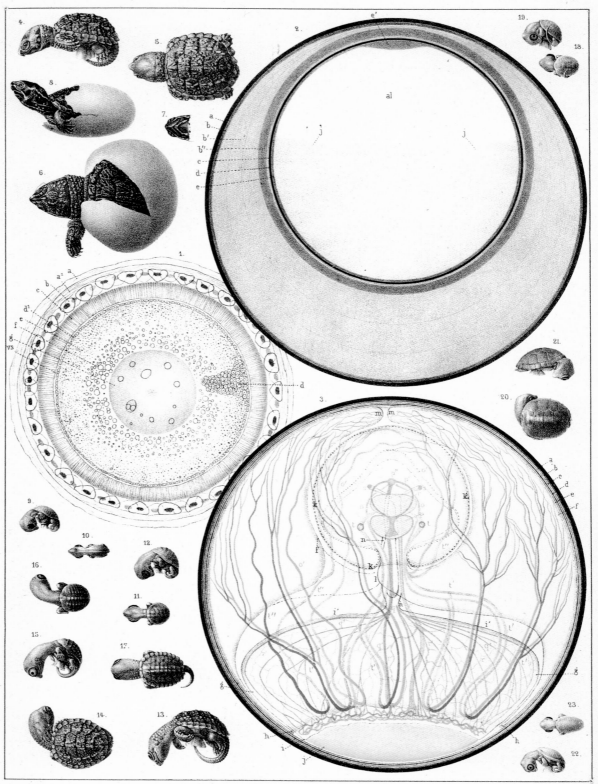

Clark & Sonrel from nat. L.H.Bradford & Co. print.

DIAGRAMS OF THE EGG & EMBRYOS OF CHELYDRA SERPENTINA, OZOTHECA ODORATA AND CHRYSEMYS PICTA.

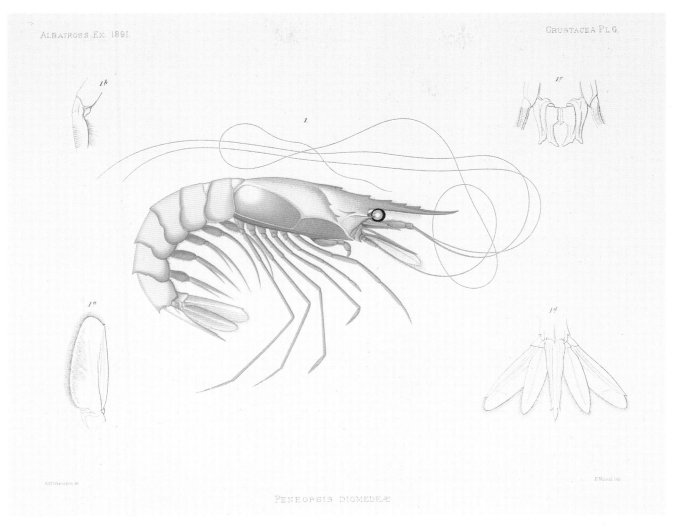

PENEOPSIS DIOMEDEÆ

Plate 54. *Peneopsis diomedeae*, female, natural size. Color lithograph by B. Meisel, from a drawing by A. Magnus Westergren, for the description of Alexander Agassiz's oceanographic collections of the *Albatross* Expedition, in Faxon, *The Stalk-Eyed Crustacea* (1895).

*On opposite page:*
Plate 53. Diagram of the egg and embryos of turtle species. Drawn by H. J. Clark and A. Sonrel from nature, printed by L. H. Bradford & Co., for Agassiz, *Contributions*, vol. 2 (1857).

245

PLATE V.

HESPERORNIS REGALIS, Marsh.

6.34 Cervical vertebrae of the toothed bird, *Hesperornis regalis* Marsh. Lithograph drawn by Frederick Berger and printed by Emil Crisand for Marsh, *Odontornithes* (1880).

PLATE VIII.

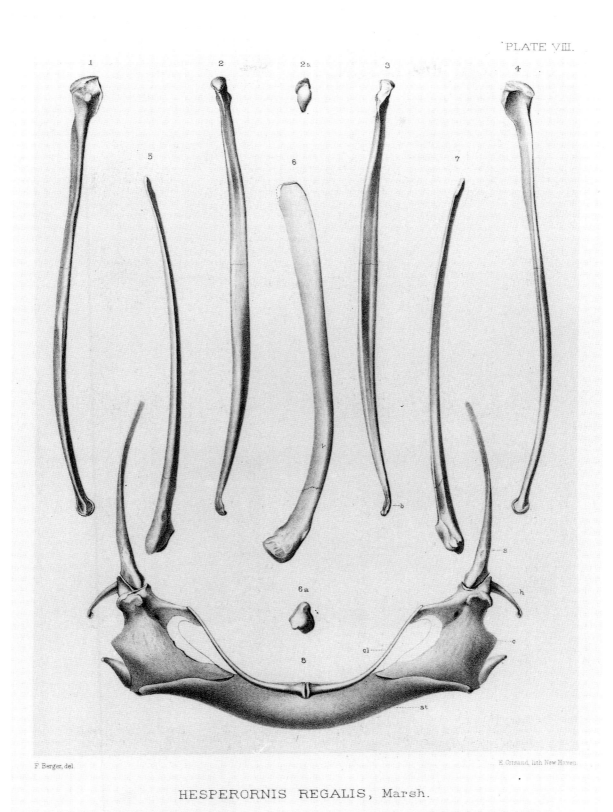

HESPERORNIS REGALIS, Marsh.

6.35 Humerus, scapula, and shoulder girdle of *Hesperornis regalis* Marsh. Lithographs drawn by Frederick Berger and printed by Emil Crisand for Marsh, *Odontornithes* (1880).

Plate 55.

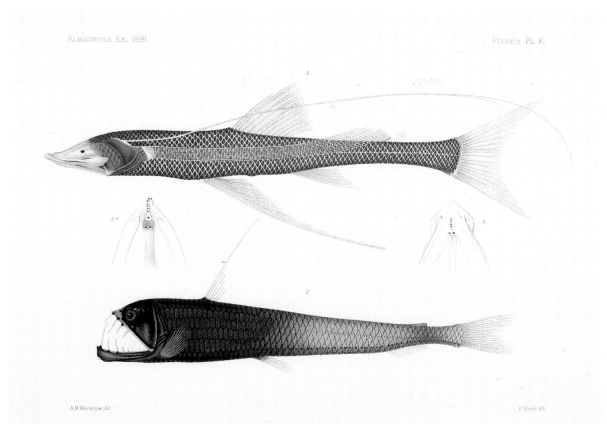

A.M.Westergren,del.                                                    B.Meisel lith.

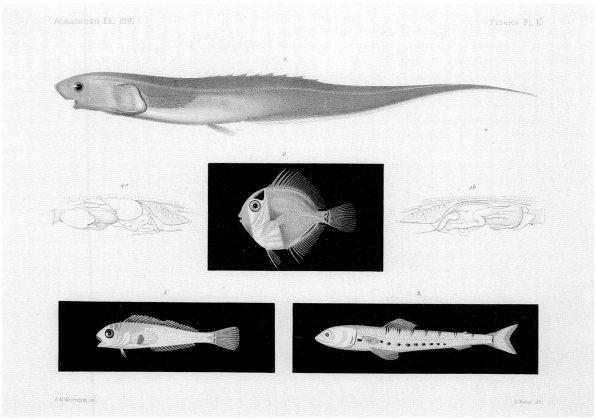

A.M.Westergren,del.                                                    B.Meisel lith.

Plates 55–57. Color lithographs by B. Meisel, from drawings by A. M. Westergren, for Garman, *The Fishes* (1899).

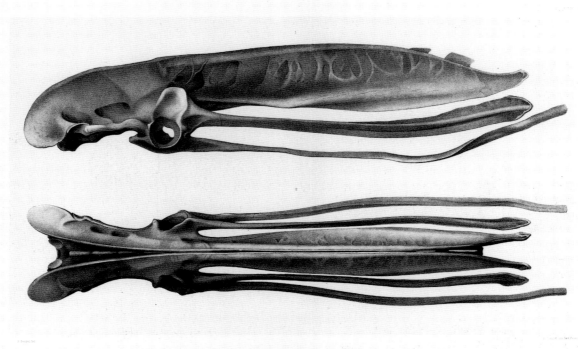

HESPERORNIS REGALIS, Marsh.

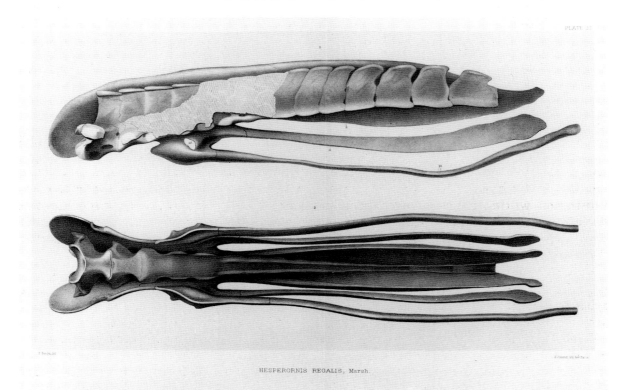

HESPERORNIS REGALIS, Marsh.

6.36–37  Lateral, superior, and inferior views of the pelvis of Marsh's *Hesperornis regalis*. Lithographs drawn by Frederick Berger and printed by Emil Crisand for Marsh, *Odontornithes* (1880).

6.38–41 Lithographs drawn by Frederick Berger and printed by Emil Crisand for Marsh, *Dinocerata* (1884).

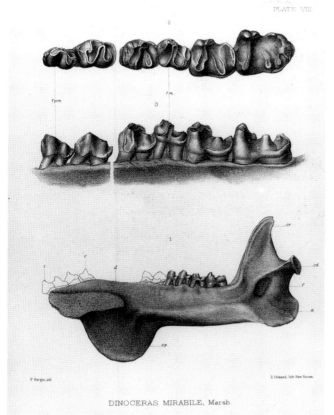

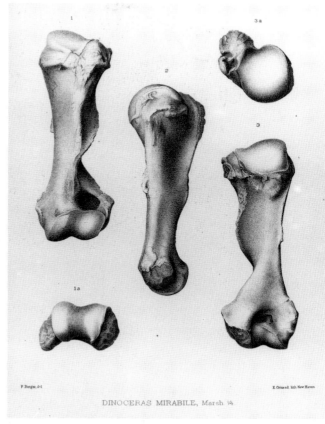

6.38 Jaw and teeth of *Dinoceras mirabile*.

6.39 Humerus of *Dinoceras mirabile*.

one reviewer wrote that the quality of the lithography "combined [with] an artistic finish which had made each plate a kind of finished picture."[49] Marsh defended his investment in illustration, writing that he intended them "to do full justice to the ample material . . . and where possible, to make the illustrations tell the main story to anatomists. . . . The text of such a Memoir may soon lose its interest, but good figures are of permanent value."[50] Fine specimens of the medium and obviously expensive, the plates broke no new stylistic territory. All the pictorial and compositional elements and devices had been formulated decades before and had become the common vocabulary of morphological and anatomical pictorial description. The illustrations of some of Amer-

ica's most important contributions in support of evolution were also some of the most conservative. The government-sponsored series of reports in which the plates appeared betrayed no hint of a dispute within the discipline over priority of description or interpretation. The style of the plates gave the work a dignified imprimatur.

The directors of institutional publication programs, private and state sponsored, maintained the established format and style of the volumes in their series, although variations on the conventions crept into the plates. James Hall steered the New York paleontological survey and publication program through abatements in legislative appropriations and through repeated storms of professional controversy

FIGURES

6.34–42

251

PLATE LIV

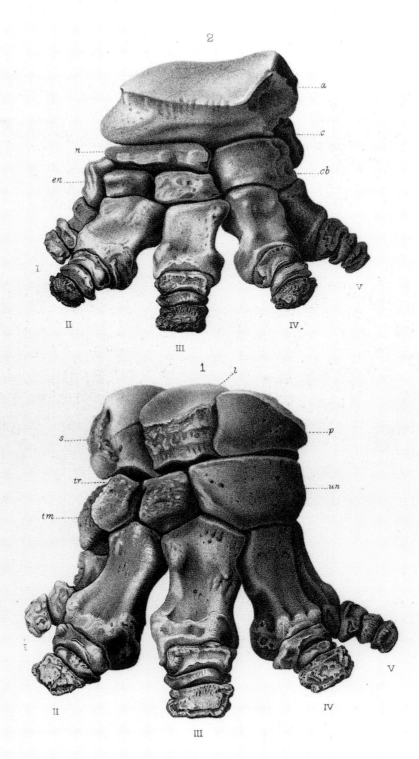

DINOCERAS MIRABILE, Marsh. ½.

6.40  Feet of *Dinoceras mirabile*.

PLATE XLI.

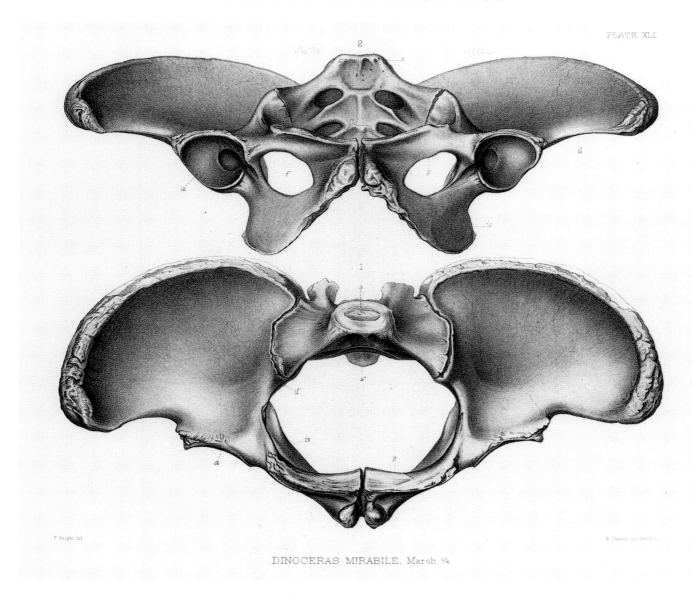

DINOCERAS MIRABILE, Marsh ¼

6.41  Pelvis of *Dinoceras mirabile*.

6.42 Frederick Berger and Emil Crisand drawing on lithographic stones in Marsh's laboratory.

over priority, loyalty, and geological interpretation. Meanwhile, the illustrations for the reports went through several important changes of personnel and print media. After an experiment with photographic reproduction, the reports of the 1880s and '90s settled back into the use of lithography and its attendant descriptive drawing style. James H. Emerton and Ebeneezer Emmons, Jr., however, brought some modifications to the standard treatment and composition of the plates. In earlier plates such as those drawn by Mrs. Hall and Mrs. Brooks, each figure stood alone in relation with the other figures; the overall plate composition preserved the sense that each specimen had been drawn separately. In later plates, the figures were arranged in unified compositions, which in turn acknowledged the defined space of the frame. In the plates produced from Emerton's drawings, lithographed by Frederick Swinton among others, the figures, densely packed into the rectangular frame, seem to have a life of their own. The Emerton plates create a startling sense of movement, in part owing to the curves characterizing the order being illustrated. The activity of these plates also resulted from abandoning a rigid arrangement in rows, in favor of a mixture of figures of different sizes into dense arrangements based on their relative shapes. Emmons's work, on the other hand, carried the basic conventions of order and symmetry in the opposite direction; his figures of fossil sponges were treated monumentally and with emphasis on the geometry of their forms. Both impulses, Emerton's toward romantic movement and Emmons's toward the classically massive and geometrical, represented individual approaches to and variants of the design constraints imposed by professional convention.

Louis Agassiz's Swiss-born son, Alexander Agassiz (1835–1910), made his fortune in min-

FIGURES
6.43–48

254

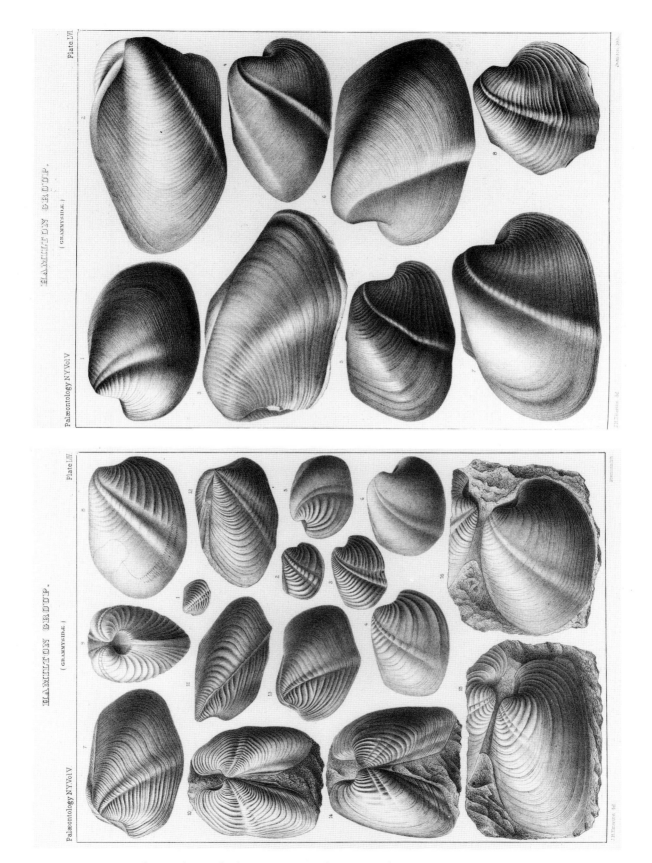

6.43–44 Grammysidae. Lithographs by Riemann and Swinton, from drawings by J. H. Emerton, for Hall, *Paleontology of New York: Lamellibranchiata. II*, vol. 5 (1885).

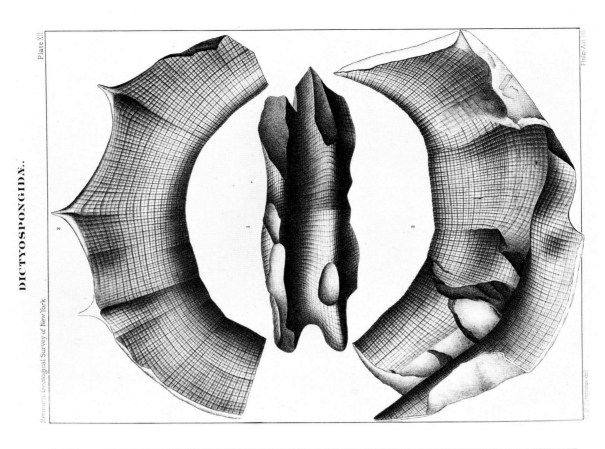

6.45–48  Fossil sponges. Lithographs by Philip Ast and C. Fausel, from drawings by Ebeneezer Emmons, for Hall and Clarke, *Memoir on the Paleozoic Reticulate Sponges* (1898).

Plate XXIX

E. Emmons del

Philip Ast lith

Plate XLVII

C. Hasse, lith.

DICTYOSPONGIDÆ.

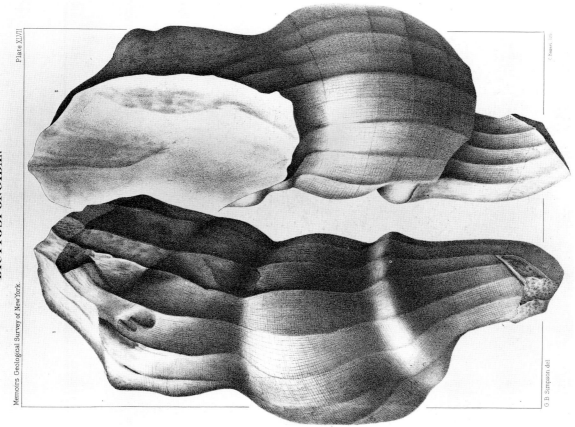

G. B. Simpson del

ing and his career in oceanography in the United States, but he maintained close ties with English and European colleagues.[51] In the younger Agassiz's adherence to the established graphic conventions into the early twentieth century, one can see both a personal conservatism and strong links with European publications in his field, marine biology. Like his father, Agassiz employed in-house illustrators, his own accomplished drawings and watercolors of fish and marine invertebrates setting a high standard of draftsmanship. Paulus Roetter, an Alsatian painter who had drawn cacti for reports of federal western surveys, drew spiney echinoderms for him.[52] During the 1890s and into the early years of the next century, Albertus Magnus Westergren, an inventor and zoologist in his own right, supervised the shipboard sketching on Agassiz's oceanographic cruises, and drew the crustacea and fishes of the collections for publication. Except for Westergren's flourishes, reflecting the mannered organicism of Art Nouveau, the treatment of the figures and plate composition of his work recalled his predecessors of the previous generation.[53]

PLATES
54–57

Again like his father, Agassiz published lithographic illustrations that advertised their costliness. The chromolithographs illustrating the memoirs on fishes published under his direction were hand finished with silver and gold details.[54] Although American chromolithography became associated with inexpensive reproductions of popular paintings and with early advertising, high-finish color lithography remained expensive. Works of natural history produced for a popular audience took advantage of the appeal of color in their illustrations, but even most commercial natural history ventures were limited by market response. (Julius Bien never completed his full-size lithographic edition of Audubon's *Birds of America*.) As ever, advanced methods viable in the realm of commercial printing cost more than many scientific publishers could afford. But with his millions in Michigan copper, Alexander Agassiz could match even royal publications. The

plates he published rivaled those produced for Prince Albert of Monaco, an amateur marine biologist and a friend of Agassiz.

Alexander Agassiz's maintenance of this high standard reflected in part his opposition to and competition with federally supported science (although he relied on the marine specimens obtained from the federal Coast Survey for his own investigations). Taking a position congruent with contemporary laissez-faire economics, Agassiz opposed the centralization of science in federal programs. He believed that individual initiative provided the ideal model for scientific investigation and that competition from the government threatened to deny private scientists their deserved priority and status in the field.[55] Agassiz attacked in particular the Geological Survey, which had grown the largest of all the federal programs with a scientific mandate.[56] In response to the federal publication programs, Agassiz put his fortune behind his defense of private science and supported an extensive list of museum serial publications with a substantial budget for illustration.

Agassiz revealed another side of his conservatism in his insistence that the new fields of zoology, such as physiology and embryology, should not displace taxonomy from its central position in the discipline. After Louis's death, Alexander Agassiz had taken up the directorship of the museum his father had founded at Harvard, the Museum of Comparative Zoology. Agassiz was reluctant to admit laboratories for physiological studies into the museum building, which he expanded and completed according to his interpretations of his father's original plan. Physiology, embryology, and, at the turn of the century, the emergent study of genetics, he insisted, belonged at the Harvard Medical School.[57] His institution existed to support zoological collections and their description. Although he permitted publication of physiological investigations in his museum's journals, his own taxonomic works, not surprisingly, adhered to the well-established conventions.

FIGURES
6.49–52

P. Roetter del                 Dougal sc.

CEREUS ENGELMANNI

6.49 *Cereus engelmanni*, a cactus. Steel engraving by W. H. Dougal from a drawing by Paulus Roetter, for Englemann, *Botany of the Boundary: Cactaceae* (1859).

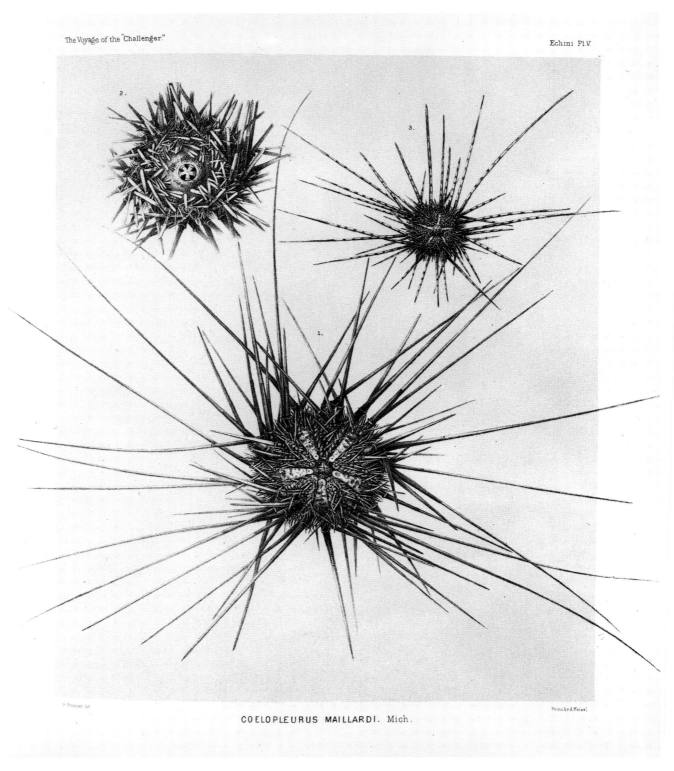

COELOPLEURUS MAILLARDI. Mich.

6.50 *Coelopleurus maillardi*, Mich. Lithograph printed by A. Meisel, from a drawing by Paulus Roetter, for Agassiz, *Report on the Echinoidea* (1881) of the *Challenger* Expedition.

1 MUNIDA PROPINQUA  2 MUNIDOPSIS VICINA
3 MUNIDOPSIS CILIATA  4 MUNIDOPSIS AGASSIZII

6.51 *Munida propinqua* Fax. and *Munidopsis* species. Lithograph by B. Meisel, from drawings by A. M. Westergren, for Faxon, *The Stalk-Eyed Crustacea* (1895).

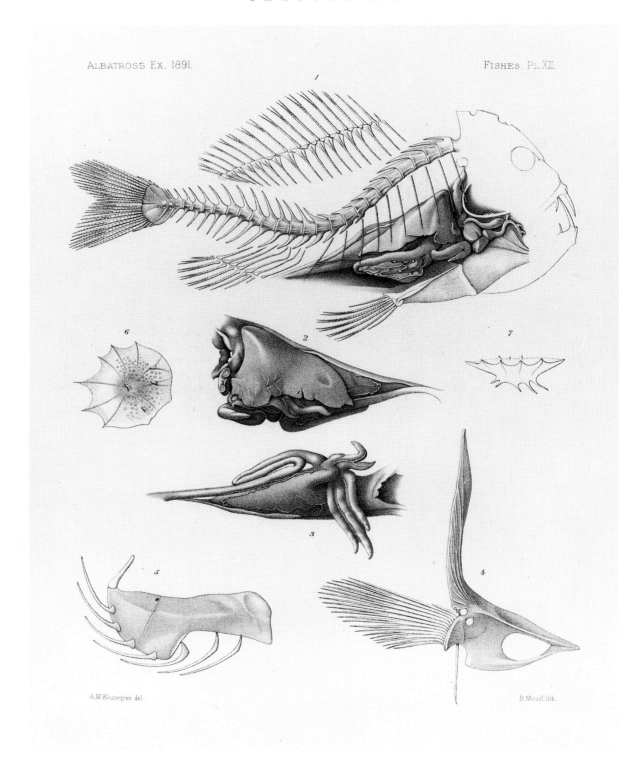

A.M.Westergren del.

B.Meisel lith.

6.52  Skeleton, viscera, and scales of *Caulolepsis subulidens* Garm. Lithograph by B. Meisel, from a drawing by A. M. Westergren, for the description of Agassiz's collection of benthic fishes made on the *Albatross* Expedition, in Garman, *The Fishes* (1899).

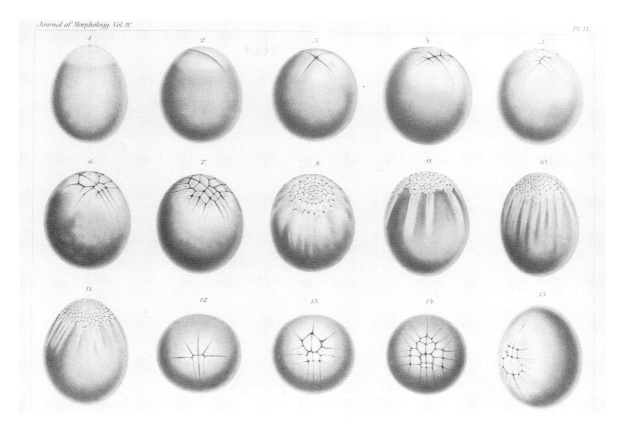

6.53  The unsegmented ovum of *Loligo Pealei*, Lesueur, and stages of its cleavage. Lithograph by Werner and Winter from drawings by S. Watase, in *Journal of Morphology* 4 (1891).

The new subdisciplines which Agassiz believed threatened the centrality of taxonomy adopted existing conventions for their own use. The lithographs reproducing the investigators' drawings in the *Journal of Morphology*, founded by Charles Otis Whitman in 1887 and "devoted principally to *embryological, anatomical* and *histological* subjects," conformed to many of the pictorial conventions associated with comparative zoology.[58] Lithographed by the Meisels of Boston and by lithographic draftsmen in Frankfurt, Germany (where color lithography could be had at a lower cost than at home), plates depicted cell division in the neat rows of figures formerly used to depict distinct species. On the premise that Western scientists would read from left to right and from top to bottom, the figures followed that order to represent sequences of change over time.[59] The late-century articles on development were not the first to use the conventions associated with comparative morphology to depict change; individual figures in the embryological illustrations of Louis Agassiz's *Contributions*, for example, showed stages of development. The new application, however, perpetuated the convention in a new interpretive context.

Plates illustrating the new fields of investigation conformed to the established conventions in part because those conventions were strongly associated with lithography. But the kinds of investigations in which the *Journal of*

FIGURES
6.53–54

Journal of Morphology. Vol. XII

Pl. XIX

6.54 Figures of egg division, from preserved specimens. Lithograph by B. Meisel, from drawings by C. O. Whitman, in *Journal of Morphology* 12 (1897).

*Morphology* specialized exerted pressure for change in the subject matter and appearance of zoological illustration. Physiology and embryology opened the organism and shifted the focus of study from external morphology to internal function. But just as the established conventions accommodated nontaxonomic subjects, they also accommodated technological innovation. Zoologists experimenting with new printing media expected the new techniques, especially photography, to accommodate existing conventions.

# 7

# The Lens and the Line: Photography and Microscopy

L OUIS AGASSIZ attached epochal importance to the application of photography and photographic reproduction to scientific illustration. In 1862 he wrote to congratulate the lithographer Lodowick Bradford:

> I have examined with the deepest interest the photolithographs you have made for Dr. Dean's microscopic preparations of the brain. They are admirable and exceed in beauty anything of the kind done thus far with which I am acquainted. If these attempts could be successfully applied to all the objects of Natural History which require to be illustrated in scientific investigations, they would inaugurate a new era in our science.[1]

From the first introduction of photography, the American scientific community was eager to harness the potential of photography, the medium that made mechanical transcriptions of nature, to its pictorial needs. Dean's photographic illustrations of sections of mammalian brains, taken through a microscope and reproduced by Bradford's own method of photolithography, marked an important advance in yoking the camera lens to the printing press. The initial interpolation of photographic steps into the illustration process, however, had little effect on the conventions that organized the appearance of the plates. Despite the spirit of experimentation in which zoologists and pale-

ontologists explored ways to exploit the lens in illustration, their first results indicated that they expected of the medium not a radically different view of nature or the natural object, but rather a more complete reproduction of the visible, guided by the established conventions of depiction and composition for their fields. Agassiz's praise of the "beauty" of Bradford's photolithographs reminds us that no term of comparison was available for early process images, except for the existing style of illustration. Later, when photography had been fully incorporated into picture printing, however, it began to modify the conventions of zoological illustration, but in ways that the first experimenters with photographic illustration would not have predicted. Instead of providing unlimited detail, photomechanical reproduction enforced a reduction to line and reinforced a growing tendency from within biology toward schematization.

FIGURE

7.1

The initial expectations that zoologists expressed for photography seemed to derive in part from the role of the camera lucida in scientific illustration. This device, invented in 1807, was smaller and more portable than the camera obscura. The camera obscura, originally a darkened room, was a box with a pinhole at one end, through which an inverted image was cast onto a flat surface. Equipped with a lens in the sixteenth century, and the subject of optical study

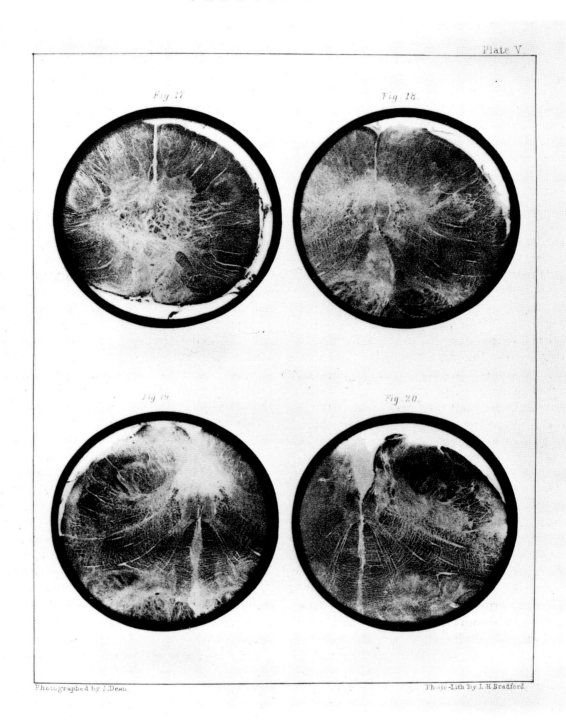

7.1 Sections of the human medulla. Photolithograph by Lodowick H. Bradford of photographs by John Dean, "The Grey Substance of the Medulla Oblongata and Trapezium," *Smithsonian Contributions to Knowledge* 16 (1864).

throughout the seventeenth and eighteenth centuries, the camera obscura had long served as a tool in perspective and landscape drawing.[2] Experiments with the camera obscura had introduced light-sensitive paper to the projection surface in an effort to make the image permanent, leading to the invention of photography.[3] While the camera obscura projected a traceable image onto a drawing surface, the camera lucida, basically a prism on a mount, created a floating image, inverted like the projection of the camera obscura. The need to hold the eye in a fixed position in relation to the prism, so as to preserve the relationship of the "image" to the drawing surface, meant that using the camera lucida required some skill, but it became standard equipment for the scientific illustrator. The scarcity of mention of the use of the camera lucida in American zoological illustration probably stems from the scarcity of documentation by the illustrators themselves, but meticulous authors such as Joseph Leidy and Samuel Scudder noted the use of the instrument in the preparation of their drawings. The increasing appearance of a continuous outline around figures for zoology suggests the widespread reliance on the camera lucida, because an initial outline "pinned" or located the unstable prism image on the drawing surface and helped the illustrator hold the eye steady while filling in the interior detail. In addition, the inversion of the image by the camera lucida would have facilitated the transcription of drawings into engraving or lithography, for which the original drawing had to be reversed in order to print the right way around. Just as the camera obscura and camera lucida transposed spatial relationships from three dimensions onto a flat surface, thus assisting the draftsman in foreshortening and coordinating the relative sizes of the objects on the paper, photography promised to eliminate the final stages, the tracing of the image and its translation into the marks of a drawing. In photography the impartial action of light rather than the interpretive habits of the eye and hand pro-

duced a permanent record of form, detail, and spatial relationships.

Shortly after the 1839 announcement of the daguerreotype, invented in France virtually contemporaneously with the English invention of the photograph, the American painter Samuel F. B. Morse introduced and promoted it in the United States.[4] Requiring long exposure times and an immobile subject, and used preeminently as a portrait medium, the daguerreotype image was produced directly on a polished metal plate covered with a light-sensitive emulsion. Unlike photography, which produced a negative from which one made positive prints, daguerreotypes could not be reproduced in multiple copies. Each image was unique. Furthermore, the metal plates were not compatible with a book format, therefore offering limited applications for publishing. The daguerreotype, however, remained the most popular photographic medium in the United States through the 1850s, to some extent holding at bay the successful commercial development of photographic negatives and prints.

The incompatibility between the daguerreotype and book printing did not prevent its use in scientific illustration. In 1852 David Dale Owen published five steel engravings based on daguerreotypes in his report on the geological surveys of Wisconsin, Iowa, Minnesota, and the Nebraska Territory.[5] In the early 1840s the French had developed a way to etch and engrave daguerreotype plates for printing, but the process had found little application.[6] Whether the printers of Owen's plates produced them by etching the actual daguerreotypes, or from engravings made by hand after the daguerreotype images, is not clear. The captions suggest the latter—"Engraved on steel from daguerreotypes of the original specimens"—but the fineness of the engraving preserves the bright and silvery tonal range of a daguerreotype, markedly different from the standard line engraving with etch that characterized most line-engraved work.[7]

In the same publication, Owen, clearly

searching for a direct transcription of the specimen to the printing plate, experimented with another method that eliminated the photographic step completely, but which shared with photography the advantage of mechanical reproduction. The remaining steel engravings were produced by the medal-ruling process, a topographical tracing directly from the surface of the fossil specimen to the plate. The process, used in France and England since the early nineteenth century, had first been applied in the United States in 1817 by the Philadelphia engraver Christian Gorbrecht, but had resulted in a distorted image. Tilting the tracer arm at forty-five degrees to the traced surface angle corrected the distortion.[8] Owen wrote of the process with enthusiasm:

> Figures of organic remains—many of them never before described—are being executed on steel. Of these there will be some twenty-four or twenty-five quarto plates highly finished, some of which are facsimiles of the originals, engraved entirely by machine. One arm of this machine traverses the specimen and guides a diamond point which rules corresponding lines through the etching-ground on the steel plate; after which the lines are deepened in the steel by the etching effect of acid. This process produces a perfect counterpart of the original, in the most silvery effect of light and shade.[9]

FIGURES 7.2–5

Owen's enthusiasm for the medal-ruling process expressed the pressure to transcend the human limitations of drawing and engraving and to attain an uninterpreted rendering of the object. His experiment attests to the power of photography to up the ante in the stakes for verisimilitude in pictorial representation for science.

FIGURE 7.6

With the introduction of the negative plate, the options for publishing photographs broadened. A posthumous publication of 1861 by James Deane, a student of fossil footprints, had individual photographic prints pasted into the volume.[10] Not only, however, were inserts expensive to publish in large editions, but photographic prints were fragile and subject to fading. The key to the use of photography for the illustration of scientific publications lay in its application to press printing.[11] In 1856 the Duc de Luynes had offered a prize "with a view of hastening the moment, so much desired, when . . . printing or . . . lithography shall permit the reproduction of the marvels of photography, without the intervention of the human hand in the design."[12] Photography, however, either in the form of the daguerreotype or the photograph, presented problems of translation in printing similar to those of other tonal media. Lithography lacked the range of tones rendered by the chemistry of photography (which in turn did not reproduce the full range of tones apprehended by the human eye). Relief printing required the resolution of the photographic grey scale into lines or dots. The challenge was clear; the double promise of photography, that of direct transcription and of streamlining the labor of picture printing, prompted constant experimentation in Europe, England, and the United States by printers eager to exploit the medium for its economy, and by authors eager to invest their illustrations of scientific subjects with the authority of the lens. Meanwhile, printers and their customers treated photographs like any other pictures, accepting engravings and lithographs made after them as reliable reproductions of the originals. That the original had been a photograph, considered a direct "copy" of nature, was sufficient for the picture to partake of the representational promise of the medium.[13]

Lodowick Bradford and a partner, James A. Cutting, began experimenting with photolithography around 1856, and patented their

*On following pages:*

7.2–3 Steel engravings from daguerreotypes for Owen, *Report on a Geological Survey of Wisconsin, Iowa, and Minnesota* (1852).

7.4–5 Medal-ruled steel engravings for Owen, *Report of a Geological Survey* (1852).

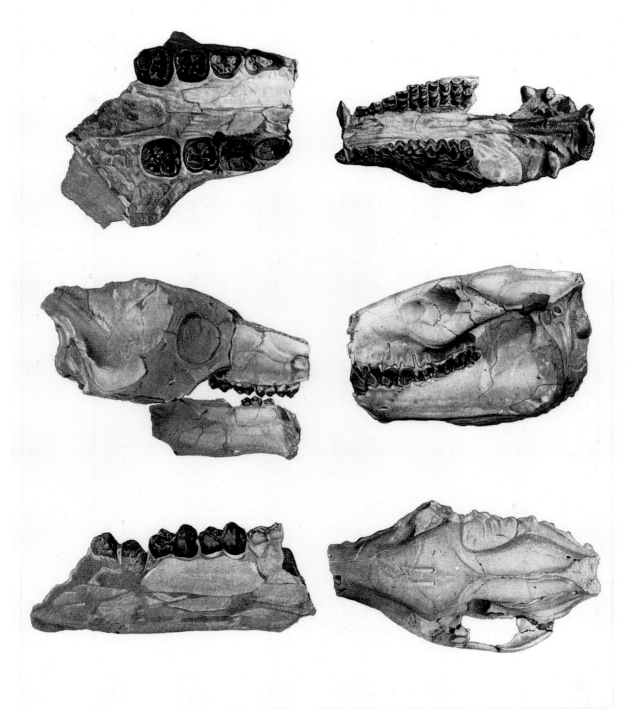

7.2 Remains of Archaeotherium and Oreodon.

Tab XII

7.3  Remains of a species of land turtle.

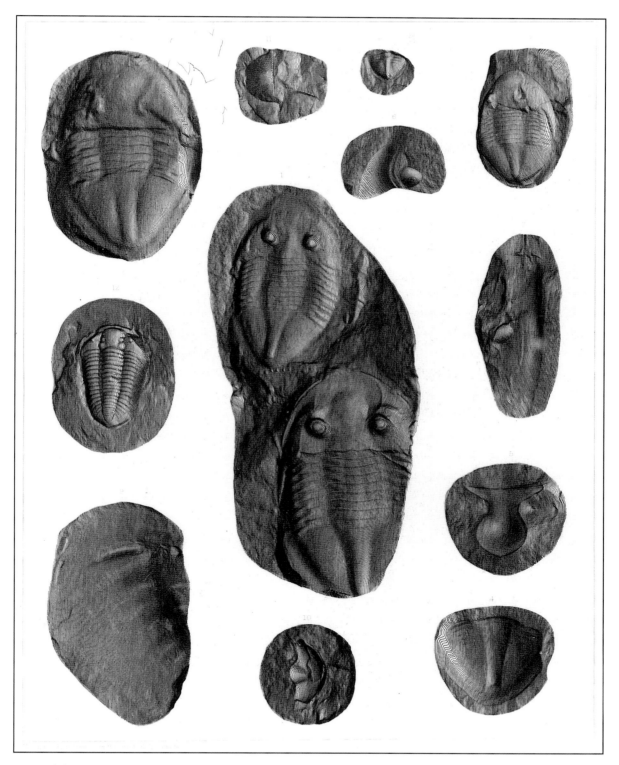

7.4  Trilobites and Brachiopod.

271

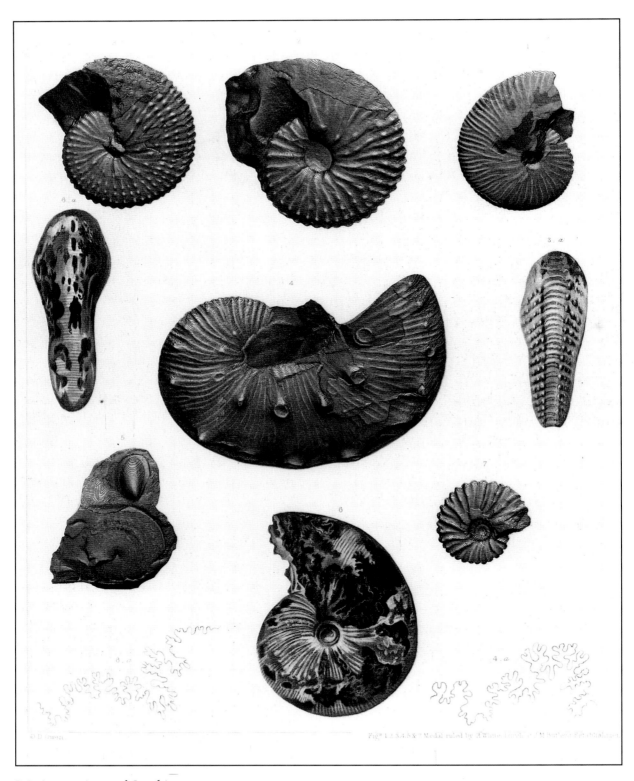

7.5 Ammonites and Scaphites.

Pl. 17

7.6 Photograph of fossil footprints by James Deane, from the posthumous publication of his *Ichnographs from the Sandstone of Connecticut River* (1861).

process two years later. They prepared a photo-sensitive mixture of gum arabic, sugar, and potassium bichromate, applied it to a lithographic stone, and exposed the stone either in the camera or under a negative.[14] Thomas Mayo Brewer (1814–1880), a Boston physician, publisher, and ornithologist, was one of the earliest customers for Bradford's process.[15] When Brewer could not obtain the services of Sonrel for illustrating his work on bird eggs, he turned to photography. Brewer's description of the method suggests that the hybrid process of photolithography had unsettled the vocabulary of rendering:

In all instances the illustrations are taken directly from the egg itself, and in none has any attempt been made to make use of drawings. It has been demonstrated by repeated trials, that no drawing of an egg, however skillful may be the draughtsman, can be obtained, in ordinary cases, which gives with sufficient accuracy of light and perspective the shape and markings of the object represented. It has been deemed advisable, therefore, to abandon entirely the original design of including among the illustrations copies of eggs of which access could not be had in the orginals.[16]

Brewer indicated indirectly that in both substance and illustration his work aspired to new standards.[17] He omitted from his publication the many drawings of eggs provided by a colleague, James Trudeau, with whom he had begun the study jointly. Having adopted photographic reproduction of the eggs, Brewer referred to Trudeau's drawings in his text, but they could not "be made use of in illustrating the present work."[18] Drawing and photography, then, represented incompatible standards of depiction. In certain cases at least, Brewer hinted, drawing belonged to outmoded and photography to corrected observation of nature.

In his introduction, Brewer gave to Bradford the place and praise traditionally reserved for the primary illustrator:

In mentioning those to whom special obligation should be acknowledged for important assistance, the author refers with pleasure to the services of Mr. L.H. Bradford, whose ingenious application of photography to the aid of lithography has given valuable results that could have been reached by no other means. The illustrations which are due to his intelligent skill are probably the most perfect representations of eggs that have ever yet been achieved. The credit of this is chiefly due to Mr. Bradford, who has thus greatly contributed to the completeness and value of the work by furnishing illustrations of unsurpassed excellence.[19]

Bradford continued to produce photolithographs into the 1860s. And Baird, with whom

273

Bradford had been reluctant to contract for printing Sonrel's fine lithographic drawing, published the printer-inventor's photolithographs of Dean's photomicrographs of the mammalian brain in the *Smithsonian Contributions to Knowledge* of 1862. No other printers, however, adopted and developed Bradford's method, and it disappeared commercially after he abandoned lithography and returned to engraving in 1868.[20]

PLATES
58–59

Without knowing that Brewer's plates were photolithographs, a viewer would find little to distinguish them from other lithographic zoological illustrations. The plates combined earlier natural history with midcentury zoological conventions.[21] As in Trudeau's watercolor drawings of individual eggs, each egg sat and cast a shadow on an individual surface, yet the figures were arranged in rows in the requisite manner of the composite plate for comparative zoology. It seems that Bradford assembled a montage of individual photographic negatives, which he then exposed on the photosensitized stone. Bradford then had to produce color separations on multiple stones. Indeed, the justification of the different impressions creates an odd effect. Although each egg is fully shaded to depict relief, there is a sharp boundary around each oval; the effect—contradicting the spatial message of the cast shadow—is as if half-eggs had been set down on the paper. The same effect appears in the color lithograph plates of eggs published by Charles Bendire in 1892, although the later figures, their long axes vertical, have lost their individual shadows and float free.[22]

PLATES
60–61

So far the plates produced by the insertion of photography into the illustration process emulated the graphic conventions of the field. Indeed, James Hall resorted to a photographic printing process in order to publish already-prepared drawings of fossils. Hall's team of illustrators had produced hundreds of drawings for publication before he had finished the accompanying text. Pressured for results by the legislature and anxious to preserve his priority, Hall decided to anticipate the completed publication on the New York Devonian collections with an 1876 edition of the plates alone, reproduced in a process known as the albertype, or artotype, invented in 1869.[23] Hall's experiment eliminated the step of translating the original drawing by hand into an engraving or lithograph. The New York City firm of Edward Bierstadt, one of the principal printers in the process, did the work.[24]

The albertype process, one of several "photogelatin" methods that stimulated great interest but proved of short-lived commercial importance, offered several advantages for scientific illustration, the chief one being its durability for large print runs. On the other hand, the albertype required an elaborate series of technical steps. To produce an albertype from a sheet of drawings, one first photographed the original, and then peeled the exposed negative film off its glass support. That negative did not, however, serve as the eventual printing surface; the printing surface was prepared by coating another piece of thick glass with a photosensitive solution, to be hardened by exposure to light. The underlayer of the solution, exposed from behind through the glass, hardened, while the upper surface remained soft. A rinse then washed away the soft layer, and another light-sensitive gel layer was applied. This last was the gel onto which the negative was exposed; it became the "type" plate for printing, after being suitably washed and fixed.[25]

A contemporary description of the process extolled its economy: "An infinite number of impressions can be taken . . . to keep as many presses in operation as may be required. One operator attending one press can produce about 200 prints a day." A border, title, and caption could all be printed at the same time as the main image. Furthermore, compared with the fragility of a photographic print, which faded when exposed to light, the albertype had the advantage that "there is no danger of the print's fading as the ink is carbon."[26]

Since the albertypes reproduced drawings, Hall made no special claims for photography as an objective medium or superior transcription of nature. The photographic experiment simply effected rapid publication of the plates, staking Hall's pictorial claim for priority for the new species. Hall did not repeat the albertype experiment. Nor did he consider a volume consisting of plates alone a completed monograph. The sheets of drawings printed as albertypes in the *Devonian Fossils* served again as the prototypes for the lithographs printed three years later to illustrate a regular volume in the *Paleontology of New York* series.[27] Hall's return to lithography for illustrating the complete—official—version of his work indicated that he, for one, had no intention of exchanging the established authority of lithographic illustration for photography.

In contrast to Hall, Alexander Agassiz appeared to agree with Brewer when he wrote in 1871: "The accuracy of photographic illustration is of course far beyond that of engraving or lithography."[28] During a recent study tour of the echinoderm collections in European museums, Agassiz had seen in Munich a way to print photographs like lithographs, "very little above the price of printing with us."[29] In a published note on the available processes for printing photographic illustration, he predicted: "As soon as a few practical difficulties of printing the separate figures of a plate at one impression are overcome, we shall be able to illustrate our memoirs accurately and economically."[30] As Bradford's experiments showed, printing separate figures in one impression posed a considerable practical problem. Glass-plate negatives were exposed in the camera one at a time. To conform with the conventions of midcentury zoological illustration, each specimen had to be photographed individually, usually from more than one direction. To conform to taxonomic practice, illustrations for many animal groups had to include details of structure, often enlarged. The pictures then would have to be assembled on a printing surface in a single composition. When produced from photographs, the resulting plate gave the effect of each specimen sitting in a separate space or box, contradicting the established convention of a uniform

space and lighting. Agassiz's early uses of photography for zoological plates preserved existing compositional conventions as much as possible, despite these "practical difficulties." Moreover, his description of the photographic examples reproduced with his article maintained the conventions used for captioning drawn illustrations. He specified whether the figure was reproduced natural size or reduced; he provided the identifying binomial of the subject; he asserted that "both negatives, of course" had been "taken from nature"; and he attributed the work to the printers and firms responsible.[31]

Agassiz had prepared many of the drawings for the lithographic illustrations of his debut monograph on the embryology of starfishes; but with the availability of photographic print media, he questioned his former assumptions about the ability of drawing to convey all he wanted. In claiming that "the accuracy of photographic illustration is of course far beyond that of engraving and lithography" he revealed the basis of his own standards of illustration. The advantage of photography lay, it seemed, not in aiding the original observation, nor in permitting an alternative vantage point for viewing the natural object. Rather, with photography, as Agassiz observed, zoologists could obtain "figures with an amount of detail which the great expense of engraving and lithography would usually make impossible, even were it mechanically practicable."[32]

Sonrel, master of detail, abandoned lithography to open a photographic studio in 1863. There the elder Agassiz frequently sat for his portrait. Jacques Burkhardt and the young Alexander Agassiz posed together at Sonrel's, and the students from the museum crossed the river to be photographed in their new uniforms before going to war.[33] Sonrel did not record his reasons for shifting media. His position as an independent lithographer in an age of large printing firms may have priced him out of the competition. On the other hand, if he was accustomed to using the camera lucida, the transition may have seemed less a change of me-

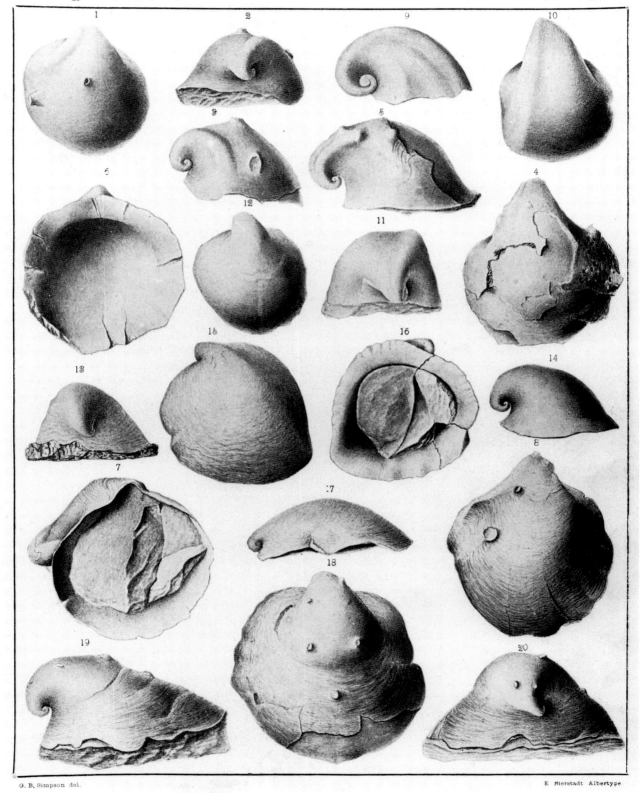

7.7 Platyceridae. Albertype by E. Bierstadt of drawings by G. B. Simpson for Hall, *Illustrations of Devonian Fossils* (1876).

7.8  Platyceridae. Lithograph by Riemann from drawings by G. B. Simpson for Hall, *Paleontology of New York*, vol. 5, part 2 (1879).

dium than a change of tools. Or, Sonrel may have taken up the camera in a personal quest for ultimate verisimilitude. Louis Agassiz's mention of Sonrel's "persevering zeal" indicated that he had taken drawing for the *Contributions* as a challenge. One of Agassiz's students characterized Sonrel as a perfectionist.[34] It is possible that in lithography Sonrel had finally confronted the limits of his own eye and hand, and turned to the lens to transcribe likeness. Yet these motives need not have been mutually exclusive. Bradford's return to engraving after a career in lithography and the invention of a method of photolithography suggests that there were multiple personal and economic reasons for printers to shift media in the rapidly changing world of commercial picture making during the 1850s and '60s.

Sonrel's move to photography did not end his work in scientific illustration. He had been photographing echinoderm specimens for Alexander Agassiz for the illustration of his 1872 memoir, *Revision of the Echini*.[35] When the Boston fire of 1872 destroyed many of the lithographic illustrations prepared for the work, Sonrel replaced lost lithographs with additional photographs. In a note of thanks to Charles Darwin for the receipt of a copy of his latest book, *The Expression of the Emotions in Man and Animals*, sometimes cited as the first scientific book to employ photographic illustration, Agassiz wrote:[36]

> I have not had a moment to look into it, . . . in part owing to the great fire which has devastated Boston. . . . I have been hit pretty hard, not in a money way, but what is worse infinitely, I have lost a year's work by the destruction of six Plates of anatomy with the original drawings, of which I have not even a sketch. They had been sent to Boston the morning of the fire to be lettered preparatory to printing. In addition I lost all the stones of the first parts of the "Revision of the Echini"; fortunately about three fifths of the edition of the Plates had been struck off and was safely housed at the Museum.[37]

For Agassiz's memoir, the individual photographs were mounted onto sheets in the standard arrangement of figures, and printed for publication by the woodburytype process, invented in 1865. The woodburytype was another photo-gelatin method that conveyed the full range of detail of the original negative and was, like lithography, a continuous tone process, producing prints with a rich, almost purple range of tones, and a surface with a slight grain but great clarity.[38] Agassiz had written enthusiastically to Darwin, "I have made pretty extensive use of the new processes of photographic printing in my book, and from what I have succeeded in obtaining trust it will hereafter be possible to supersede the old lithographic processes, which are wasteful in time and money, and not half as accurate."[39]

Alexander Agassiz also turned to photographic printing processes to reproduce his own original drawings.[40] This application of photography, together with Hall's excursion into photographic printing of drawings, suggests that the advantage of photography in eliminating steps in the reproductive process was as important as its status as an ultimate standard of representation. Photography might have privileged representational status, but it did not discredit skilled illustration; the two could supplement one another. Sometimes a zoologist wanted the selective and analytical possibilities available through drawing. Alexander Agassiz's reproduction of his own drawings by a photographic printing process demonstrated his assumption that a drawing had authority when it represented direct observation by a scientist of standing. The step in which that authority could be lost was in the translation of the drawing to the printed plate under the hand of a draftsman. The availability of an alternative—photography—widened the gap between scientist and draftsman.

In the fine arts, on the other hand, the availability of the alternative drew attention to artistic value in the interpretive skills of the hand-engraver and lithographer. Sylvester Rosa

Koehler, curator of prints at the Boston Museum of Fine Arts, expressed regret over the purported "objectivity" of photography in an exhibition catalog of 1892. "The old processes"—engraving and lithography—"in their highest development, are artistic and give free scope to the personal element."[41] In contrast, "the modern photomechanical processes . . . are scientific, and seek to eliminate the personal element." They are not interpretations, they are "servants, whose merit is measured by the degree to which they find it possible to repress their own individuality."[42] In science the process of translation by the engraver or lithographic draftsman had to be tightly controlled either by direct supervision or by a professional consensus establishing graphic conventions and imposing constraints. Clearly, for the descriptive purposes of zoology and paleontology, a history of working relationships between authors and printers—such as Baird's with Sinclair, or Hall's and the New York Lyceum's with Gavit and Duthie—had established an acceptable standard of detail and manner of drawing, one that balanced the capabilities of commercial engraving and lithography with the practices and requirements of the discipline. As for the "individuality" of an original zoological drawing, the established practices of the discipline defined drawing by the scientist as an act of original observation; despite the incorporation of photography into the practices of observation and pictorial depiction, drawing retained that status. A lithograph or engraving made after such a drawing—provided the conditions and context of reproduction met standard practices—continued to convey the same authority. This persistence of practice and production speaks to the capacity of the discipline to absorb photography to its own purposes. In zoological illustration the greatest pressure for change came not from the establishment of new and preemptive standards of representation, but from the insertion of photography into commercial print media.

Photography played an important role in the development of wood engraving as a major commercial medium. During the 1860s printers, especially the popular illustrated weekly newspapers engaged in heated competition for readership, moved from pasting an original drawing onto the surface of the engraving block to coating that surface with a photosensitive emulsion and exposing a negative of the drawing. In either case, the engraver then used hand tools to translate the image into line work.[43]

To the scientific author, wood engraving offered two great advantages. The first dated from the fifteenth century. The boxwood blocks, or in the nineteenth century their electrotype facsimiles, could be made the same height as the metal type, set with the text in the same press bed, and printed simultaneously to combine text and illustration on a single page. Even before the introduction of photography to wood engraving, the medium had become increasingly important for zoological illustration, especially in publications of octavo-sized editions. The growing use of wood engraving for text figures and diagrams printed on the same page with the taxonomic descriptions helped it transcend its associations with popular natural history illustration, often crude copies taken from other popular sources.

The second advantage was wood engraving's relative cheapness compared to lithography or steel engraving. In contrast to metal-plate engraving, in which the incised lines retained the ink after the plate surface was wiped clean, in wood engraving, as in letterpress, the raised surface of the hard woodblock took the ink. Unlike metal engraving, which required a paper stock heavy and strong enough to withstand being pressed into the inked incisions, or lithography, which required a stouter paper than letterpress, wood engraving could be printed on cheap paper, although fine line work required paper of a better quality. For regular, parallel lines, the dominant manner of engraving on wood in the commercial press, cheap paper sufficed. In the growth of mass publishing, the mechanization of paper manufacture and the

A. SONREL, Photo.

American Photo-Relief Co., Printers, Phila.

7.9 Echinoderms. Woodburytype printed by the American Photo-Relief Company, from photographs by A. Sonrel, for Agassiz, *Revision of the Echini* (1872–1874).

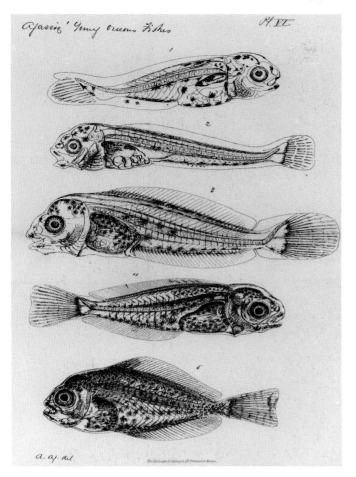

7.10 Stages of butterfish development. Heliotype Printing Co. reproductions of drawings by Alexander Agassiz, in Alexander Agassiz, "On the Young Stages of Some Osseous Fishes," *Proceedings of the American Academy of Arts and Science* 17 (1881–1882).

introduction of wood-based rather than linen rag paper played a central role.[44] Many zoologists of the mid- to late nineteenth century wrote textbooks and popular editions of their studies that were marketed successfully in inexpensive editions.

With the expanded use of wood engraving to reproduce originals in all varieties of mediums, its practitioners developed variations on the standard technique of parallel lines and cross

hatching. Writing to instruct in the techniques of printmaking and printing processes, James Hodson, an English wood engraver, explained: "Where the drawing is made wholly in pencil, the work of the engraver is called 'fac-simile,' because every line and mark of the pencil drawing has to be carefully preserved. When 'wash' is used to show the tone and shadows, the engraver has to make use of such lines and tinting as will produce upon the block the proportion of colour or tone indicated by the wash."[45] The white-line wood engraving technique, in which the lines incised into the surface of the inked block described the figures, came to rival in popularity the black-line style, in which the surface, or inked lines, constructed the figure. Henry Marsh, a wood engraver whose treatment of scientific subjects attained great influence in the United States, had achieved microscopically fine textures and detail in his white-line engravings of drawings by Sonrel and Burkhardt for the 1862 edition of *A Treatise on Some Insects Injurious to Vegetation* by the Boston entomologist, Thaddeus William Harris.[46] Marsh's work was the initial inspiration for Cornell entomologist John Henry Comstock to study insects, and for his wife, Anna Botsford Comstock (1854–1930) to learn wood engraving.[47]

In the winter of 1885, when Anna Comstock traveled to New York to study wood engraving with her friend, Susanna Phelps Gage, wife of Cornell anatomist and microscopist Simon Henry Gage, they represented a small but growing number of women entering the field of illustration. Anna Comstock had taught herself to draw and engrave insects for her husband's work, and Susanna Gage had been taking a correspondence course with wood engraver John P. Davis at the Cooper Union. Now they attended classes in Davis's studio.[48]

Davis decorated his studio in New York with masterpieces of engraving for the inspiration of his students. A member of the school of wood engravers who had broken with the formal manner of black-line engraving, Davis taught

FIGURES
7.11–17

281

7.11–12 Wood engravings from drawings by Alexander Agassiz for Alexander Agassiz and Elizabeth Cary Agassiz, *Sea-Side Studies in Natural History* (1865).

The same Actinia (Metridium marginatum) fully expanded ; natural size

7.11 *Metridium marginatum*, fully expanded.

the new and controversial white-line technique. In 1880 Davis had written about the older manner: "A certain kind of line, it was held, should be used to represent ground; another kind to represent foliage; another . . . sky . . . flesh . . . drapery, and so on. Each sort of line was the orthodox symbol for a certain form."[49] These rigid formulas had developed in part to facilitate the rapid translation of drawings and photographs into line work by corps of staff engravers under pressure to meet newspaper deadlines. The large-format newspaper illustrations engraved on wood were actually produced in sections and then bolted together for finishing and printing. A standardized technique provided some insurance that the portions would cohere stylistically on reassembly. Davis claimed that the use of photography to print the original picture onto the engraver's block, introduced in 1858 but not in general use until the mid 1870s, had liberated the engraver from

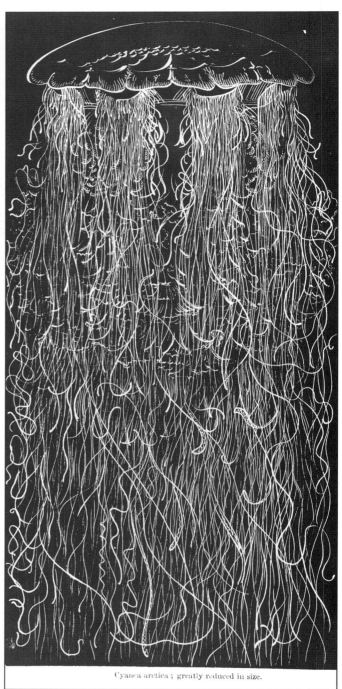

Cyanea arctica ; greatly reduced in size.

7.12 *Cyanea arctica.*

282

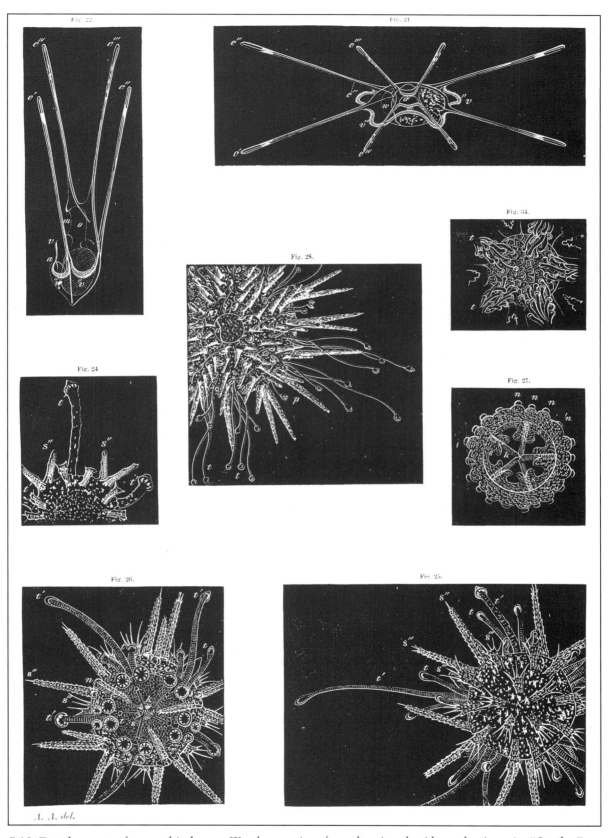

7.13 Development of sea urchin larvae. Wood engravings from drawings by Alexander Agassiz, "On the Embryology of Echinoderms" (1864).

Fig. 106. — White-headed or Bald Eagle.

7.14 Eagle, after Wilson.

7.14–15 Wood-engraved illustrations from Sanborn Tenney and Abby A. Tenney, *Natural History of Animals* (1875).

do not hatch medusæ like the parent, but each hatches a little hydroid which is first free, then afterwards becomes attached to a shell, sea-weed, or stone, and from this

Fig. 484. — Portuguese Man-of-War.

7.15 Portuguese Man-of-War.

little hydroid others branch till a little community of hydroids has grown up, as in Figure 478. From these hydroids bud again the Coryne, Figure 479.

In some kinds, as Tubularia, Figure 481, the hydroid

7.16–17 Wood engravings by Henry Marsh for Harris, *A Treatise on Some Insects Injurious to Vegetation* (1862).

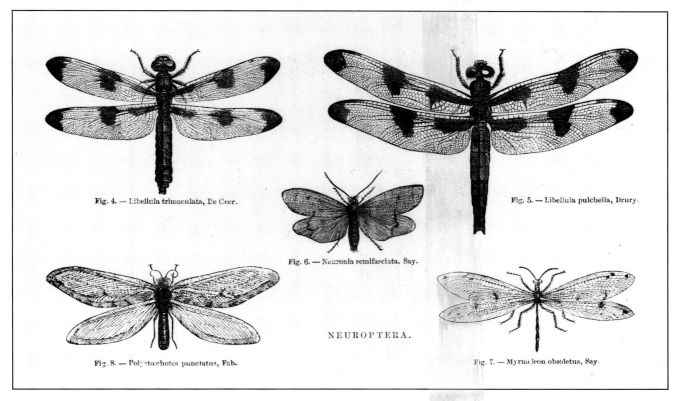

7.16 Neuroptera.

the constraints of standardization.[50] "Instead of merely symbolizing the work of the artist, the engraver now makes use of all methods by which he can fix on the block, as accurately and perfectly as possible, the original picture. . . . This abandonment of conventional recipes, this enlarged liberty with respect to means, is the distinguishing characteristic of the new school."[51]

The English wood engraver W. J. Linton, Davis's predecessor at the Cooper Union, was a vocal opponent of the new style and deplored what he viewed as the corruption of wood engraving. He considerd its widespread commercial use "a return to the old mechanism of the plank-cutters"; he believed that the practitioners of interpretive wood engraving had cultivated "artificiality, softness, and polish" rather than valuing the "freshness, originality, and strength of the earlier day."[52] While Linton argued that the "engraver had to become an artist, to understand an artist's drawing" he advocated a style that avoided "foolishness of over-refinement."[53] Linton saw the use of photography on the block not as a source of freedom but as reducing the engraver's task to mindlessness. "Some years since," he wrote, "I found in the lowest depth a deeper still,—the employment of 'engravers' on no longer even drawings, but on photographs of drawings, not drawn even so as to be suitable for the purpose."[54] He likened such work to "cutting (as rats might gnaw) portions of something not understood by them."[55] Arguing for a return to the conventions, Linton insisted that the technique he advocated emulated the material represented in the original drawing as well as conveying the intention of the artist. Horizontal

Fig. 145.

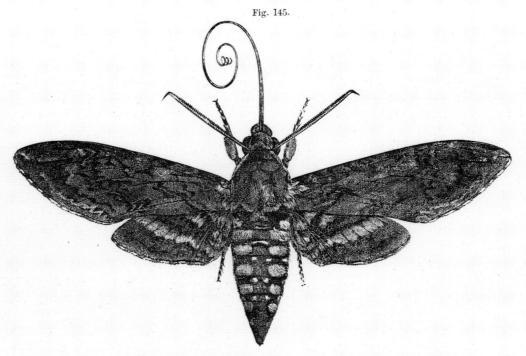

founded with the Carolina Sphinx (*Sphinx Carolina* of Lin-
næus, Fig. 145, Fig. 146, larva, Fig. 147, pupa), which it

Fig. 146.

closely resembles. It measures across the wings about five
inches; is of a gray color, variegated with blackish lines
and bands; and on each
side of the body there are
five round, orange-colored
spots encircled with black.
Hence it is called by Eng-
lish entomologists *Sphinx quinquemaculatus* (Fig. 144), the
five-spotted Sphinx. Its tongue can be unrolled to the

Fig. 147.

7.17 *Sphinx carolina.*

lines, for example, should render level water. If the engraver, whose mission he likened to that of "a translator, not a copier," observed the proprieties of materials and their differences, then "hair looks like hair; the flesh is flesh."[56] Davis, in contrast, asserted that the new method reproduced originals more accurately than the formulas of the old school. "Exactly to reproduce—that is the present aim of the engraver on wood."[57] Proponents of interpretive wood engraving aimed for a style that was virtually photographic.

When Davis's pupil Anna Comstock engraved for scientific and for popular natural history illustration, she either worked directly from the specimen or from a photograph on the block. She seems to have considered the specimen and its photograph as equivalent, for she did not differentiate between her technique for engraving directly from the insect and that for engraving from a picture. For both, she interpreted in line a nonlinear original. Like other members of the new school of wood engraving, Comstock claimed that the interpretive technique of fine incised or "white" lines translated not only tones but color and texture. Especially in her engravings of Lepidoptera, Comstock's style rendered an impressive interpretation of wing texture (the hard, shiny chiton of beetles and ants did not lend itself as readily to the white-line technique). Comstock carried into the twentieth century the engraving style for insect illustration pioneered by Marsh, whose work had impressed even Linton. Of Marsh, Linton wrote: "It is work not only of patience and remarkable eyesight, but also of true artistic skill; showing, too, . . . that there are powers of expression in wood which cannot be equalled by the rival [steel engraving] process."[58] Comstock considered wood engraving an art. "The study of color and texture by our engravers had the inevitable result of changing them from engravers to artists," she wrote in her autobiography.[59]

Anna Comstock led a public life. Long before 1913, when she received an appointment as assistant professor at Cornell, she engaged in university activities, traveled widely with her husband, published his books and her own, and lectured in her own right. Among her many public honors, she took special pride in being the third woman elected to the American Society of Wood Engravers. Her mission to educate the public about wood engraving and her educational work in natural history were intimately linked.[60] Through her natural history books for grade school use, through her own lecturing, and through her exhibitions of her own wood engravings and the work of others, Comstock's work reached a wide audience.

The wood-engraving career of Anna Comstock moved between the realms of scientific and popular natural history illustration and also helped cultivate an active intermediate terrain, that of the school and college textbook. Wood engravings illustrating school and college textbooks established a middle stylistic ground between the text diagrams of taxonomic publications and the often sentimental or overly dramatic illustrations for popular natural history. Textbook illustrations and works produced for a general audience drew on similar sources; bird illustrations, for example, reproduced repeatedly the familiar figures of Wilson and Audubon. But textbook authors like those of college professors Sanborn Tenney and James Orton, whose works enjoyed many printings and editions, exercised restraint in illustration, usually restricting the figures to the animal itself, or, when providing some background, limiting it to a modest vignette.[61]

In contrast, some authors or publishers of popular natural history of the 1870s and '80s interpreted the evolutionary contest as a mandate for illustration that resembled the dramatic scenes of the literature of adventure. In a work such as The Standard Natural History, full-page engraved illustrations borrowed from the German natural history of Alfred Brehm's Thierleben, published in multiple editions in Leipzig (one of the centers where American biologists went for advanced study) depicted vividly the violence of competition—the survival of the fittest.[62] At the same time, the scientifi-

FIGURES 7.18–21

287

ENGRAVED FROM NATURE, BY ANNA BOTSFORD COMSTOCK.

7.18 Moths. Engraved from nature by Anna Botsford Comstock for John Henry Comstock, *Evolution and Taxonomy* (1893).

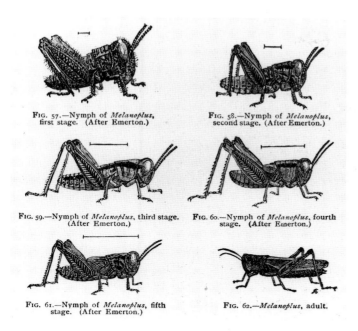

FIG. 57.—Nymph of *Melanoplus*, first stage. (After Emerton.)

FIG. 58.—Nymph of *Melanoplus*, second stage. (After Emerton.)

FIG. 59.—Nymph of *Melanoplus*, third stage. (After Emerton.)

FIG. 60.—Nymph of *Melanoplus*, fourth stage. (After Emerton.)

FIG. 61.—Nymph of *Melanoplus*, fifth stage. (After Emerton.)

FIG. 62.—*Melanoplus*, adult.

7.19 Illustrations of Hexapoda metamorphosis. Wood engravings by Anna Botsford Comstock for the Comstocks' *A Manual for the Study of Insects* (1895).

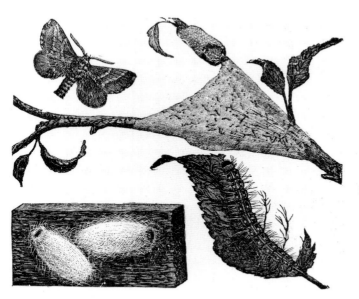

7.20 *Clisiocampa americana*, eggs, tent, larva, cocoons, and adult. Wood engraving by Anna Botsford Comstock for the Comstocks' *A Manual for the Study of Insects* (1895).

632　　　　*THE STUDY OF INSECTS.*

only about twice the length of the head and thorax. This sex is very rare; it can be recognized by the venation of the wings, which is similar to that of the female. Nothing

FIG. 762.—*Pelecinus polyturator*, female.

is known regarding the habits of this species, but it is supposed to be parasitic, like the Ichneumon-flies.

### Family CHRYSIDIDÆ (Chry-sid'i-dæ).

#### *The Cuckoo-flies.*

The cuckoo-flies are wonderfully beautiful creatures, being usually a brilliant metallic green in color. The species are of moderate size, the largest being only about a half inch in length. They can be distinguished from other  Hymenoptera by the form of the abdomen, in which there are only three or four visible segments (Fig. 763), except in the male of a single genus (*Cleptes*), where there are five. The abdomen is convex above and flat or

FIG. 763.—*Chrysis nitidula.* concave below, so that it can be readily turned under the thorax and closely applied to it. In this way a cuckoo-fly rolls itself into a ball when attacked, leaving only its wings exposed.

Although these insects are handsome, they have very ugly morals, resembling those of the bird whose name has been applied to them. A cuckoo-fly seeks until it finds one of the digger-wasps, or a solitary true wasp, or a solitary

7.21 Hymenoptera. Wood engravings by Anna Botsford Comstock for the Comstocks' *A Manual for the Study of Insects* (1895).

289

Figure 7.22.

7.22–23 Dramatic depictions of the survival of the fittest: *Carcinus moenas*, Green Crabs, dashed against a rocky shore by high surf (7.22, *opposite*); and *Platydactylus mauritanicus*, Tarente (7.23, *above*). One of the geckos has taken a bite out of the other's neck. Wood engravings by K. Jahrmargt for Kingsley, *The Standard Natural History*, vol. 2, *Crustacea and Insects* (1884), and vol. 3, *Lower Vertebrates* (1885).

cally up-to-date contributions to *The Standard Natural History* by eminent zoologists were reflected in other more sober figures in the manner of mainstream scientific publishing. The influence of Darwinian theory appeared not only in the thrilling scenes of natural violence, but in the increased interest in the invertebrate phyla. Evolutionary theory had stood the old chain of being, formerly top down from the Creator to the lowest infusoria, on its head. The pre-Darwinian texts in the popular literature had invariably started with Man, devoted

hundreds of pages and illustrations to domesticated animals and exotic mammals, as many to birds, and increasingly fewer to fish, reptiles, insects, and mollusks.[63] The *Standard Natural History*, adopting the bottom-up perspective of evolutionary development, devoted entire volumes to invertebrates; the human species at the end of the story was described in the terms of the new science of anthropology. For the youthful middle-class readers of such works, the dramatic action and exotic settings depicted in the wood-engraved illustrations were an important medium through which they developed their view of nature. If then some readers turned to educational literature for serious study, they found figures, although more restrained, printed in a medium similar to those that had originally kindled their enthusiasm.

In time, the photomechanical processes induced a change in the rendering of the original drawings for zoological illustration. Many of the picture-printing methods incorporating photography were oriented toward the reproduction of line rather than continuous tone. In response, zoological illustrators adapted their work from shaded or tonal drawing to line or line and stipple drawing. While the proliferation of photolithographic, electrotype, and screen-printing techniques would encourage a wide diversity of drawing styles in popular illustration, it contributed to the reverse effect in zoological illustration; it accelerated the consolidation of a new convention in zoological drawing.[64] Concurrently, influenced by the rising importance of embryology and physiology, taxonomic diagnosis and its accompanying illustrations focused increasingly on organs requiring dissection or microscopic analysis.[65] Together these developments established the basis for the development of a linear drawing style for zoological illustration.

Initially, experimentation and innovation resulted in a flourishing diversity of technique and style, with bits and pieces of old conventions mingling with new ones. While some publications adhered to illustration with crayon lithographs and maintained the conventions

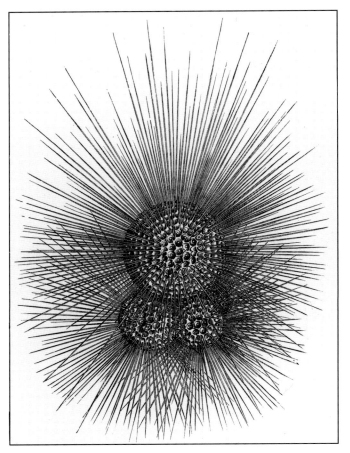 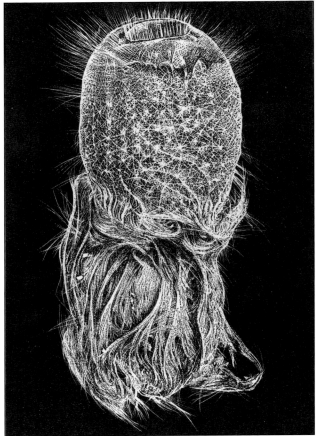

7.24–25 Wood-engraved figures. *Globigerina bulloides*, greatly enlarged, a species of foraminifera (7.24); and a sponge species, *Holtenia carpenteria* (7.25). From John S. Kingsley, *The Standard Natural History*, vol. 1, *Lower Invertebrates* (1885).

established around that medium, others adopted at least in part the new print media, with a resulting heterogeneous appearance in the illustrations. Edward S. Morse, who had been the illustrator among Louis Agassiz's first group of students, kept a file of examples of new picture-printing processes in an effort to keep abreast of the changing technological scene.[66] Sylvester Rosa Koehler mounted an exhibition and wrote an extensive catalog explaining the diversity of media, some lasting and some ephemeral.

Some publications incorporated new diagnostic techniques and new media but also maintained the ordering conventions associated with midcentury lithographic illustration. Samuel Scudder's *The Butterflies of the Eastern United States and Canada*, published privately in 1889, was such a work.[67] Scudder's preparations for illustration as well as his printed plates provide a historical overview of his discipline, its drawing conventions, and the print media available to him. Scudder had first conceived of his study on butterflies while a student of Louis Agassiz. As had Agassiz, Scudder

Plates 58–59. Birds' eggs. Photolithographs by Lodowick H. Bradford, for Brewer, *North American Oölogy* (1857).

Plates 60–61. Color lithographs by Ketterlinus for Bendire, "Life Histories of North American Birds," *Smithsonian Contributions to Knowledge* 28 and 32 (1892 and 1895).

OOLOGY OF NORTH AMERICA.

Plate 58.

OOLOGY OF NORTH AMERICA.

Plate 59.

Plate 60.  Hawk eggs.

Plate 61.  Cuckoo, Jay, Crow, and Lark eggs.

THE DE VINNE PRESS.

7.26 "Butterflies in black—Nymphalidae (Nymphalinae)." Printed by the De Vinne Press from new electro-types, from original woodcuts engraved by Henry Marsh. From Samuel Hubbard Scudder, *Butterflies of the Eastern United States and Canada* (1889).

THE UNIVERSITY PRESS.

7.27 Nymphalidae and Papilionidae. Wood engravings reproduced from different sources. From Scudder, *Butterflies of the Eastern United States and Canada* (1889).

Plate 62.

emphasized difference; he distinguished more species than his colleagues.[68] But for Scudder, the individual specimen was not sacrosanct. He instructed his illustrators to "borrow" parts of one specimen in composing a pictorial representative of the species for the plate.[69] Some of the figures in his plates dated from the 1860s and had been drawn by Jacques Burkhardt, whose work would have been available to him only through Agassiz. Other figures harked back to the period of the naturalist and were copied from original drawings by John Abbot.[70] The majority of Scudder's illustrators, however, belonged to a later generation, notably James Henry Blake (another former Agassiz student), James H. Emerton, who studied and published on spiders in addition to illustrating zoological subjects, and Mary Peart, one of several women who drew for Scudder.[71] The figures that these and additional illustrators prepared included the larvae and caterpillar stages of the life cycle, insect parasites of butterflies, maps of geographical distribution of New England species, microscopic examinations of wings and scales, dissections, and histological studies.

FIGURES
7.26–35

The color lithographs opening the volume and depicting butterfly species set the tone and style for the rest of the plates, although the new printing processes permitted and, indeed, enforced, a very different graphic treatment of the subject. Thomas Sinclair, still willing to bid for a zoological contract although he knew well that the author would be particular about quality, printed the majority of the chromolithographs, requiring from eight to fifteen stones.[72] Sinclair's color lithographs were followed by electrotyped reproductions of wood engravings by Henry Marsh; photogravures of pencil drawings by James Henry Blake; photographic relief plates of ink drawings of wing venation by Emerton; gelatin process photographic reproductions of pencil drawings. The Meisels of Boston, specializing in technical lithography, produced some delicate color work and fine line lithography for the illustrations of wing venation, internal anatomy, embryology, and miscroscopic slide sections of tissue.[73] When the plates depicted drawings made with the aid of a camera lucida or a microscope, Scudder noted the size of the microscope objective or approximated the enlargement; when the fig-

PLATES
62–64

*On opposite page:*

Plate 62. Scudder's captions gave complete identifications, attributions, and descriptions of media. The information he provided for his fourth plate read: "Butterflies in color.—Nymphalidae, especially Argynnidi. Printed in color from ten stones by Thos. Sinclair and Son, after drawings by L. Trouvelot (figs. 1–3, 5–8) and G. A. Poujade (fig. 4). Natural size. Where both surfaces are given, the detached wings show the under surface." From Scudder, *Butterflies of the Eastern United States and Canada* (1889).

*On following pages:*

7.28 Hesperidae. Photogravure printed by the Boston Photogravure Co., from india-ink drawings by James Henry Blake. From Scudder, *Butterflies of the Eastern United States and Canada* (1889).

7.29 Male abdominal appendages, Nymphalidae. Lithography by B. Meisel, from drawings by Edward Burgess and J. H. Emerton. From Scudder, *Butterflies of the Eastern United States and Canada* (1889).

7.30 Butterfly wings. Photographic relief plate prepared by the Boston Photogravure Co., from ink drawings by J. H. Emerton. From Scudder, *Butterflies of the Eastern United States and Canada* (1889).

7.31 Wing patches and folds found in male butterflies. Pencil drawings by J. H. Emerton, reproduced photographically by the gelatin process by the Boston Photogravure Co. From Scudder, *Butterflies of the Eastern United States and Canada* (1889).

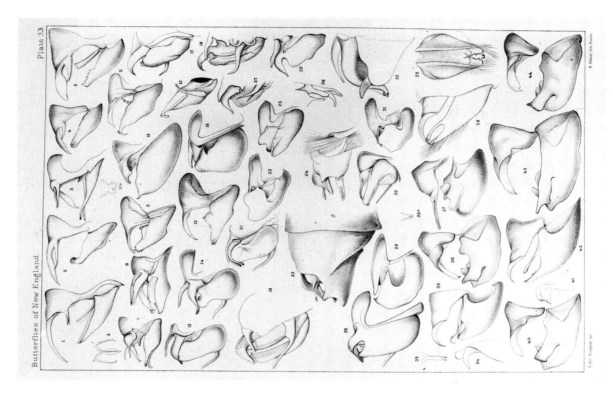

Butterflies of New England    Plate 33

Figure 7.29.

Butterflies of New England    Plate 17.

Figure 7.28.

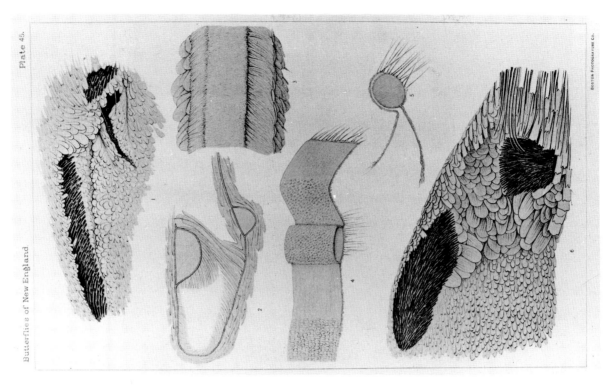

Plate 45.

Figure 7.31.

Plate 40.

Figure 7.30.

Plate 7.

Plate 63. Pierinae. Chromo-lithograph from twelve stones by Thomas Sinclair and Son, after drawings by J. H. Blake, Sidney L. Smith, and G. A. Poujade. From Scudder, *Butterflies of the Eastern United States and Canada* (1889).

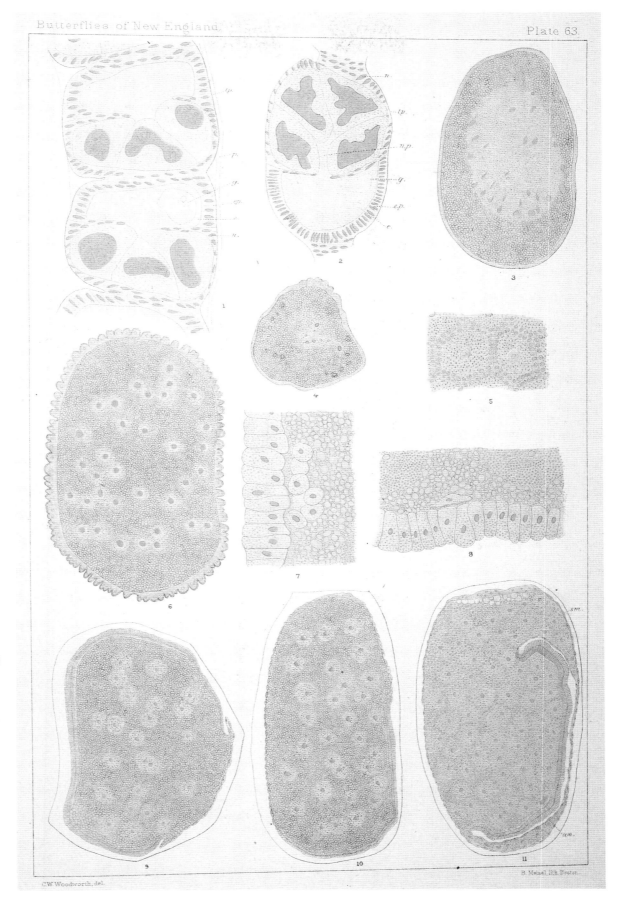

Plate 64.
Embryology
of *Euvanessa
antiopa.*
Three-color
lithograph by
B. Meisel,
from drawings
made with a
camera lucida
by C. W.
Woodworth,
from sections
about .00025-
inch thick of
organs and
eggs. From
Scudder,
*Butterflies of
the Eastern
United States
and Canada*
(1889).

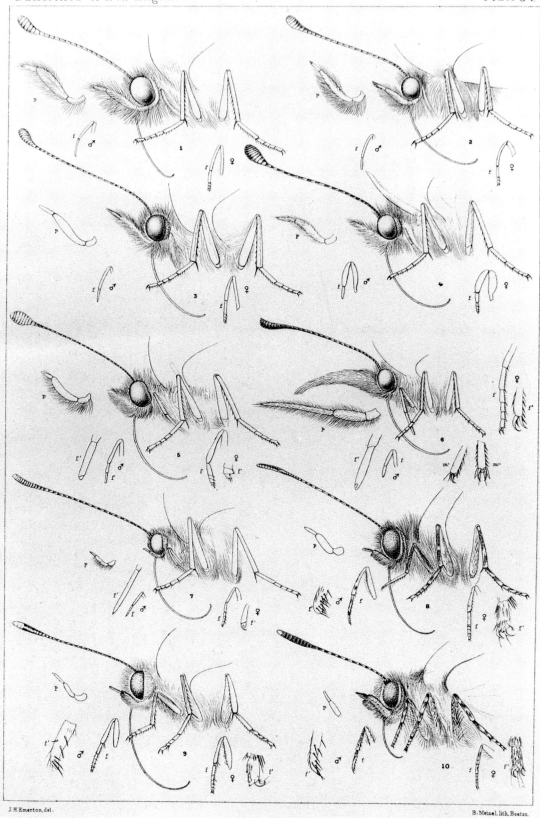

J.H.Emerton,del.

B. Meisel, lith, Boston.

7.32 Eyes, tongue, antennae, palpi, and legs of Imago. Nymphalidae and Lycaenidae. Lithograph by B. Meisel from drawings by J. H. Emerton and Henri Metzger. From Scudder, *Butterflies of the Eastern United States and Canada* (1889).

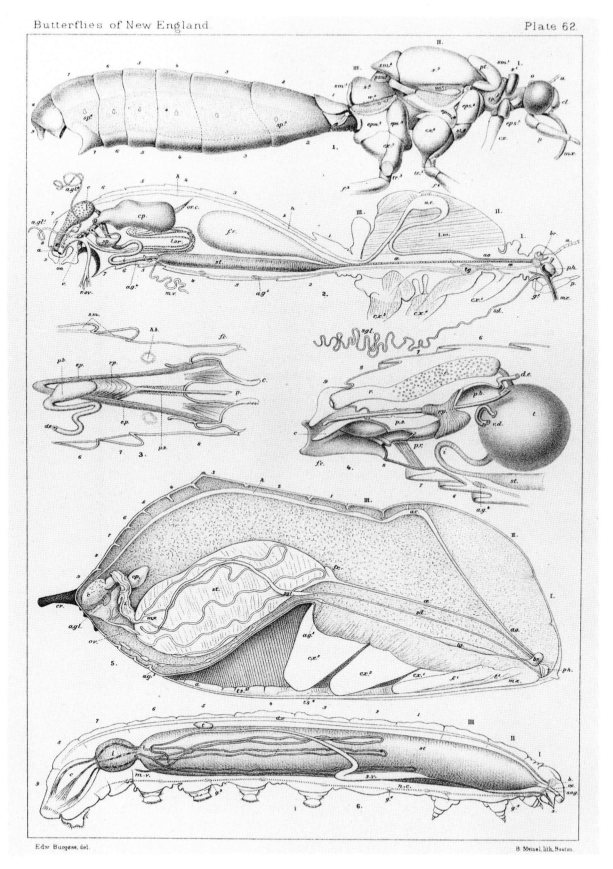

Edw Burgess, del.                                                    B Meisel lith. Boston.

7.33 Anatomy of *Anosia plexippus*. Lithograph by B. Meisel from drawings of dissections made by Edward Burgess and J. H. Emerton. From Scudder, *Butterflies of the Eastern United States and Canada* (1889).

Plate 68.

B Meisel.lith.Boston.

Figure 7.34.

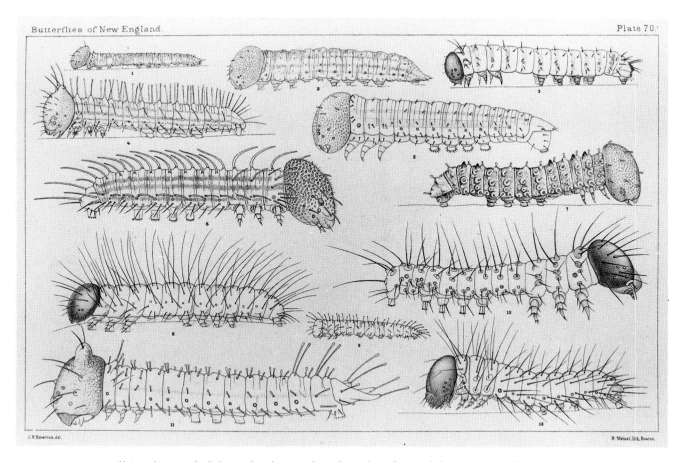

7.35 Caterpillars of Nymphalidae at birth, greatly enlarged. Lithograph by B. Meisel, from drawings by J. H. Emerton, J. H. Blake, and Mary Peart. From Scudder, *Butterflies of the Eastern United States and Canada* (1889).

ure represented a thin section, he noted its thickness. The variety of print media was disciplined or unified by Scudder's insistence on adherence to the conventions of symmetrical plate composition. The work as a whole presents itself as classically ordered, concealing the trouble Scudder went to in achieving his effects. After returning Sinclair's work twice

*On opposite page:*
7.34 Eggs. Lithograph by B. Meisel from drawings of highly magnified eggs by A. Assmann, J. H. Emerton, J. H. Blake, and Mary Peart. From Scudder, *Butterflies of the Eastern United States and Canada* (1889).

for revision, Scudder accepted a lower bid for the remaining work from Julius Bien, who had printed the lithographic edition of Audubon's birds. Months later, however, Bien admitted that he could not print the plates, forcing Scudder to find yet a third firm to complete the job. To compound Scudder's problems, the photogravure company also submitted unacceptable work.[74] Diversity in printing only increased the necessity for constant vigilance on the part of the author.

Like Scudder's plates, the illustrations in the joint work of zoologist Addison E. Verrill, also a former student of Louis Agassiz's, and illustrator James H. Emerton reflect in part the va-

305

riety of approaches in Verrill's study of the organism, in part the borrowing of figures from previous publications, and in part the diversity of print media reproducing the original drawings. Unlike Scudder's work, that of Verrill and Emerton produces a hodgepodge effect; but its very heterogeneity offers an entry into the changes in practice that influenced the changing conventions in zoological illustration.

In 1879 and 1880 Verrill published a study of the cephalopods—squids and octopuses—of the northeastern coast of America, illustrated mostly with Emerton's drawings.[75] Verrill's discussion of the species included both amateur reports and his own extensive field collecting and studies of behavior. He provided at once a historical review of the literature on the groups, and an updated taxonomic diagnosis. His descriptions included details of his study of the animals' development and life cycles. There were also, mixed within Verrill's scientific analysis of the giant squid species, elements of the long tradition of popular fascination with the animal; he reminded his audience that Victor Hugo's description of the "Poulpe" was entirely fantastic, confounding the octopus and the polyp.

Verrill also felt compelled to distinguish his method of obtaining illustrations from those of the illustrated popular accounts. A note on *Architeuthis Harveyi*, a squid species attaining the impressive length of over thirty feet, explained the steps through which Verrill constructed his illustration of the animal. First Paulus Roetter, illustrator at the Museum of Comparative Zoology, had made a drawing based on photographs of a specimen caught in a net off Newfoundland in 1873. After the specimen, badly mutilated by its struggle with the net, arrived at the museum, Verrill corrected the drawing and his own diagnosis of the species by comparing the specimen with the photographs and with the original measurements. In this way, he produced a pictorial reconstruction of the animal, which he distinguished from other wood engravings made from the

two original photographs "published in several magazines and newspapers, but . . . engraved with too little attention to details to be of much use in the discrimination of specific differences."[76] His picture, "prepared . . . with the greatest care possible," illustrated the taxonomic identification.[77]

Emerton's illustrations for Verrill's report contrast markedly with the work he did for Hall and the New York Survey. For Hall, Emerton drew fossil specimens in the midcentury style for reproduction in crayon lithography. For Verrill, Emerton's drawing techniques matched the diversity of available print media as well as the diversity of Verrill's approaches to description, reflecting both fieldwork and laboratory study, and including microscopic examinations. Like Dean's photographs of mammalian brain sections in 1862, Emerton's schematic rendering of development was depicted within a circle, to represent the microscope's field of vision. When Emerton drew several of the smaller species from specimens kept alive in aquaria, he portrayed their characteristic motions. Other illustrations provide a vivid account of handling the dead specimen in the laboratory. One plate shows the giant squid suspended on a bar over a large dish, the tentacles draped and looped over the bar to exhibit the arrangement and number of the suckers. For an illustration to accompany a standard taxonomic description of an octopus, Emerton drew the animal apparently in a jar of alcohol, its arms arranged to display their characteristic features.

Emerton's drawing technique corresponded to illustrative purpose. The plates illustrating the whole animal and corresponding to species descriptions portrayed contour, either shaded in crayon lithography or drawn in line and stipple and reproduced by photolithography; line and stipple photolithographs predominated over crayon lithographs. For illustrations of development and internal anatomy and in contrast to the contour drawings for the taxonomic descriptions, Emerton used line and stipple,

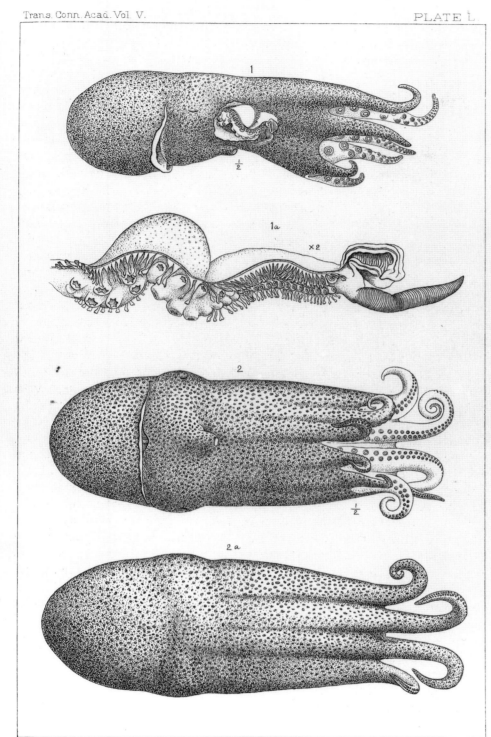

J H Emerton from nature.

Photo Lith Punderson&Crisand New Haven Ct.

ALLOPOSUS MOLLIS, VERRII.

*At left and on following pages*:

7.36–40
Squid and octopus.
Photolithographs
and lithographs by
the New Haven firm
of Punderson and
Crisand, from
drawings by
J. H. Emerton,
for Addison
Emory Verrill,
*Transactions of
the Connecticut
Academy of Science*
5 (1879–1882).

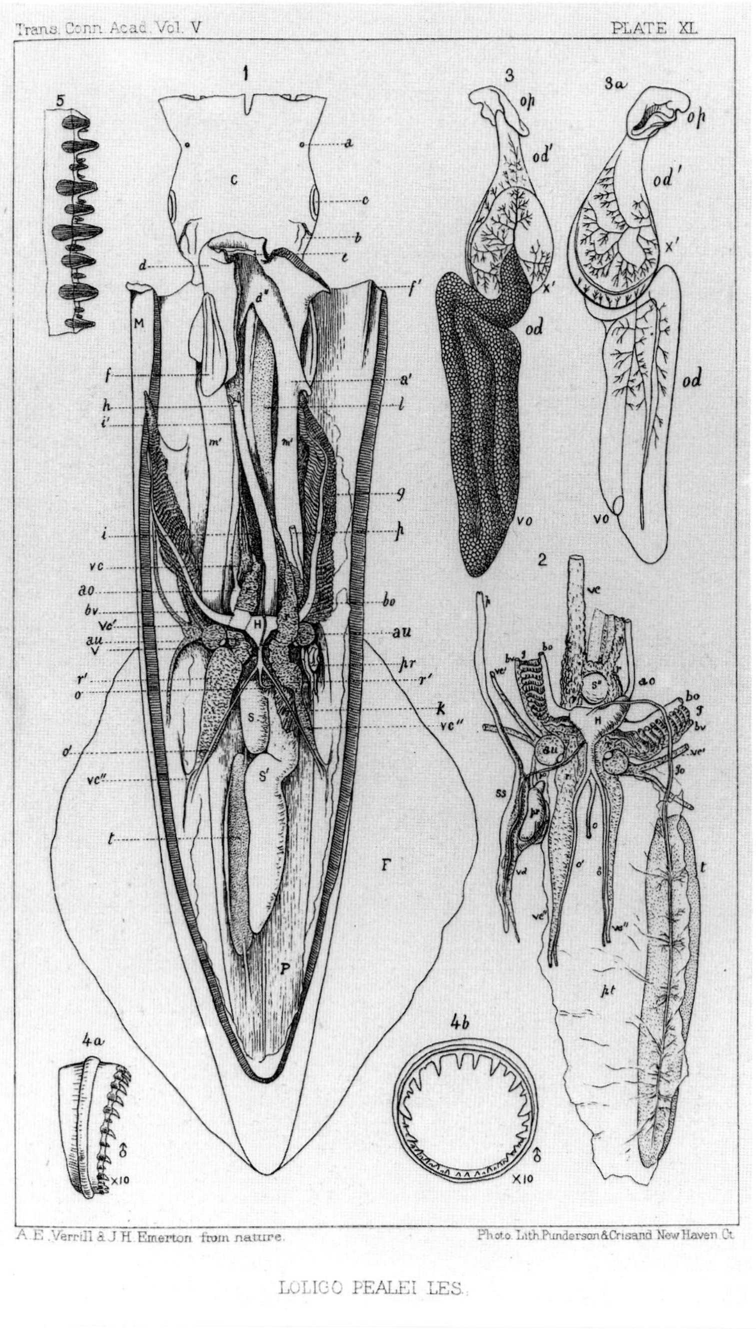

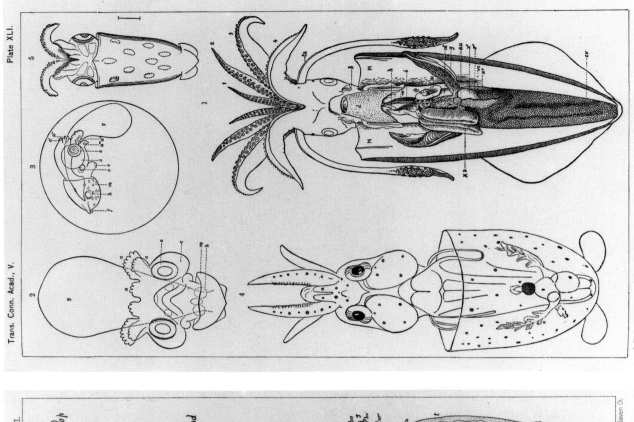

Trans. Conn. Acad. Vol. V
PLATE XL
A. E. Verrill & J. H. Emerton from nature.
Photo. Lith. Punderson & Crisand New Haven Ct.
LOLIGO PEALEI LES.

Figure 7.37.

Trans. Conn. Acad., V.
Plate XLI.
J. H. Emerton and A. E. Verrill, from nature.
Loligo Pealei Les.

Figure 7.38.

Figure 7.40.

Figure 7.39.

but predominately line. This technique simplified and flattened the animals' organs and bodies. In the cut-away views, parallel hatching indicating the body wall emulated centuries-old conventions of technological illustration.[78] Emerton unified these techniques by giving both the contoured and the more schematic drawings a rigid symmetry. In one plate he aligned the octopus figures in profile, neatly one above the other, arms tightly curled. In the plate of the bottled octopus, the arrangement of the arms invoked the preserved specimen of museum taxonomy, but Emerton brought into alignment the constraints of the straight-sided glass jar with the most basic of the compositional constraints, the rectangular page. In the drawings of developing squids, Emerton's reduction of features to a minimum and his emphasis on symmetry transformed the fluid bodies of the animals into schematic models. In contrast to Emerton's effort to evoke a live animal in the plates corresponding to the descriptions, the drawings illustrating development subordinated verisimilitude to legibility and rational arrangement of the organs and layers of tissue.

What, if anything, did the range of Emerton's illustrations for Verrill's study indicate? At one extreme, he drew the messy arrangement of the giant squid over a dish; at the other, he reduced the young animals to schemata. Emerton's almost playful recording of laboratory practices in some of his cephalopod illustrations testifies to the fact that those practices, although taken for granted among professionals, authenticated the illustration of a scientific document. The scientific context of Verrill's and Emerton's work—laboratory study and publication in a scientific journal—compensated for the increasing abbreviation of their pictorial description.

FIGURES 7.36–40

Many of the conventions developed in the context of the decades-long predominance of crayon lithography were no longer practicable in a competitive industry. Despite the vast and changing array of photomechanical printing processes, most of them had one thing in common: they reproduced line work best. So, to line work zoological illustrators turned, constructing, during the last quarter of the century, the basis for a new style of taxonomic illustration. Line and stipple drawings maintained their clarity when reproduced in photo-lithography. Although stippling—dots massed or dispersed to render relief—was laborious, stipple drawings were prepared oversize and then reduced photomechanically in reproduction. Not only did stipple maintain clarity when reduced, but reduction intensified the effect of the stippling to approach the effect of crayon lithography. Increasingly, as in Emerton's work, a continuous outline bounded the stippled figure. While the outline may have represented the use of the camera lucida, it also provided a clear definition of edge and contained the areas of dots. As stippled illustration began to predominate in the literature, the outline began to develop into a convention in its own right. Indeed, many drawings dispensed with stippling and the illusion of contour and consisted of a series of outlines of internal parts, as in Emerton's drawings of spider genitalia that illustrated his own work.

FIGURES 7.41–44

The emerging linear conventions were redefining pictorial description for taxonomic illustration. Increasingly, for taxonomic analysis, the whole animal was reduced to its diagnostic parts; the representation of the animal was reduced to outline. The diagnostic approach to taxonomic illustration merged with the shift in printing technology to line media, and together they contributed to the schematization of the animal in zoological illustration.

The schematization of the animal emphasized diagnostic and analytical disciplinary practice. There may have been, for Emerton and other illustrators working in the new manner, a connection between symmetry, simplification and schematization, and scientific objectivity. Schematic rendering also brought zoology into congruence with other sciences. Since midcentury, zoological illustration had shared conventions with technical illustration, among them the single light source coming from the

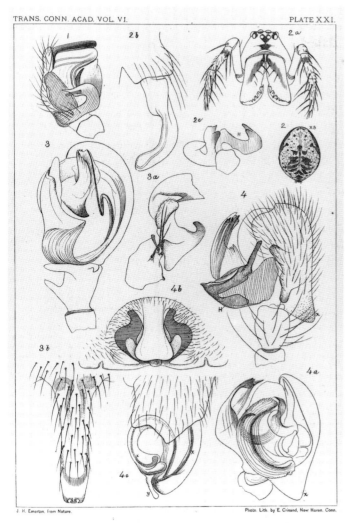

TRANS. CONN. ACAD. VOL. VI.    PLATE XXI.

J. H. Emerton, from Nature.   Photo. Lith. by E. Crisand, New Haven. Conn.

DRAPETISCA, &c.

7.41 Spider genitalia. Photolithograph by E. Crisand, from drawings by J. H. Emerton, for his study of New England spiders, in *Transactions of the Connecticut Academy of Science* 6 (1882).

upper left-hand corner at a forty-five-degree angle.[79] Emerton's use of striation to indicate the cutaway body wall also emulated technical drawing. Such an overlap of illustration conventions heightened connotations of exactitude and measurement in zoological pictures. These values perhaps compensated for reduction of detail to schemata and created the basis for new criteria of "accuracy." Drawings such

as Emerton's developing cephalopods or spider genitalia lacked the kind of detail conveyed in crayon lithography or the photo-gelatin media on which Alexander Agassiz founded his standard of photographic accuracy in the early 1870s. Yet Verrill asserted that his illustrations conveyed the requisite detail for science. Agassiz wanted comprehensive detail; Verrill, selective detail.

Initially, photography had seemed to promise scientific authors comprehensive detail; its influence on print technology reinforced a standard of selective detail.[80] For microscopists, the tension between the two was played out in an acute consciousness of the processes of observation and representation in science; they came down on the side of the selective powers of the eye and hand to control the comprehensiveness of the camera lens.

Advances in optics and chemistry, and in industrial technology, contributed to rapid developments in microscopy. Improved lenses and staining techniques, more precise microtomes for serial sectioning, and milling machines for precision threads and attachments, opened new realms of scale for investigation. Yet the microscope as a tool for observation had limitations that made users aware of their own visual limitations and of their psychological projections and interpretations. From microscopists came a critical discussion of seeing and representing. Their awareness of the deceptions of vision, the tricks of the lens, and the traps of optical artifacts contrasted markedly to the predominant unspoken assumptions of picturing for science, especially in the almost universal assumption of the transparency of photography. Microscopists were aware that their perceptions depended on myriad factors: the quality of their equipment; their skill in preparing their material; their experience in differentiating optical from biological phenomena; and their understanding of the effects of staining media on different tissues.[81] Specialists urged that all microscropists should know the laws of optics. Nevertheless, published misinterpretations of optical artifacts caused controversy and some

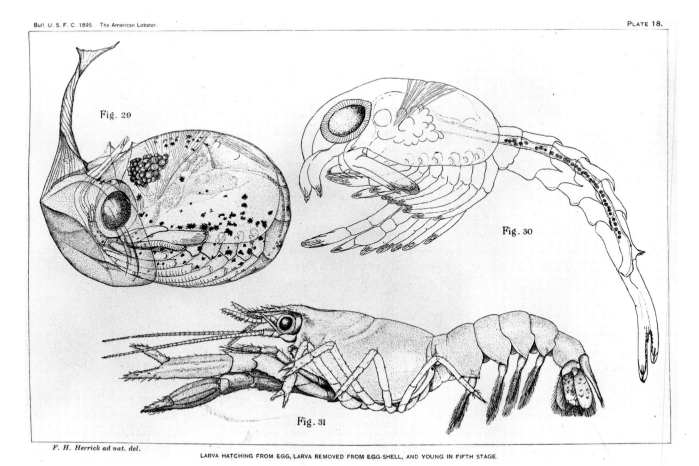

Fig. 29

Fig. 30

Fig. 31

F. H. Herrick ad nat. del.

LARVA HATCHING FROM EGG, LARVA REMOVED FROM EGG-SHELL, AND YOUNG IN FIFTH STAGE.

7.42  Lobster larvae. Photomechanical reproduction of line-and-stipple drawings by Francis Hobart Herrick for his *The American Lobster* (1895).

embarrassment to the profession.[82] In the first part of the nineteenth century, many anatomists distrusted the distorting apparatus of the compound microscope. As practitioners learned to distinguish optical artifacts from biological phenomena, they also spoke and wrote with some skepticism about the process of scientific observation.

The critical approach of the miscroscopists entered zoology through teaching and popular manuals of microscopy.[83] In 1892 Simon Henry Gage, author of one of the most popular teaching manuals of microscopy, reminded his readers that despite the "greatness of the microscope, and the truly splendid service it has ren-

dered . . . it is after all, only an aid to the eye of the observer, only a means of getting a larger image on the retina than would be possible without it."[84] Gage warned that the interpretation of the retinal image depended on "the character and training of the . . . brain behind the eye."[85] The increasing precision of the instruments meant nothing if observers did not correctly interpret what they saw through the lens.

Microscopists emphasized the importance of drawing in the service of observation. Because they relied on drawing for recording and organizing the solitary and piecemeal act of observation, they continued to rely on lithography to

FIGURE
7.45

312

transcribe their observations into print. One British microscopist, Lionel Beale, whose manual enjoyed extensive circulation in the United States—Alexander Agassiz owned the 1869 edition—included instructions for preparing woodblocks for drawing and engraving and even for drawing on lithographic stones directly from the microscope. Beale anticipated readers' objections that a draftsman should make the drawings rather than the microscopist himself by countering that the draftsman would have to become a microscopist to know what to draw: "It will, I know, be said that these processes, above described [wood engraving, lithography, tracing], are of a nature which any intelligent draughtsman can perform, and hardly worth the labour which a microscopical observer, who wishes to carry them out, must be content to bestow."[86] But, he warned, "It is . . . impossible to obtain a good representation of any microscopic object without long and careful study, as it is to produce any other object in nature; and surely it is hard to expect a draughtsman, who is engaged in copying various subjects to spend hours in looking at specimens through the microscope, observing things which he neither knows nor perhaps desires to know anything about."[87]

Microscopists' statements about observation in their instruction manuals, in speeches to their societies, and in their journal articles reiterated what had become unspoken truisms in science at large. They were motivated to express these truths and to discuss seeing in science as other biologists were not. Manuals of microscopy, revised every few years to incorporate innovations, were used as college texts and by amateurs, and thus served as introductions to the field; this broader readership may have prompted the cautionary passages. Along with instructions for staining and making sections and with descriptions of the advantages of direct and indirect lighting and other methods and techniques, the manuals included introductions into the basic principles of scientific illustration. Zoological illustrators were typi-

cally either recruited from other drawing professions or were trained on the job. Zoologists who needed to learn to draw had available in the latter half of the nineteenth century an increasing number of popular manuals for sketching landscapes from nature. But only in the literature of microscopy did one find expressed the professional consensus of how and what to draw for the science. Problems of magnification and relative scale, rules for drawing directly from the object, suggestions for the use of the camera lucida and for devising different and consistent marks for different tissues received detailed description. Despite advances in photography through the lens, for most microscopists communication of their observations depended on their drawing skills.

The microscopists addressed repeated warnings on seeing for science not only to the students for whom they wrote textbooks, but also to their professional colleagues. At the 1883 meeting of the American Association of Microscopists, the president spoke on "The Verification of Microscopic Observations."[88] "The votary of the tube must master a technique complex and difficult."[89] Even if the microscopist knew the laws of optics, the nature of prisms and mirrors, and refraction and polarization, the eye could be deceived. Even with improved techniques in bleaching, staining, cutting, and mounting, or with knowledge of the chemical effects of media, the observer must take heed: "Things seen must not too readily be taken to be the . . . realities. The eye . . . needs more than the other senses, to be trained to see aright."[90] As a first step in verification, the speaker recommended verification of one's observations by others. "Even the well trained eye is in danger of projecting the mental preconceptions . . . into the focal plane of the objective, and seeing . . . not what is really there, but what some theory demands shall be." A second opinion could help dissolve these spectral appearances.[91]

Drawing could confirm observation. Drawing with the aid of the camera lucida preserved

313

GOODE AND BEAN.—OCEANIC ICHTHYOLOGY.

PLATE LVI.

206. STEPHANOBERYX GILLII. (p. 187.)     207. TRACHICHTHYS DARWINII. (p. 188.)
208. HOPLOSTETHUS MEDITERRANEUS. (p. 189.)

Figure 7.43.

PLATE LXXVIII.

272. Lycodes Esmarkii. (p. 303.)
274. Lycodes Frigidus. (p. 305.)

273. Lycodes Reticulatus. (p. 305.)
275. Lycodes Mucosus. (p. 306.)

7.43–44 Species of the genera Stephanoberyx and Trochichthys (7.43, *opposite*), and the genus Lycodes (7.44, *above*). Photomechanical relief print reproducing line-and-stipple drawings, from George B. Goode and Tarleton H. Bean, "Oceanic Ichthyology," *Smithsonian Contributions to Knowledge* 31 (1895).

315

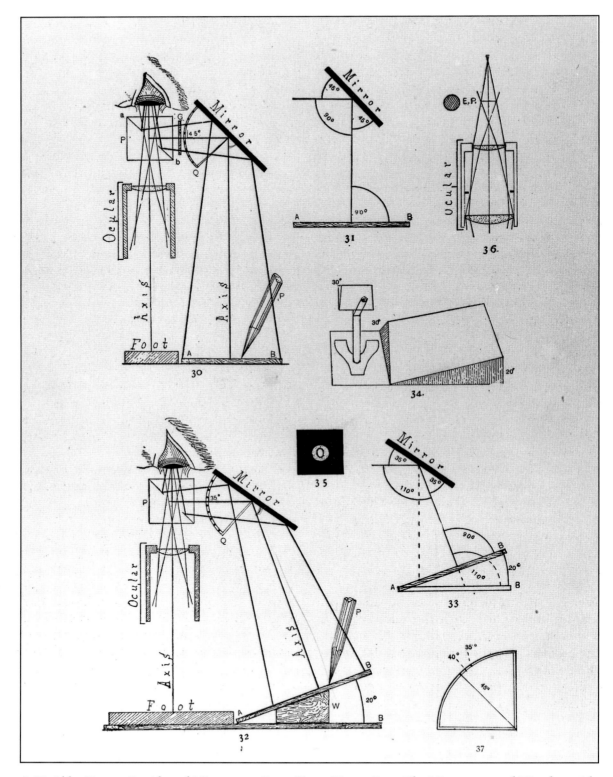

7.45  Abbe Camera Lucida and Microscope. From Simon Henry Gage, *The Microscope and Histology*, 4th ed. (1892).

a record of the observation but, more important, "by the very act of drawing, our attention is quickened and our recollection made more clear."[92] And although photomicrography produced "a record that is almost entirely free from the fallibility inherent in mere drawing," microscopists, having learned caution from one kind of lens, were more skeptical than their contemporaries about the truthfulness of photography.[93]

The speaker's qualifications of the powers of photography in microscopy contrasted to the confidence of earlier commentators on photography, who had taken the view that the camera, replacing the human eye and hand with the processes of physics and chemistry, provided privileged scientific records. The author of the article on photography in the 1870 edition of *The New American Cyclopedia* believed that the camera could serve not only as a means of representing human observations but also as an observer of things otherwise inaccessible. He anticipated a dual role for photography in science, "the instant recording of ephemeral phenomena," and the representation of objects in detail "unattainable by the hand of man." "The general popularity which photography has . . . attained is founded not only upon the realized perfection with which it can perpetuate external forms, but also upon the anticipated advantages hereafter to accrue from it in several of the higher departments of science."[94] (By "higher" the writer meant pure research, as opposed to scientific research with a commercial or industrial application.) He cited the use of photography to document solar and lunar eclipses and its use with the microscope. In that capacity, according to the article's author, photographs supplied evidence superior to human observation: "In this manner questions of the utmost importance in physiology and the sciences of organization, which have long been

in dispute, have received a final solution, and permanent representations have been obtained of transient phenomena occurring in living organisms."[95]

The 1883 speaker to the Association of Microscopists cautioned that although photography eliminated errors of hand, eye, and judgment, it "has errors of its own."[96] He reminded his audience that the camera cannot focus itself, nor differentiate diffraction images from negative or dioptric ones. "Hence its own record needs careful interpretation."[97] Nor could the camera distinguish details near the limits of visibility. Still, a photograph could preserve a record so that later, when science asked different questions about the subject, a photograph would show what a drawing, as a deliberate selection of detail, might have missed. And where the human eye reached its own limits, especially in the twilight of high magnifications, although photographs "may not be absolute proofs," long camera exposures could record what the eye could not see.[98] Here, the speaker moved beyond a comparison and contrast between the advantages of drawing versus photography, into a realm of technology-dependent perception.

Photography may have offered new possibilities in extending vision, but the speaker put drawing first: "A drawing . . . serves . . . a double purpose. It preserves for us a transcript of what we saw . . . and it serves as a ready means of interchange. . . with others, outweighing, often, many a page of description."[99] Perhaps by 1883 such an assertion seemed an overstatement of the obvious, but more likely, it reaffirmed the value of drawing in what was becoming a mechanized and photographic age. Moreover, although the conventions of pictorial transcription were subject to change, a drawing's role in interchange and description remained central to scientific practice.

CHAPTER

$8$

# The Zoologist's Province

IN the last decades of the nineteenth century, the prose and pictorial conventions of professional zoology and popular natural history seemed irrevocably segregated. The popular literature celebrated the observer's emotional relationship with nature. The heroic ideal of the artist-naturalist had been thoroughly domesticated. All men, women, and children who loved nature could participate in the deeply satisfying activity of observing animals and plants in city parks, at the seashore, in the garden, or along country lanes; sketching from nature enriched the enjoyment of observation. Many popular works emphasized the local; their illustrations depicted common species going about their daily business. On the other hand, initiation to the profession required formal education capped by study in Europe, training in museum or laboratory technique, adoption of an impersonal ethos. Animals in most zoological plates looked different from other pictures of animals; anatomical parts rather than the whole animal were becoming the subject of illustration. The viewer needed special knowledge to read such pictures. Even so, the boundaries between the two realms were not as firm as some professionals might have wished.

Professional practice endorsed one element of the naturalist-illustrator tradition; drawing remained identified with observation. But institutional hierarchy distinguished zoologist from illustrator, although the trend toward

schematic representation in illustration made mastery of the appropriate techniques accessible to more zoologists. Moreover, photomechanical reproduction meant that in many instances zoologists no longer required an illustrator to transform working drawings into publishable plates. Where illustrators were retained, the new conventions limited the range of permissible drawing techniques. Lithographic plates such as those published by Alexander Agassiz continued to appear, but they also followed the trend toward a flattening of the animal.

The illustrations in Alfred Goldsborough Mayer's monograph, *Medusae of the World*, published in 1910, exemplified these professional developments.[1] Mayer, who undertook his study at the suggestion of Alexander Agassiz, had his own drawings from living specimens reproduced in chromolithography by A. Hoen and Company of Baltimore. Mayer clearly modeled his own illustrations on Sonrel's medusae plates for Louis Agassiz, but Mayer's treatment of his subject differed markedly from Sonrel's. Sonrel rendered his subjects as rounded forms; the later plates depict the animals as flat designs. To the staff illustrator at the Tortugas Marine Laboratory of the Carnegie Institution, Mayer assigned the task of tracing previously published illustrations for reproduction in line as text figures.[2]

The ever-growing literature of popular natural history, on the other hand, continued to

PLATE

65

318

meet and cultivate a taste for prose and pictorial description of the whole animal in nature. First-person narratives placed the author, often the author-illustrator, at the center of the description. The works of William Hamilton Gibson (1850–1896), writer and illustrator of popular natural history, exemplified this fashion.[3] Gibson combined descriptions of rural characters with descriptions of natural history and landscape and produced illustrations that integrated text and image into a single narrative.[4] In contrast to the standardization of line and stipple style in specialized zoological illustrations, popular natural history illustration like Gibson's took full advantage of type-compatible line media. The illustrations of many popular works reproduced lively pen work with a fresh and informal quality, as if sketched rapidly on the spot. Gibson also called attention to the rich variety of engraving styles used to interpret his drawings for reproduction; in one of his books, he identified the engraver of each illustration.[5] Among them was Henry Marsh, whose engravings had been the inspiration of Gibson's youth.

FIGURE 8.1

Illustrations characterized by movement and by the reinsertion of the observer into the textual and pictorial narrative shared few of the conventions of zoological illustration, and vice versa. Some zoologists, however, crossed the disciplinary boundaries. Henry Christopher McCook, Presbyterian minister, entomologist, and vice-president of the Philadelphia Academy of Natural Sciences, wrote in both genres. McCook's popular work of 1885, *Tenants of an Old Farm*, combines natural history with rural nostalgia, complete with a dialect-speaking Mammy. The volume includes the natural history illustrations of Edward Shepard and Frank Stout, and humorous interpretations—cartoons—of the insect life cycle by Daniel Beard. The interpretive sketching style and strong narrative element that predominate in McCook's natural history-fiction also runs through the line drawings illustrating his treatise on orb-weaving spiders, *American Spiders*

*and Their Spinning Work*, published in 1889.[6] An institutional imprint prevented the narrative quality of McCook's illustrations from discrediting the scholarly content of the work. Similarly, publication in a federal fisheries report gave official endorsement to Henry Wood Elliott's distinctly idiosyncratic portraits of fur seals, drawn in Alaska in 1873 as part of a government study.[7]

FIGURES 8.2–6

Ornithology, perhaps more than any other field, was under constant pressure from its popular hinterland to accommodate the old natural history values. The spirit of Wilson and Audubon animated the ever-growing ranks of amateur bird-watchers and devotés. A flourishing popular literature, to which women were important contributors, offered its readership prose and pictorial narrative description of bird life.[8] In the late decades of the nineteenth century, scientific ornithology—an exclusively male preserve—centered on museum study and epitomized the movement toward strictly codified description and the tight control of illustration convention. Louis Agassiz's former student, Joel Asaph Allen, oversaw the professional boundaries of description and debate as chief editor of *The Auk*, the journal of the American Ornithologists' Union (A.O.U.), founded in 1883. Competition over claims of true appreciation of birds existed between what Allen called the "science" and the "sentiment" of ornithology.[9] Vigilance against sentiment associated with the popular literature was all the more necessary because professional ornithology included many sophisticated amateurs, some of whom held semiofficial positions in major museums, to which they donated valuable collections.

Many contributions to *The Auk* consisted largely of narratives of ornithologists' expeditions in search of rare or interesting species; the male authorship, descriptions of hunting and shooting, and the emphasis on collecting made these essays acceptable to the professional literature. Much of the popular literature, in contrast, emphasized the domestic side

319

Plate 65. *Cyanea capillata* var. *versicolor*. Chromolithograph by A. Hoen and Co. of a drawing from life by Alfred Goldsborough Mayer, for *Medusae of the World*, vol. 3 (1910).

PLATE XX.

Plate 66. Finches. Hand-colored lithograph from Baird, Brewer and Ridgway, *A History of North American Birds: Land Birds* (1874).

victim securely, and held him to await assistance. It came. The entire neighborhood had been apprised of the battle, and in less than five minutes the ground swarmed with an army of re-enforcements. They came from all directions; they pitched upon that hornet with terrible ferocity, and his complete destruction was now only a question of moments. I experienced a sort of malevolent delight at such a fitting expiation for a life of rapine and murder. Already a dozen pairs of teeth were working at the joints of his wings, and those members had soon been severed from the body had I left him to his fate; but there was a problem of engineering skill connected with his capture which I wished to solve, and I concluded to come to his rescue, and

STRATEGY VERSUS STRENGTH.

even spare his life if need be, in an interesting experiment. I therefore dislodged all the ants excepting the two original assailants. The overwhelming attack upon the hornet had made him furious, but these pugnacious little fellows were even now more than his match, and still held him as before. No sooner, however, did I remove from their grasp those

8.1 Strategy versus strength. A page with a reproduction of a black-line engraving by Henry Marsh, from a drawing by William Hamilton Gibson for Gibson's *Highways and Byways* (1882).

of bird behavior and bird-watching by describing birds observable from the country or suburban home and garden. Ornithologists resisted having to pay attention to publications outside the discipline. William Brewster, curator of birds at the Museum of Comparative Zoology, reacted with annoyance when he discovered that Harriet Mann Miller had preempted his priority in the description of a new species. Why, he asked, had she not published it "in some accredited scientific journal," instead of in a popular book with a title like *Little Brothers of the Air*.[10] Miller responded that she, who had no training in scientific ornithology, had only wanted to share with others her delight in birds. The interchange and the tension between amateur birders and ornithologists was reflected in the illustrations of their respective literatures. On the one hand, the dominance of taxonomy in American ornithology introduced the technical style of illustration into works intended for a mixed audience of professionals and advanced amateurs, such as Elliott Coues's *Key to North American Birds* of 1872.[11] On the other hand, the discipline never eliminated vestiges of natural history landscape conventions from its professional illustration.

FIGURES
8.7–9

The illustrations of Baird's protégé, Robert Ridgway (1850–1929), appointed curator of the Department of Ornithology at the United States National Museum in 1880, represented one extreme of the professional stance. Ridgway led the discipline in the standardization of nomenclature. He urged in particular the recognition of geographic varieties through the adoption of trinomial nomenclature.[12] Baird had warmly encouraged Ridgway, an enthusiastic boy-naturalist, by corresponding with him about birds and collecting, and sending him paper, paints, and tips on drawing birds.[13] It was on Baird's invitation that Ridgway had come to Washington. On entering professional life, Ridgway adopted a style of illustration in marked contrast to his youthful watercolors of birds in nature. In the early 1870s, Baird, Thomas Mayo Brewer, and Ridgway collaborated on a compre-

hensive work on American birds, to which the junior author contributed illustrations of bird and plant groups, as well as many drawings of birds' heads.[14] Thereafter, Ridgway's professional publications were characterized by a strict style of taxonomic illustration that depicted not the whole bird but instead focused on the parts or characters defining genera—the head, beak, wing, tail, and feet—reduced to their linear minumum.[15]

PLATES
66–67

FIGURES
8.10–13

Ridgway, a thoroughgoing taxonomist, also contributed to the standardization of ornithological description through his *Nomenclature of Colors for Naturalists* of 1886.[16] Ridgway intended his work to adapt to the needs of zoology, botany, and mineralogy the technical discussion of color and the standardization of color nomenclature for industry, such as Wilhelm von Bezold's *Theory of Color in Its Relation to Art and Art-Industry*, published in an American edition in 1876.[17] He commented that "popular and even technical natural history demands a nomenclature which shall fix a standard for the numerous hues, tints, and shades which are currently adopted, and now form part of the language of descriptive natural history."[18] Ever the systematist, Ridgway noted that the introduction of aniline dyes and pigments had spawned a confusing synonymy: "An inspection of the sample-books of manufacturers of various fancy goods (such as em-

*On following page:*

Plate 67. Woodpeckers. Hand-colored lithograph from Baird, Brewer, and Ridgway, *A History of North American Birds: Land Birds* (1874).

Plate 68. The purples. Hand-colored plate from Ridgway, *Nomenclature of Colors* (1886).

Plate 69. "Illustrations of mimicry between members of the 'Sylvanus' group of the genus Heliconius and various Melinaeas, etc." Color lithograph by B. Meisel from drawings by A. G. Mayer, in Mayer's article, "On Color and Color-Patterns in Moths and Butterflies," *Bulletin of the Museum of Comparative Zoology* 30 (1897).

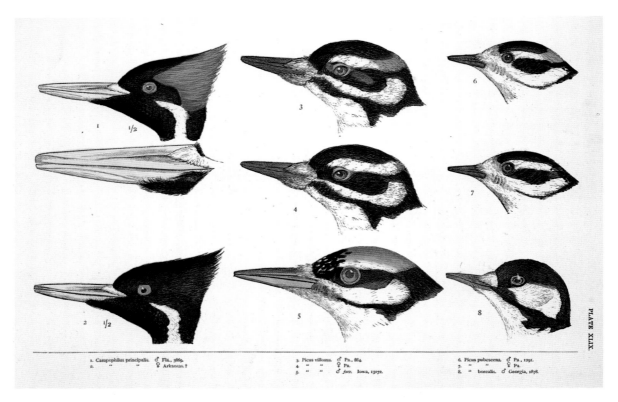

Plate 67.

Plate 68.

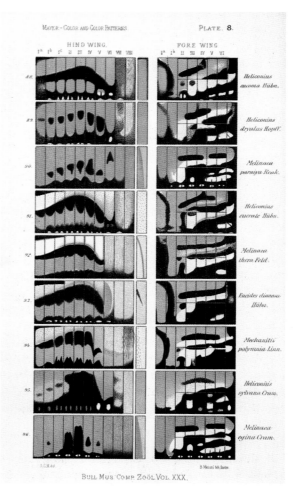

Plate 69.

Plate 70. "Peacock in the Woods." From a painting by Abbot H. Thayer, assisted by Richard S. Meryman. Color lithograph by A. Hoen, Baltimore, from Thayer, *Concealing Coloration in the Animal Kingdom* (1909).

An example or two of moulting as seen in special individuals will serve to define more clearly the above general description. A nearly mature female of Argiope cophinaria was observed (August 19th) in

**Argiope Moulting.** the final stage of moulting. When first seen she was suspended head downward to the central shield of her snare, as represented in Fig. 58. The cephalothorax had already escaped from the shell, and the dorsal part of the moult still clung to the pedicle and stood straight out at right angles to the body. The abdomen was just ready to escape, and, indeed, slipped out of the shell as I approached, and the skin lay in a rumpled mass at the end of the thread by which the creature was suspended.

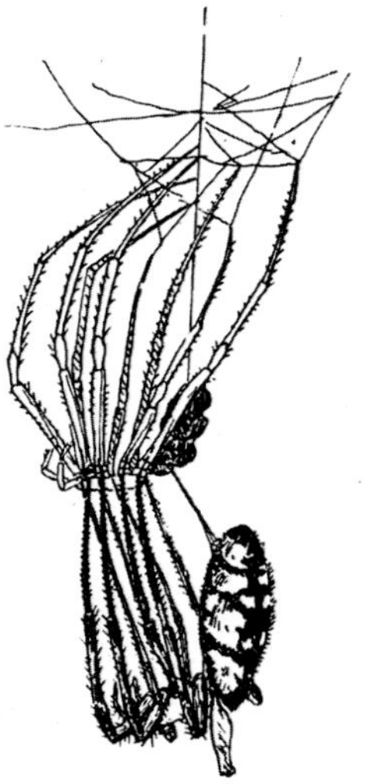

FIG. 58. 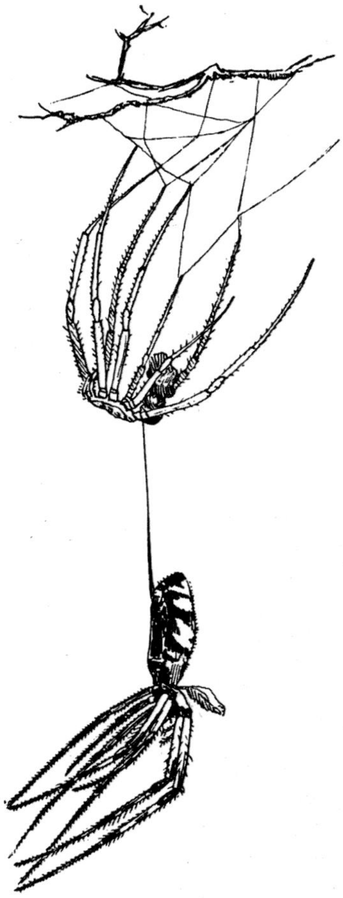          FIG. 59. 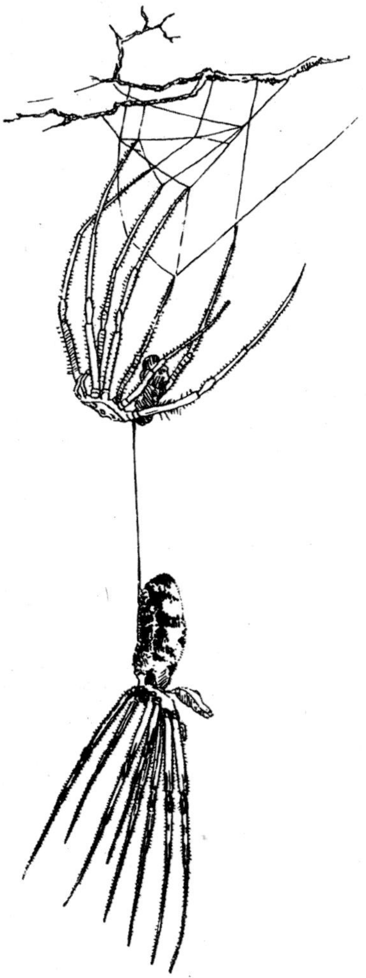                    FIG. 60.

FIG. 58. Argiope in the last act of moulting. FIGS. 59 and 60. Argiope stretching her legs just after moulting.

The body was bent upwards in a horseshoe shape, and the legs were partly freed from their moult. A few paroxysms occurred by which the legs were forced further and yet further out of the skin; then, first escaped the first pair, then, in a very brief space thereafter, the two second legs; immediately the third pair followed, and in brief succession the fourth pair. The spider's body dropped downward, and she stretched herself as

Figure 8.2.

8.2–5  Pages from McCook, *American Spiders and Their Spinning Work* (1889).

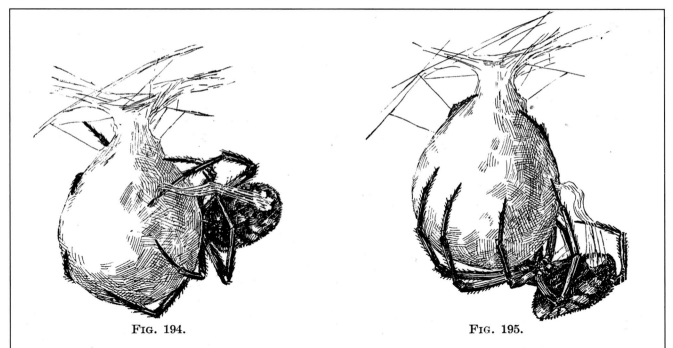

FIG. 194.                    FIG. 195.

FIGS. 194, 195.   The action of Argiope in drawing out silk with the spinning legs.   The alternation of the legs appears by comparing the figures.

8.2–3  Scenes from the life cycle of Argiope.

FIG. 275.   Ballooning Lycosids ascending from a fence post, and floating before the wind.

8.4  Ballooning Lycosids.

PLATE II

Plate 71. "Male Ruffed Grouse in the Forest." From a painting by Gerald H. Thayer. Color lithograph by A. Hoen, Baltimore, from Thayer, *Concealing Coloration in the Animal Kingdom* (1909).

PLATE IV

Plate 72. "Male Wood Duck on Shallow Water," from a sketch by Richard S. Meryman; and "Male Wood Duck in a Forest Pool," from a painting by Abbot H. Thayer, assisted by Richard S. Meryman. Color lithograph by A. Hoen, Baltimore, from Thayer, *Concealing Coloration in the Animal Kingdom* (1909).

FIG. 303 (upper figure).  A Tarantula rampant, just before striking.
FIG. 304 (lower figure).  Tarantula in act of striking.

8.5  Tarantulas in action.

THE COUNTENANCE OF CALLORHINUS.—A LIFE STUDY OF AN ADULT MALE FUR-SEAL.
(Full face of old male, profile and under view of female heads.)
Drawing by Henry W. Elliott, North Rookery, Pribylov Group, July 5, 1873. (p. 75.)

PLATE 26.

8.6 Life-study of an adult male fur seal. Reproduction of a drawing by Henry W. Elliott, 1873, in George B. Goode, *The Fisheries and Fishery Industries of the United States* (1884–1887).

broidery silks and crewels) is sufficient to show the absolute want of system of classification which prevails, thus rendering these names peculiarly unavailable for the purposes of science, where absolute fixity of the nomenclature is even more necessary than its simplification."[19]

Ridgway found, however, that colors, like birds, did not submit readily to a single linear system. His experience with ornithological taxonomy helped him find the solution, albeit one with limitations similar again to zoological taxonomy: "In order to make comparison of allied shades and tints more easy, it has been endeavored to place all belonging to a particular class together on one plate; but in not a few cases it has been a difficult matter to decide

PLATE
68

PLATE XII

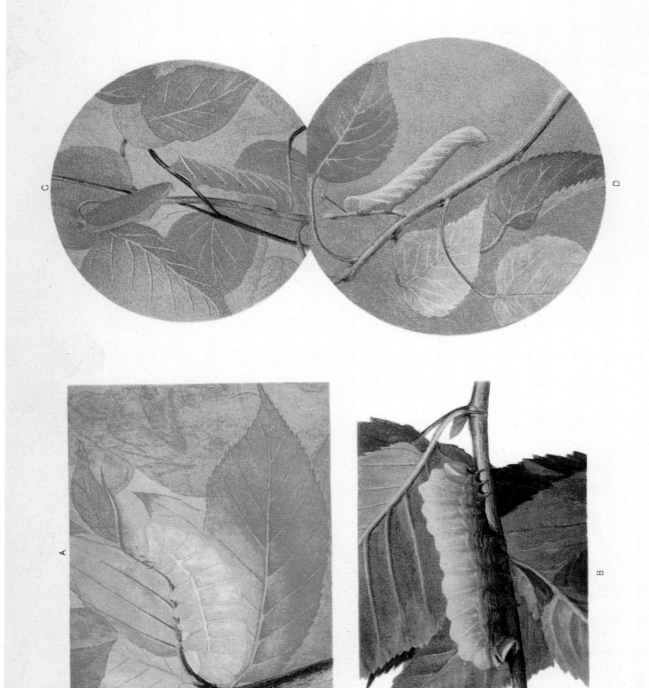

Plate 74. "White Flamingoes, In morning or evening sunlight." "Red Flamingoes, In morning or evening sunlight." "The Skies they Simulate. The bright parts consisting of the very same color-notes as those in the sketch of the birds themselves." Reproduced from a painting by Abbot H. Thayer. Color lithograph by A. Hoen, Baltimore, from Thayer, *Concealing Coloration in the Animal Kingdom* (1909).

*On opposite page:*

Plate 73. Luna and Sphinx Moth caterpillars, upright and inverted, to demonstrate their concealing coloration. Figs. A, C, and D, painted from life by Gerald H. Thayer, the leaves painted by M. E. and E. B. Thayer. Fig. B, painted from life by Louis A. Fuertes. Color lithograph by A. Hoen, Baltimore, from Thayer, *Concealing Coloration in the Animal Kingdom* (1909).

333

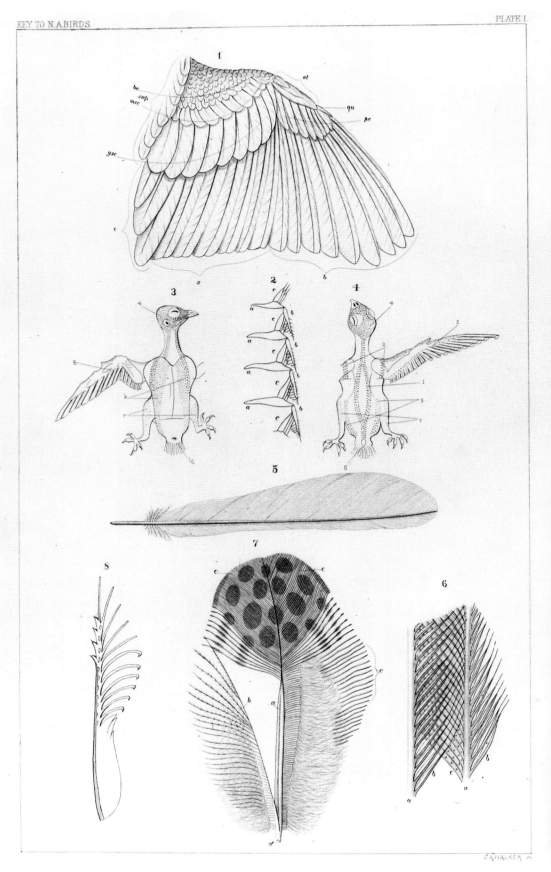

Figure 8.7.

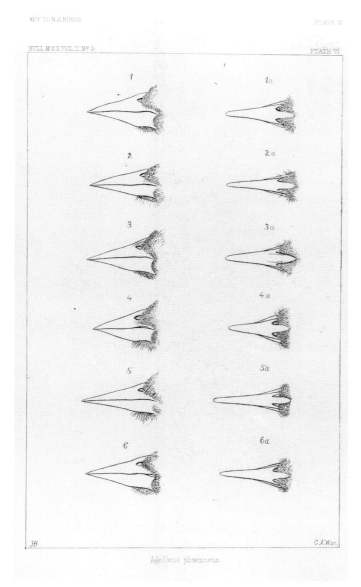

8.7–8 Bird anatomy and variations of beaks. Steel-engraved frontispiece and plate from Coues, *Key to North American Birds* (1872).

upon which plate a certain one should be put, the decision being in some instances almost purely arbitrary."[20]

One of Ridgway's aims in standardizing color nomenclature was to train the rising generation in its systematic application. To assist the beginner in the rudiments of proper taxonomic practice, he included with his presentation of color terminology an introduction to ornithological terms and a diagram of bird anatomy. Ridgway's book provided a sort of kit. Added together, accurate color nomenclature and anatomical terminology, correct use of the tables of standard systems of measurement, combined with attributions based on the plates representing feather patterns and egg shapes would result in a professional, standardized ornithological description. Ridgway's color manual addressed the need to groom recruits from the popular phalanx of bird-watchers into ornithologists. A youth intent on entering the discipline had to make a transition from the narrative, descriptive prose and illustration of popular natural history to the formal, diagnostic, and analytical style of taxonomy.[21]

FIGURES
8.14–15

But the large, devoted, and articulate ranks of amateur birders would not accept without dispute the narrow definition of ornithology propounded by the likes of Ridgway. Popular authors and the general public took part in the great debates of the day, such as whether the introduced English sparrow should be eliminated; whether amateur collecting, especially of eggs and nests, should be discouraged; whether legislation should be passed to protect birds from casual hunting and commercial predation for the millinery trade.[22] The groundswell of the bird protection movement could not be ignored.

The A.O.U. shared with the amateurs the goal of bird protection legislation. The professional organization harnessed the strength of the popular bird protection movement through its encouragement of the state Audubon societies organized, largely by women, in the late 1890s.[23] The Audubon societies soon founded their own publication, *Bird Lore*, which published articles by leading ornithologists and by popular writers, along with reports from the state chapters. The official acknowledgment of popular ornithology by the profession in effect reendorsed an illustration style that included

8.9 Great Blue Heron. Wood-engraved text figure after Audubon, from Coues, *Key to North American Birds.*

the elements familiar since the time of Wilson—an emphasis on the living bird in nature and a tendency toward bird portraiture. In exchange for its accommodation of the popular movement and taste, however, the profession exacted—tacitly—the authority to rule on standards of description and illustration.

One way to control the quality of bird illustration in the hybrid literature was to select illustrators for semiofficial endorsement, as the discipline endorsed acceptable women authors for associate membership in the A.O.U. Frank M. Chapman, Allen's assistant at the American Museum of Natural History and as-

sistant editor of *The Auk*, played an important role as intermediary between the professional and popular branches of ornithology. Chapman promoted Ernest Thompson Seton, the Canadian artist-naturalist who wrote for a popular audience. Professional approval both bolstered Seton's reputation and also exercised a certain check on his tendency to anthropomorphize the behavior of wild animals in his storytelling.

The young Louis Agassiz Fuertes (1874–1927), named after but not related to Louis Agassiz, produced pictures of birds that incorporated many of the natural history conventions.[24] Contemporaries called his work bird portraiture. Fuertes launched his career illustrating books by women, Florence Merriam's *A-Birding on a Bronco* of 1896, followed by Mabel Osgood Wright's *Citizen Bird*, a book for children.[25] Merriam's reputation as one of the most vigorous and scientifically correct of the women birders, Wright's status as one of the few women associates of the A.O.U., as well as the professional eminence of her co-author, Elliott Coues, helped Fuertes avoid being designated an illustrator of women's and children's books. Introduced by Coues, Fuertes also attended, participated in, and exhibited his work at the meetings of the A.O.U.[26] Moreover, he made a specialty of depicting birds of prey. The discipline united in hailing Fuertes as the heir of Audubon.

In asserting its authority to rule on standards of illustration, the profession also exercised its authority to dictate the appropriate terms by which a painter could contribute to the discipline. It was one thing to produce bird portraits, but quite another when the painter Abbott Handerson Thayer asserted that his powers of observation and interpretation exceeded those of ornithologists. The controversy over Thayer's theory of concealing coloration engaged influ-

*On following pages:*
8.10–12 Illustrations from Baird, Brewer, and Ridgway, *A History of North American Birds: Land Birds* (1874).

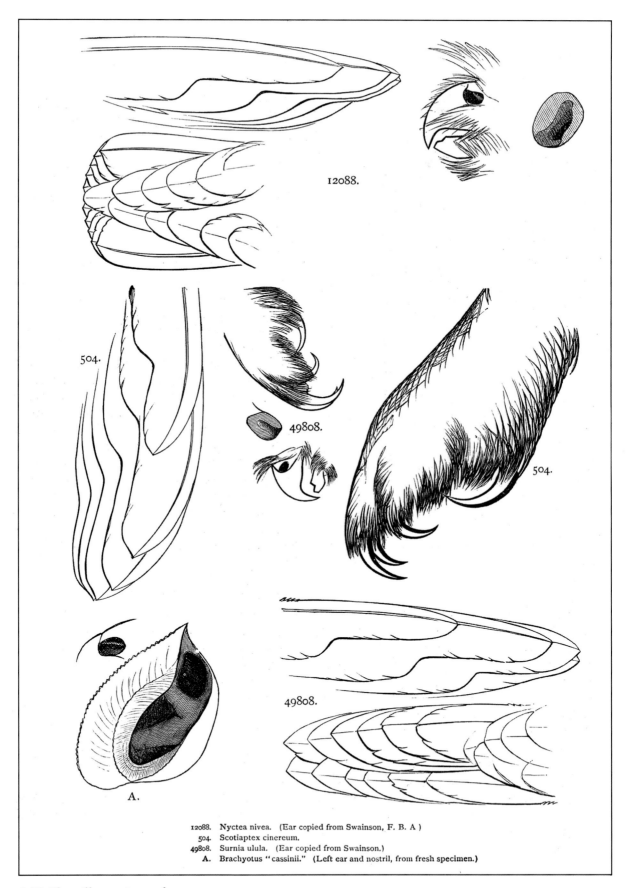

12088.

504.

49808.

504.

49808.

A.

12088.  Nyctea nivea.  (Ear copied from Swainson, F. B. A )
504.  Scotiaptex cinereum.
49808.  Surnia ulula.  (Ear copied from Swainson.)
A.  Brachyotus "cassinii."  (Left ear and nostril, from fresh specimen.)

8.10  Plate illustrating owl anatomy.

*Syrnium cinereum.*

8.11 *Syrnium cinereum.* Wood-engraved text figure.

ential members of the extended turn-of-the-century ornithological community, professionals and amateurs alike. Theodore Roosevelt became Thayer's principal antagonist. Ultimately, the profession rejected Thayer's claims.

In 1896, when Thayer first introduced his experiments and articles on animal countershading, they were well within the acceptable standards of argument, and they were well received in ornithological circles.[27] He demonstrated how the darker markings on the top of an animal's body and paler colors underneath helped obscure the animal when light fell from above and cast the lower parts of the body in shadow. Chapman recalled Thayer's flair for dramatic demonstration. "One rainy day in the spring of

338

*Nyctea scandiaca.*

8.12 *Nyctea scandiaca.* Wood-engraved text figure.

1896, wearing an old suit and rubber boots, Thayer came into my office at the Museum and said, 'Come out into the square; I've something to show you.'"[28] Once outside, Thayer pointed to a patch of ground in shadow and demanded: "'How many decoys do you see?' . . . 'Two,' I replied and described them as brownish, about six inches long and elliptical in shape."[29]

Closer examination revealed "in fact, four decoys . . . the two nearly invisible ones were painted pure white on the lower half; whereas the conspicuous decoys were the same color throughout."[30] When Thayer demonstrated his theory at the A.O.U. meetings in Cambridge that fall, he again convinced his audience. At the same meetings, he met Fuertes and, accord-

Plate CXII.

Man. N. Am. B.

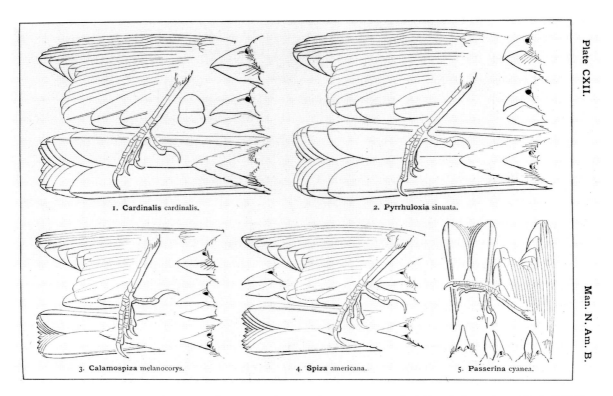

1. **Cardinalis** cardinalis.

2. **Pyrrhuloxia** sinuata.

3. **Calamospiza** melanocorys.

4. **Spiza** americana.

5. **Passerina** cyanea.

8.13 Outline illustration of generic characters, from Ridgway's *Manual of North American Birds* (1887).

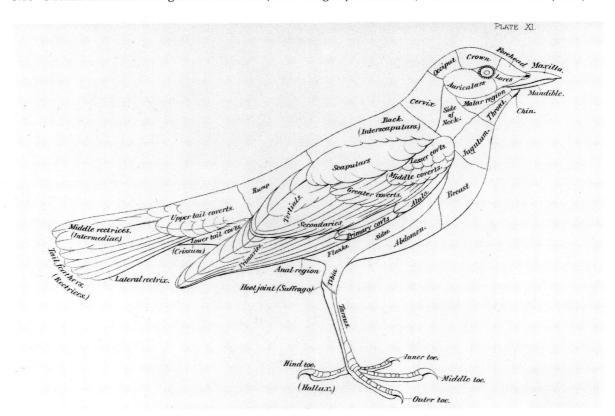

8.14 Diagram of bird anatomy, from Ridgway's *Nomenclature of Colors* (1886).

PLATE XIV.

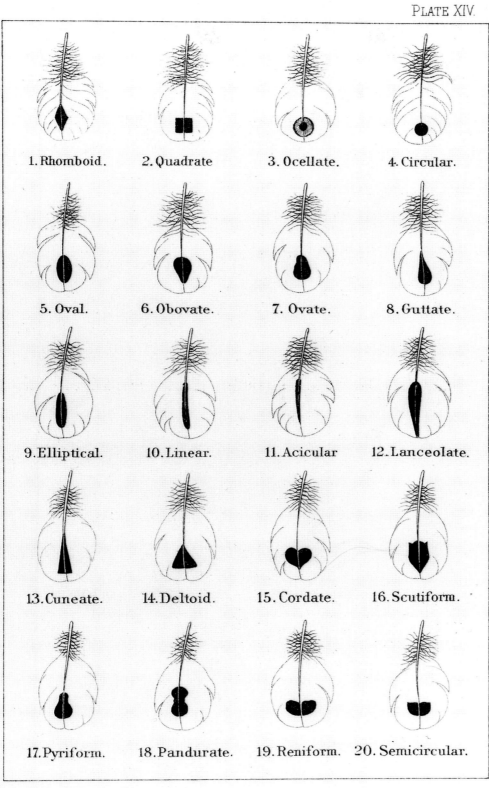

8.15 Feather types, from Ridgway's *Nomenclature of Colors* (1886).

ing to Chapman's account, "recognizing the genius of the young bird artist" took him into the Thayer family to train him in painting. Chapman considered that "this association was of incalculable value" to the younger artist.[31]

Not content with having explained countershading, Thayer expanded his initial theory into a universal law: all markings and patterns in the animal kingdom protected their wearers, and thus were adaptive for survival. Other naturalists working on explanations of coloration within an evolutionary framework also emphasized utility, but took a more pluralistic view. Colors sometimes served as warnings, sometimes as identification in courtship, and sometimes as mimicry, a kind of protection.[32] Alfred Goldsborough Mayer conducted a study to analyze color and patterning in butterfly wings.[33] Like the painter, Mayer sought to formulate laws of coloration and focused on the role of color and pattern in avoiding predation. But like the English naturalist, Frank Beddard, Mayer cautiously postponed conclusions until science had accumulated more information.[34]

PLATE 69

Thayer saw pluralism as inconsistency. He believed that the law of natural selection dictated that all animal coloration prove useful. For him, predation and avoidance of predation were the principal factors in survival and thus in nature. Even conspicuous colors, he argued, could be interpreted as concealing and protective, because animal colors and patterns quoted those of the environment and effectively hid the individual. Thayer offered the example of zebras hidden by dappled light in high grass. The color frontispiece of Thayer's 1909 treatise on his expanded theory, *Concealing Coloration in the Animal Kingdom*, written with his son Gerald, depicted the peacock hidden in the woods. "The Peacock's splendor is the effect of a marvelous combination of 'obliterative' designs, in forest-colors and patterns," wrote Thayer. All imaginable forest tones could be found in the male bird's feathers so that the bird disappeared among the leaves "to a degree past human analysis."[35] Thayer chose some

PLATE 70

challenging examples to demonstrate the universality of his explanation. His insistence on the concealing coloration of the peacock, zebra, and flamingo exhausted his credibility among ornithologists.

Thayer countered resistance with the assertion that artists were better qualified than scientists to analyze the mechanisms of concealment. "The entire matter has been in the hands of the wrong custodians," he insisted. "Appertaining solely to animals, it has naturally been considered part of the zoologist's province," leading to fragmented interpretations and false hypotheses. The problem of animal coloration, Thayer continued, "properly belongs to the realm of *pictorial art*, and can be interpreted only by painters. For it deals wholly with optical illusion, and this is the very gist of a painter's life."[36]

Thayer's emphasis on illusion reflected the influence of the French impressionists, who, since the 1870s, had shifted the concerns of painting to the laws of optics and the illusions of light and color. Thayer argued that the artist is born with "a sense" of understanding illusion: "From his cradle to his grave, his eyes, wherever they turn, are unceasingly at work on it,—and his pictures live by it. What wonder, then, if it was for him alone to discover that the very art he practices is *at full*—beyond the most delicate precision of human powers—on almost all animals?"[37] It remained to the artist, a "specialist" in the visible, to reveal the "beautiful *things discovered* . . . as palpable and indisputable as radium and X-rays" to the appreciation of "all men." Lest his audience miss the point, he reiterated, in italics: "*Naturalists have not understood the principles of objects' distinguishability.*"[38]

Chapman described having to submit to Thayer's aid in selecting the correct blue crayon to match the color of snow in shadow. But Chapman insisted that illusion and color were not the only factors in determining the limits of Thayer's theory. "I admit the correctness of your observations as a color expert, but

PLATES
71–74

as an expert in Blue Jays I cannot see how they have any bearing on the relation between a Blue Jay and its environment in winter," the ornithologist protested. While the blue jay might "disappear" against shadowed snow—a white sheet upon which the painter Childe Hassam threw his shadow in Thayer's Central Park demonstration—several species of blue jays did not, claimed Chapman, winter in snowy climates, nor "as an arboreal species" did they often feed on the ground. This knowledge rendered Thayer's superior color sense irrelevant. "Thayer looked at me sadly," Chapman related, "And I fear from that day he regarded me as a backslider from the ranks of his earliest converts."[39] Thayer believed that the principles of color overrode the ornithologist's observations of bird behavior. Chapman had only proved Thayer's contention that naturalists lacked "the full color sense necessary as a basis for studying obliterative coloration."[40] In a note appended to the introduction to his treatise, Thayer wrote: "If, like a multitude of people, one cannot see that shadows on an open field of snow, or on a white sheet, under a blue sky, are *bright blue* like the sky overhead, one will probably prove more or less defective in all color-perceptions."[41] Eventually even the genial Chapman was forced to preserve his professional reputation by opposing Thayer's exaggerations of the principle of obliterative coloration.

Even if Thayer had not stretched his claims for concealing coloration, zoologists would probably have remained unconvinced by his expanded argument. Professional resistance to his theory turned on several points. Foremost among them was that Thayer based his explanation of concealing coloration on the mechanism of natural selection, one aspect of evolutionary theory which American zoologists had not embraced.[42] Also important, however, were Thayer's use of visual evidence and his insistence on the painter's privileged vision.

Neither Thayer's illustrations nor his method conformed to professional conventions. Abbott and Gerald Thayer painted in oils and watercolors the pictures reproduced as color lithographs in their book. Although diagrams and photographs supplemented the reproductions of paintings, none of the illustrations spoke in the pictorial idiom of science. Thayer also derived his illustrations differently from the way zoologists made theirs; he formulated his law of concealing coloration and then devised his demonstrations and painted his illustrations.[43] In contrast, Ridgway's approach to color, that of a systematic arrangement and nomenclature, was congruent with the central tenets of the discipline. Zoologists used illustrations to document the specimens on which they based their descriptions; only after the patient accumulation of examples did the specialist attempt to formulate a generalization.

Thayer took refuge from his critics in his insistence on the artist's inborn, superior visual understanding. Animal coloration "belonged to the realm of *pictorial art*"; illusion was "the very gist of a painter's life." Those who remained unconvinced were "defective in all color-perceptions." We have seen how the development of institutions and practices internal to natural history and the discipline of zoology interacted with changing print technology to shape and alter the appearance of zoological illustration and distance it from the fine arts. At first glance, the contrast between the role of the Peales and that of the Thayers in the natural history communities of their day suggests that science and art, and professional zoologists and amateur naturalists, had become divided by unbridgeable gulfs. At the beginning of the nineteenth century, the painter Rembrandt Peale declared that artists were more qualified than anatomists to understand form. A century later, painter Abbott Thayer asserted that he understood illusion better than scientists did. The Peales' concept of observation and representation had been congruent with that of the naturalists of their community. Thayer's was not. Rembrandt Peale had insisted on his equality with anatomists. Thayer insisted on a fun-

343

damental difference between artists and scientists. And whereas the Peales enjoyed a central place in the development of American natural history, Thayer failed in his attempt to persuade his critics—although he succeeded in stimulating debate and further experiment.[44] Contrary to Thayer's claims, however, professional resistance to his claims had less to do with the methodological and pictorial divide between science and art (although Thayer had many supporters among painters) than with a diversity of scientific approaches to observation.[45]

The profession itself was split three ways. There were zoologists who based their practice on taxonomy, those who based theirs on observation of the animal in nature, and those who favored actual experiments. Yet all three approaches were considered professional practice. Thayer's method was not one or the others. He attempted to prove his theory to skeptics with specially contrived artifacts such as painted decoys, cardboard cutouts, and on one occasion, a "painting" of a sunrise made out of a flamingo skin.[46] His critics argued that his demonstrations did not replicate natural conditions. Nor did his demonstrations meet the professional criteria of experimentation; Thayer conducted his demonstrations not to test animal behavior in nature but to convince a human audience.

Despite Thayer's failure to convince zoologists, the controversy he provoked revealed some important constants in American zoology. At the beginning of the century, the Peales, like their naturalist colleagues, had placed observation over theory. Thayer proposed a universal law—a theory—to explain observable patterns in nature; his zoologist audience preferred statements based on observation. Despite decades of dominance within the discipline by museum-based taxonomy, natural history remained an important strand of zoology. The debate about Thayer's law of concealing coloration was a natural history debate, that is, the questions turned on the meaning of the colors and behaviors of live animals in nature. More-

over, although specialists ultimately set the terms of the discipline, the Thayer episode engaged not only the professional community, but the wider amateur community. Observation as method and pictorial representation as evidence were central to the controversy.

In a way, the Thayer controversy constituted a near miss. Thayer attempted to influence the professional discipline because he, like so many other American amateurs of natural history, espoused its founding values and their national cultural legacy—the visual emphasis bequeathed by the early naturalists, the ethos of the artist-naturalist, the enduring link between observation and pictorial representation. But Thayer and his critics did not give these values the same meaning. Both sides in the controversy resorted to insisting on privileged, mutually exclusive professional practice. Their arguments of last resort demonstrated once again that within the zoologist's province, consensus about the accuracy of pictorial representation had less to do with content or technique than with consensus about the practices behind its creation.

American naturalists of the early nineteenth century emphasized the value of personal eyewitness in their study of American nature. The unity of observation and representation advocated and practiced by the early naturalists had been short-lived but profoundly influential in what became the divergent specialist and popular realms of nature study. The enduring stamp of the naturalists' credo on popular natural history is easy to identify: nearly a century after Wilson and Audubon, pictures inspired by and resembling theirs illustrated the popular literature of natural history. The influence of the early naturalists on zoology was refracted by the lens of change. As the discipline differentiated itself from popular natural history, it institutionalized pictorial conventions that signaled that distinction. Naturalists of James Dwight Dana's generation believed it was necessary to reject the narrative elements of their predecessors and to establish an impersonal, analytical identity. At midcentury, the field of zoology redefined the elements of description by eliminating narrative and by defining the animal in a scientific, as opposed to natural, environment—the museum, the laboratory, the shallow space of the page. The personal narrative of the artist-naturalist was banished from the prose of the professional zoologist, but narrative persisted in illustration. Ornithology never completely discarded elements developed by the naturalist-illustrators. Physiological studies reintroduced a narrative meaning to the convention of the composite plates. Moreover, other founding values persisted. Even as zoology codified its practice and consolidated its pictorial conventions, zoologists maintained a close identification between the quality of illustration and the quality of the work. In the 1870s Othneil C. Marsh echoed Alexander Wilson in asserting that illustations were the most enduring contribution he could make to his discipline.

The early naturalists had insisted on their authority to contribute to natural history on an equal basis with Europeans. They also insisted that their foundation in direct observation and representation of living nature distinguished American natural history from European. Many of the most distinguished early practitioners, however, were not native-born. The developing scientific institutions found one road to achieving parity with European science in the emulation of English and continental styles of illustration and publication. The arrival of European lithographers after 1848 helped to raise the quality of commercial lithographic draftsmanship and technique, both important components in the consolidation of conventions in midcentury American zoological illustration. What then, if anything, distinguished American zoological illustration as American? The early insistence on difference and republi-

can values had been largely silenced by the entry of American science into the international arena. That entry, however, could not have been achieved without the development of public natural history institutions; arguments for their founding emphasized the public utility of natural history. Much of the grist for the midcentury mill of zoological publication came from the state and territorial surveys—public natural history on an unprecedented scale. Native animals remained the primary subjects for study and illustration. And when ornithology accommodated its amateur branch—to work together in preserving native species—the resulting hybrid organizations adopted the name of Audubon, the national symbol of the artist-naturalist.

The illustrator and draftsman fared less well than the ideal of the artist-naturalist. Increased division of labor in scientific institutions concentrated status at the upper end of the hierarchy of production and obscured the contributions of assistants such as laboratory technicians, preparators, and illustrators. In taxonomic zoology, the movement toward a stricter codification of taxonomic description was matched by the effort to control pictorial representation within a narrow range of permissible conventions, while the incorporation of photography into picture-printing technologies encouraged linear treatment of scientific subjects. These trends reduced the illustrator to a technician who produced pictures to conform to precise specifications within a narrower range of permissible drawing techniques than had been available to the midcentury museum illustrator or lithographer. Photomechanical reproduction limited the role the printer played in affecting the appearance of the finished plate. But just as microscopy reaffirmed the practice of the drawing observer, the development of a schematic illustration convention simplified the preparation of illustrations, reviving the identity of observation and representation.

345

# Notes

———— ❈ ————

PREFACE

1. Throughout this book, "American" refers to the United States only.

INTRODUCTION
ANIMAL PICTURES AND NATURAL HISTORY

1. The principal sources on which I base my discussions of pictorial conventions of drawing and print media are E. H. Gombrich, *Art and Illusion: A Study in the Psychology of Pictorial Representation* (Princeton, N.J.: Princeton University Press, 1969); William Ivins, *Prints and Visual Communication* (New York: Da Capo Press, 1969); Estelle Jussim, *Visual Communication and the Graphic Arts: Photographic Technologies in the Nineteenth Century* (New York: R. R. Bowker, 1974). There are aspects of each of these works, however, that require a critical approach. The work of Gombrich has become a central part of the canon of theory on scientific illustration without a full accounting of either the influence of Karl Popper on Gombrich's ideas about scientific illustration or Gombrich's use of information theory. See, however, W.J.T. Mitchell, *Iconology: Image, Text, Ideology* (Chicago: University of Chicago Press, 1986) pp. 37–38. Jussim relies heavily on information theory. It permits her, however, to analyze photomechanical printmaking, which Ivins, with his faith in the transparency of photographic representation, credits with offering a truly objective pictorial medium. Mitchell's analysis of rhetoric about images also throws into question Ivins's, Gombrich's, and Jussim's use of metaphors of language to discuss pictures; see especially Mitchell, *Iconology*, pp. 80–82.

2. For an analysis of science as a series of translations, see Bruno Latour, *Science in Action: How to Follow Scientists and Engineers through Society* (Cambridge, Mass.: Harvard University Press, 1987), especially chapter 6, "Centres of Calculation," pp. 215–257.

3. Edward R. Tufte, in *Visual Display of Quantitative Information* (Cheshire, Conn.: Graphics Press, 1983), discusses the development of graphic conventions for charts, graphs, and diagrams and provides historical and comtemporary examples. For a preliminary discussion of the use of diagrams in biology, see also Peter J. Taylor and Ann S. Blum, "Pictorial Representation in Biology," in *Biology and Philosophy* 6, no. 2 (April, 1991): 125–134.

4. A large literature now argues and supports these positions. See, for example, Samuel Y. Edgerton, Jr., *The Renaissance Rediscovery of Linear Perspective* (New York: Harper and Row, 1975); Elizabeth L. Eisenstein, *The Printing Press as an Agent of Change: Communications and Cultural Transformations in Early-Modern Europe*, 2 vols. (New York: Cambridge University Press, 1980); Latour, *Science in Action*. Latour has stated that "if scientists were looking at nature, at economies, at stars, at organs, they would not *see* anything. . . . Scientists start seeing something once they stop looking at nature and look exclusively and obsessively at prints and flat inscriptions." Latour, "Visualization and Cognition: Thinking with Eyes and Hands," *Knowledge and Society: Studies in the Sociology of Culture Past and Present* 6 (1986): 16.

5. Many studies have discussed and reproduced American natural history and zoological illustration, among them S. Peter Dance, *The Art of Natural History: Animal Illustrators and Their Work* (Woodstock, N.Y.: Overlook Press, 1978); Martina Norelli, *American Wildlife Painting* (New York: Watson-Guptill, in association with the National Collection of Fine Arts, Smithsonian Institution, Washington, D.C., 1975).

6. Martin J. S. Rudwick, in "The Emergence of a Visual Language for Geological Science, 1760–1840," *History of Science* 14 (1976): 149–195, provides a model for seeking a relationship between the consolidation of a discipline and its use and development of illustration conventions. His study, like Barbara M. Stafford's *Voyage into Substance: Art, Science, Nature and the Illustrated Travel Account, 1760–1840* (Cambridge, Mass.: MIT Press, 1984), stops at a crucial juncture. Once the conventions have been established, how does the discipline make them work? What is the relationship between changes in practice *within* an established discipline and its interaction with, for example, the printing

industry, a changing audience, and other kinds of social developments? I will address these issues in subsequent chapters.

7. For a social history of animals in early modern England, with important implications for Europe and America, see Keith Thomas, *Man and the Natural World: Changing Attitudes in England, 1500–1800* (London: Allen Lane, 1983). Thomas's work incorporates a rich variety of primary sources, documenting the complex interconnections of animals and society. The development of a discipline of natural history and the later culture of science in no way meant an end to the construction and changing social meanings of animals. Thomas's study covers the developments leading up to the urban and industrializing nineteenth century, for which John Berger's "Why Look at Animals," in *About Looking* (New York: Pantheon Books, 1979), explores the meanings of zoological gardens, urban pet-keeping, and children's toys which formed part of the resonant social background for zoological science, museums, and scientific pictures of animals. David Elliston Allen's *The Naturalist in Britain: A Social History* (London: Allen Lane, 1976) amplifies the view of the related developments of natural history as a field of scholarship and as a popular leisure pursuit from the late seventeenth to the late nineteenth centuries. The work of all three writers helps explain how natural history's two distinct audiences, the scholarly and the popular, developed, and suggests ways in which the professional and the popular first distinguished themselves and then interacted throughout the nineteenth century, both in Britain and the United States.

8. Examples are reproduced in Hugh Honour, *The New Golden Land: European Images of America from the Discoveries to the Present Time* (New York: Pantheon Books, 1975).

9. Again, see Thomas, *Man and the Natural World*, pp. 254–269. See also James Turner, *The Politics of Landscape: Rural Scenery and Society in English Poetry, 1630–1660* (Cambridge, Mass.: Harvard University Press, 1979), for a discussion of the elimination of work from the poetry of landscape in the seventeenth century, a process that continued in literature and painting into the nineteenth century.

10. The painter George Stubbs worked in many of these genres: horse portraiture, landscape with farming activity, anatomical illustration, romantic depictions of wild animals. See Basil Taylor, *Stubbs* (London: Phaidon Press Limited, 1971). See also John Berger's comments on Gainsborough in *Ways of Seeing* (London: British Broadcasting Corporation and Penguin Books, 1972), pp. 106–108.

11. Hal Opperman, *J.-B. Oudry, 1686–1755* (Fort Worth, Texas: Kimbell Art Museum, 1983).

12. See, for example, Fritz Koreny, *Albrecht Dürer, and the Animal and Plant Studies of the Renaissance*, A New York Graphic Society Book (Boston: Little, Brown and Company, 1985). Juxtaposition of Dürer's preparatory drawings with copies by other painters gives a striking demonstration of how representational conventions arose and were perpetuated.

13. See Latour, "Visualization and Cognition." In this article, as elsewhere, Latour draws on William Ivins's analysis of linear perspective as providing "optical constancy," constitutive in the development of science. W. M. Ivins, *The Rationalization of Sight* (New York: Plenum Press, 1973).

14. "The best index to the hegemony of the artificial perspective is the way it denies its own artificiality and lays claims to being a 'natural' representation of 'the way things look,' 'the way we see,' or . . . 'the way things really are.' Aided by the political and economic ascendance of Western Europe, artificial perspective conquered the world of representation under the banner of reason, science, and objectivity." Mitchell, *Iconology*, p. 37.

15. Historians of anatomical illustration have ransacked the history of painting and sculpture for particular models for illustrations. See Ludwig Choulant, *History and Bibliography of Anatomic Illustration*, translated and annotated by Mortimer Frank (New York: Hafner Publishing Company, 1962). Although anatomists partook and even led the way in the use of printed illustration in relation to new practices of observation, the human body and its pictorial representation constituted a separate category from the rest of natural history and natural history illustration. The human was and is invested with meanings that elicited from anatomists special kinds of claims to truth. See Thomas Laqueur, *Making Sex: Body and Gender from the Greeks to Freud* (Cambridge, Mass.: Harvard University Press, 1990), and Londa Schiebinger, "Skeletons in the Closet: The First Illustrations of the Female Skeleton in Eighteenth-Century Anatomy," *Representations* 14 (Spring 1986): 42–82.

16. Eisenstein, "From a Hearing Public to a Reading Public: Some Unevenly Phased Social and Psychological Changes," in *The Printing Press*, pp. 129–136; and Thomas, *Man and the Natural World*, pp. 70–81, on the displacement of popular by erudite knowledge of nature.

17. See Eisenstein, *The Printing Press*, especially part 1, chapter 2, "Defining the Initial Shift; Some Features of Print Culture," and part 3, "The Book of

Nature Transformed." See Ivins, *Prints*, pp. 31–46, on the introduction of the exactly repeatable image and its implications for science. Eisenstein notes, however, that "Ivins's analysis . . . tends to detach the fate of printed pictures from that of printed books. His treatment implies that the novel effects of repeatability were confined to pictorial statements. Yet these effects were by no means confined to pictures, or, for that matter, to pictures and words. Mathematical tables, for example, were also transformed." Eisenstein follows George Sarton in defining "the shift from script to print" as "'a double invention; typography for text, engravings for the images'"; Eisenstein, *The Printing Press*, pp. 54–55.

18. Pliny the Elder, *Natural History*, chapter 4, book 25, quoted in Ivins, *Prints*, p. 14.

19. Pliny's statement also highlights the importance of color in plant identification in late antiquity. Eighteenth-century Linnaean botany established the discipline on the emphasis of structure over color, considered a variable. That preserved specimens in collections tended to lose color also contributed to the emphasis on structure. But it is possible that the role of printing in the emergence of empirical studies of nature may also have contributed to the separation of color from structure by separating the two stages of production in illustration. Not only were the two steps of production separated, but the application of color added to the expense.

20. Stafford, *Voyage*, chapter 1, "The Scientific Gaze," pp. 31–56.

21. J. B. de C. M. Saunders and Charles D. O'Malley, *The Antomical Drawings of Andreas Vesalius* (New York: Bonanza Books, 1982), p. 16.

22. Karen Reeds, "Publishing Scholarly Books in the Sixteenth Century," *Scholarly Publishing* 14, no. 3 (April 1983): 258–274. Reeds also discusses the difficulty of authors in persuading publishers to print their illustrated treatises at full size compared to the ease of publishing, at lower cost, smaller-format textbooks, reissues of the classics, and pirated editions.

23. On the movement of meaning between text and illustration in twentieth-century science, see Françoise Bastide, "Iconographie des textes scientifiques: Principes d'analyse," *Culture Technique* 14 (June 1985): 133–151.

24. Ivins, *Prints*, chapter 2, "The Road Block Broken: The Fifteenth Century," pp. 21–50.

25. See Reeds, "Publishing Scholarly Books," for the reactions of sixteenth-century authors to pirated editions.

26. Vesalius acknowledged the contribution of Jan Stefan van Kalkar, of the studio of Titian, to the *Epitome* by "paying" him in reproduction rights, which gave him the virtual status of joint author; see Saunders and O'Malley, *Anatomical Drawings*, p. 26. Botanist Leonard Fuchs, in a unique recognition of his illustrators, printed portraits of the artists who drew the plants and of the artists who cut the woodblocks. Bernhard Siegfried Albinus (1697–1770), author of several works of anatomy including *Tabulae sceleti et musculorum corporis humani* (1747), defended the work of his artist, Jan Wandelaer, against criticisms of innaccuracy; see Choulant, *History*, pp. 276–283.

27. On Le Moyne's travels in North America, his natural history drawings, and questions concerning his authorship of a published natural history, see Elsa Guerdrum Allen, "The History of American Ornithology before Audubon," *Transactions of the American Philosophical Society*, n.s., 41, part 3 (1951): 440–443.

28. Clifford S. Ackley, *Printmaking in the Age of Rembrandt* (exhibition catalog) (Boston: Boston Museum of Fine Arts, 1981). See also Gombrich on repeated copying of popular whale engravings, giving rise to distortions similar to preprint hand-copied illustrations. Gombrich reproduces an anonymous Italian print of a whale with ears, and the Dutch original, also eared; *Art and Illusion*, pp. 79–81, figs. 57 and 58, p. 80.

29. Elizabeth Mongan, *Wenceslaus Hollar, 1607–1677: Birds, Beasts and Grotesques* (exhibition catalog) (Gloucester, Mass.: The Hammond Museum, 1979). On Dürer's influence on Renaissance animal and plant studies, see Koreny, *Albrecht Dürer*.

30. See Wilfred Blunt and William T. Stearn, *The Art of Botanical Illustration* (New York: Charles Scribner's Sons, 1951), for discussion and examples of early botanical illustration.

31. Ackely, *Printmaking*, pp. 94–96.

32. Mongan, *Hollar*, n.p.

33. Svetlana Alpers, *The Art of Describing: Dutch Art in the Seventeenth Century* (Chicago: University of Chicago Press, 1983), p. 71. The influence of Dutch painting on French and English animal painting and natural history illustration needs further exploration. Oudry and Hollar studied with Dutch painters, and other eighteenth-century draftsmen who drew for natural history came from Holland.

34. For a discussion of the seventeenth-century linguistic reforms in science, see Michel Foucault, *The Order of Things: An Archaeology of the Human Sciences* (a translation of *Les Mots et Les Choses*) (New York: Pantheon Books, 1971). See also Stafford, *Voyage*, chapter 1, "The Scientific Gaze," for a

nuanced discussion of the influence of these reforms on the visual representation of nature in scientific travel accounts, with similar implications for natural history illustration. I have based my brief sketch on Stafford's account.

35. Bishop Thomas Sprat, *History of the Royal Society of London, For the Improving of Natural Knowledge* (1667), quoted in Stafford, *Voyage*, p. 35. See also Peter Dear, "*Totius in verba*: Rhetoric and Authority in the Early Royal Society," *Isis* 76, no. 282 (June 1985): 145–161.

36. See Evelyn Fox Keller, *Reflections on Gender and Science* (New Haven, Conn.: Yale University Press, 1985).

37. See Alpers, *Art of Describing*, chapter 2, " 'Ut pictura, ita visio': Kepler's Model of the Eye and the Nature of Picturing in the North," and chapter 3, " 'With a Sincere Hand and a Faithful Eye': The Craft of Representation," for connections between optics and picturing, and between seeing and craft in Dutch seventeenth-century painting.

38. Robert Hooke, *Micrographia, or Some Physiological Descriptions of Minute Bodies Made by Magnifying Glasses, with Observations and Inquiries thereupon* (1665) (New York: Dover Publications, 1961).

39. Ibid., n.p.

40. Both Svetlana Alpers and Geoffrey Lapage, *Art and the Scientist* (Bristol, U.K.: John Wright and Sons Ltd., 1961), refer to Hooke's preface, but to very different ends. Alpers, in chapter 3 of *The Art of Describing*, uses Hooke's invocation of a "sincere Hand, and a faithful Eye" to connect early scientific practice with the descriptive practices of Dutch drawing and painting. Lapage based his definition of the difference between scientific illustration and "art" on the issue of control by the scientist, and cites Hooke in an attempt to prove that Hooke himself produced all the original drawings for the work. I wish to emphasize the pragmatic incorporation of existing engraving technique and compositional conventions into early empirical practice.

41. This acceptance of printing techniques would change, however, when new technologies offered new options. During the nineteenth century the invention of photography, combined with the drive toward a completely mechanized printing process, prompted zoologists to work with printers to experiment with and invent methods to bypass the manual translation of drawings into prints. See chapter 7.

42. "A depiction is never just an illustration. It is the material representation, the apparently stabilised product of a process of work. And it is the site of the construction and depiction of social difference. To understand a visualisation is thus to inquire into its provenance and into the social work that it does. It is to note its principles of exclusion and inclusion, to detect the roles that it makes available, to understand the way in which they are distributed, and to decode the hierarchies and differences that it naturalises. And it is also to analyse the ways in which authorship is constructed or concealed and the sense of audience is realised." Gordon Fyfe and John Law, "Introduction: On the Invisibility of the Visual," in Gordon Fyfe and John Law, eds., *Picturing Power: Visual Depictions of Social Relations* (London: Routledge, 1988), p. 1.

43. Gombrich, *Art and Illusion*, p. 110.

44. "*Anything* that will accelerate the mobility of the traces that a location may obtain about another place, or *anything* that will allow these traces to move without transformation from one place to another, will be favored: geometry, projection, perspective, bookkeeping, paper making, aqua forte, coinage, new ships." Latour, "Visualization and Cognition," p. 13.

45. Ivins, *Prints*, chapter 3, "Symbolism and Syntax: A Rule of the Road, The Sixteenth Century," pp. 51–70.

46. See Charles Coulston Gillispie, ed., *Dictionary of Scientific Biography* (New York: Charles Scribner's Sons, 1970–1990). Jacques Roger, "Georges-Louis Leclerc, Comte de Buffon," vol. 2 (1970), pp. 576–582; Sten Lindroth, "Carl Linnaeus (or von Linné)," vol. 8 (1973), pp. 374–381.

47. A. J. Cain, *Animal Species and Their Evolution* (London: Hutchinson's University Library, 1954), pp. 29–31.

48. Ernst Mayr, *The Growth of Biological Thought: Diversity, Evolution, and Inheritance* (Cambridge, Mass.: The Belknap Press of Harvard University Press, 1982), p. 174.

49. Ibid., pp. 171–180.

50. Charles Webster, "John Ray," in *Dictionary of Scientific Biography*, ed. C. Gillispie, vol. 11 (1975), pp. 313–318.

51. Mayr, *Biological Thought*, p. 176.

52. Ray had come to similar conclusions about a reproductive definition of species and about the need to explain transmutation; see Webster, "John Ray," p. 315.

53. See Jacques Roger, *Buffon, un philosophe au jardin du roi* (Paris: Fayard, 1989). For excerpts from Buffon's writings and discussions of his work in the context of his time, see John Lyon and Phillip R. Sloan, eds., *From Natural History to the History of Nature: Readings from Buffon and his Critics* (Notre Dame, Indiana: University of Notre Dame Press, 1981).

54. Roger, "Georges-Louis Leclerc," pp. 576–582.

55. Ibid., p. 580.

56. Quoted in ibid., p. 581.

57. David Allen, *Naturalist*, pp. 39–40.

58. Mayr, *Biological Thought*, p. 181.

59. David Allen, *Naturalist*, pp. 43–51.

60. Donald Worster, *Nature's Economy: The Roots of Ecology* (Garden City, N.Y.: Anchor Press, Doubleday, 1979), chapter 2, "The Empire of Reason," pp. 26–55. See also Raymond Williams, "Ideas of Nature," in *Problems in Materialism and Culture* (London: Verso, 1980), pp. 67–85.

61. Worster, *Economy*, chapter 1, "Science in Arcadia," pp. 3–25.

62. Thomas Bewick, *A History of British Birds*, vol. 1 (Newcastle, U.K.: Beilby and Bewick, 1797), p. iv.

63. In Charlotte Brontë's *Jane Eyre*, the lonely orphan comforted herself with an afternoon spent surreptitiously reading Bewick. Thomas Bewick, *A General History of Quadrupeds* (Newcastle, U.K.: R. Beilby and T. Bewick, 1791), and *A History of British Birds*, 2 vols.

64. See Stafford, *Voyage*, p. 41.

65. Elsa Allen, "History of American Ornithology before Audubon," p. 490.

66. Stafford, *Voyage*, p. 36.

67. Elsa Allen, "History of American Ornithology before Audubon," p. 481. On Catesby's work, see also Alan Feduccia, *Catesby's Birds of Colonial America* (Chapel Hill: University of North Carolina Press, 1985).

68. Elsa Allen, "History of American Ornithology before Audubon," pp. 543–549. Abbot, dependent on his English patrons for his living as a natural history collector and illustrator, was a loyalist.

69. In recognition of the primacy of text and verbal description, Stafford adds: "An illustration, therefore, is a picture of the thing world inserted into a verbal text, and represents a gesture towards semiotic wholeness." Stafford, *Voyage*, p. 51. The literatures of scientific travel and natural history and the development of their illustration conventions were intimately connected. In this study, however, I concentrate on the natural history plates of the travel and voyage reports.

70. On John Bartram and John Abbot, see Elsa Allen, "History of American Ornithology before Audubon," pp. 536–549. See also Joseph Ewan, "The Natural History of John Abbot: Influences and Some Questions," *Bartonia* 51 (1985): 37–45.

71. Thomas Bewick, *A General History of Quadrupeds. The figures engraved on wood, chiefly copied from the original of T. Bewick, By A. Anderson, first American edition, with an appendix, containing some animals not hitherto described* (New York: Printed by G. and R. Waite, 1804). The added animals were the hamster of Georgia, the wild sheep of California, the mammoth of New York, and the viviparous shark of Long Island.

## CHAPTER 1
### THE NATURALIST-ILLUSTRATOR

1. Several studies have described and reproduced the work of naturalist-illustrators and artist-naturalists, among them Norelli, *American Wildlife Painting*, and Dance, *The Art of Natural History*.

2. Cited in John C. Greene, *American Science in the Age of Jefferson* (Ames: Iowa State University Press, 1984), p. 417.

3. Bartram's drawings at the British Museum (Natural History) are reproduced in Joseph Ewan, ed., *William Bartram: Botanical and Zoological Drawings, 1756–1788* (Philadelphia: American Philosophical Society, 1968).

4. William Bartram, *Travels through North and South Carolina, Georgia, East and West Florida, the Cherokee country, the extensive territories of the Muscogulges, or Creek Confederacy, and the country of the Chactaws; containing an account of the soil and natural productions of those regions, together with observations on the manners of the Indians* (Philadelphia: James and Johnson, 1791).

5. Elsa Allen, "History of American Ornithology before Audubon," pp. 536–543.

6. By 1816, the Peale museum boasted a public attendance of nearly 48,000 visitors annually. Its success inspired imitations, although the imitators rarely presented as systematically organized a display as did the Peales. See Charlotte M. Porter, *The Eagle's Nest: Natural History and American Ideas, 1812–1842* (Tuscaloosa: University of Alabama Press, 1986), p. 29.

7. Greene, *American Science*, p. 282.

8. Even after the acceptance of James Smithson's bequest, permitting the establishment of the Smithsonian Institution, Congress continued to debate whether the Constitution permitted federal funding of science. See chapters 4 and 5.

9. On the Peale museum, see Charles Coleman Sellars, *Mr. Peale's Museum: Charles Willson Peale and the First Popular Museum of Natural Science and Art* (New York: W. W. Norton, 1980). See also Edgar P. Richardson, Brooke Hindle, and Lillian B. Miller, *Charles Willson Peale and His World* (New York: Harry N. Abrams, Inc., 1983).

10. The species eventually received the name *Megalonyx jeffersoni*, in recognition of Jefferson's priority of description, but the analysis of the fossil

351

was a long-term, collaborative process, initiated by Cuvier and Jefferson, augmented by José Garriga, who described the Paraguayan specimen in Madrid, and later consolidated by Anselme Demarest in 1822. See Greene, *American Science*, pp. 33–34, 282–285.

11. Ibid., pp. 284–288. On the debate over the interpretation of fossil remains, see Martin J.S. Rudwick, *The Meaning of Fossil: Episodes in the History of Paleontology*, 2d ed. (Chicago: University of Chicago Press, 1985), chapter 3, "Life's Revolutions," pp. 101–163. Rudwick's discussion shows that it is possible to write a history of Cuvier's paleontology without mentioning the Peales or Jefferson. Greene, on the other hand, demonstrates the importance to the American naturalists that their specimens were central actors in the elaboration of the new principles. For a discussion of the role of fossils in distinguishing natural from artificial imagery, see Barbara M. Stafford, "Characters in Stones, Marks on Paper: Enlightenment Discourse on Natural and Artificial Taches," *Art Journal*, Fall 1984, pp. 233–240.

12. Rudwick, *Meaning*, pp. 103–112.

13. Rembrandt Peale, *Historical Disquisition on the Mammoth, or Great American Incognitum, An Extinct, Immense, Carnivorous Animal, whose Fossil Remains Have Been Found in North America* (London, 1803). Peale began his *Disquisition*: "The revolutions which have happened on our earth, by which its original appearance has been successively changed, have, at all times, commanded the attention of the learned." He continued: "From an examination of the various strata, as discovered in mines or exposed in cliffs, we have been taught that the surface of the earth has at times been violently agitated." To corroborate his statement, Peale cited Cuvier: "The celebrated Cuvier, in his Memoir on Fossil Bones, thus commences his observations: 'It is now universally known that the globe which we inhabit, on every side presents irresistible proofs of the greatest revolutions'" (pp. 1–2).

14. Ibid., p. 76.

15. Jefferson launched the refutation of Buffon's theory of degeneration in his *Notes on the State of Virginia* of 1781. But because a theory of vigor was necessary in his case against Buffon, Jefferson could never accept the theory of extinction. Instead, he postulated that mastodons had migrated into the continent's interior and might be found by exploration. See Porter, *Eagle's Nest*, pp. 16–18, 39. In the report of the 1819–1820 government expedition to the Rocky Mountains, the author and expedition botanist, Edwin James, recalled Jefferson's theory—with an affectionate irony—when he described how

the mirage effect on the plains made distant animals appear as large as mastodons. Edwin James, *Account of an expedition from Pittsburgh . . . 1819–20* (Philadelphia: Carey and Lea, 1822–1823), p. 419.

16. Peale, *Historical Disquisition*, p. 72.

17. Ibid., pp. 38–39. Peale insisted the mastodon had been carnivorous, despite the weight of specialist opinion against that conclusion.

18. See Basil Taylor, *Stubbs* (London: Phaidon Press Limited, 1971).

19. Peale, *Historical Disquisition*, p. 39.

20. Alexander Wilson, *American Ornithology*, vol. 3, p. 4. Following Jefferson, when American naturalists referred to "vain philosophy," they meant Buffon. See Porter, *Eagle's Nest*, p. 20.

21. Robert Cantwell's *Alexander Wilson: Naturalist and Pioneer* (Philadelphia: J. B. Lippincott Company, 1961), provides a detailed biography of Wilson. Other important biographical sources are Elsa Allen, "History of American Ornithology before Audubon," and Clark Hunter, ed., *The Life and Letters of Alexander Wilson* (Philadelphia: American Philosophical Society, 1983). Hunter's biographical research and documentation are the most complete, and correct several of Cantwell's points about Wilson's trial for his political poetry, the impetus for his emigration to the United States.

22. Wilson, *American Ornithology*, vol. 3 (1811), p. ix.

23. Wilson, *American Ornithology*, vol. 1 (1808), p. 2.

24. Ibid., p. 4.

25. Ibid., p. 2.

26. The definition of "autodidact" was relative rather than absolute. In Europe, despite the existence of both private and state collections, libraries, and museums, and a few universities where students could earn degrees in natural history (Linnaeus encouraged his students to obtain doctorates), the study of birds had yet to become the profession of ornithology. For the history of that process, see Paul Lawrence Farber, *The Emergence of Ornithology as a Scientific Discipline: 1760–1850*, Studies in the History of Modern Science, vol. 12 (Dordrecht, Holland: D. Reidel, 1982). Farber characterizes Wilson and his successor John James Audubon as "colonial" contributors to ornithology. Wilson, of course, believed that he established the basis for breaking the colonial relationship of American natural history. But in the sense that the work of Wilson and Audubon was appropriated by the metropolitan establishment, Farber is correct. English and especially Scottish ornithologists made something of a cult of Wilson's work after his death, and French ornithol-

ogy considers Audubon one of its own. See René Ronsil, "L'art français dans le livre d'oiseaux: Elements d'une iconographie ornithologique française," *Mémoires de la Société Ornithologique de France et de l'Union Française*, no. 6, suppl. to *L'Oiseau et la Revue Française d'Ornithologie* 27, 4 (Paris, 1957).

27. See Cantwell, *Wilson*, p. 120-124.

28. Alexander Wilson to William Bartram, 16 August 1804, Thayer Collection, Museum of Comparative Zoology, at Houghton Library, Harvard University, reprinted in, Hunter, *Life and Letters*, p. 219.

29. Wilson, "Prospectus," 1807, reprinted in ibid., p. 271.

30. Wilson to Bartram, 15 December 1804, reprinted in ibid., pp. 225–226.

31. Wilson to Bartram, 4 March 1805, reprinted in ibid., pp. 231–232. Courtesy of Mrs. Willa Funck.

32. Wilson to William Duncan, 26 March 1805, reprinted in ibid., p. 235.

33. Bartram to Thomas Jefferson, 18 March 1805, Library of Congress, reprinted in ibid., p. 233.

34. Jefferson to Wilson, 7 April 1805, Library of Congress, reprinted in ibid., pp. 236–238. See Porter, *Eagle's Nest*, pp. 165–166, on the influence of this drawing on later American painting and engraving.

35. Wilson to Bartram, 2 July 1805, in Hunter, *Life and Letters*, pp. 243–244.

36. Wilson to Bartram, 29 November 1805, reprinted in ibid., p. 246.

37. Wilson to Bartram, 4 January 1806, reprinted in ibid.

38. Wilson to Jefferson, 6 February 1806, reprinted in ibid., p. 249.

39. Ibid.

40. Wilson, *Ornithology*, vol. 1, p. 7.

41. Ibid.

42. Typescript biography of Alexander Lawson, Manuscript Collection 79a, Academy of Natural Sciences of Philadelphia. See also Allen Johnson et al., eds., *Dictionary of American Biography*, under the auspices of the American Council of Learned Societies (New York: Charles Scribner's Sons, 1928–1955), vol. 11 (1933), p. 56.

43. Ivins, *Prints*, p. 85.

44. Cantwell, *Wilson*, p. 125.

45. Tebbel, *A History of Book Publishing in the United States*, 4 vols. (New York: R. R. Bowker Co., 1972), p. 109. Oliver Goldsmith, *An History of the Earth and Animated Nature*, 4 vols. (Philadelphia: Mathew Carey, 1795). Three of the engravers were Seymour, Scot, and Allardice.

46. Wilson to Bartram, 22 April 1806, in Hunter, *Life and Letters*, p. 254. William Bradford, one of

Samuel's forebears, had begun printing in Philadelphia in 1658; Tebbel, *History*, vol. 1, p. 164.

47. Bradford was expanding his business at that time, buying up other printers and publishers and shifting his focus from political printing to literature and arts. Cantwell, *Wilson*, pp. 136–138.

48. Tebbel also gives the number of illustrations as six hundred, in *History*, vol. 1, p. 116.

49. Ibid., p. 174.

50. Ibid., p. 116. The encyclopedia would ruin Bradford. The project had cost $200,000 by 1818, and its publication was not complete until 1824. With Bradford's bankruptcy, "the *Cyclopedia* became the property of a syndicate of engravers who had worked on the project. In order to sell the last of it, they had to get permission from the Pennsylvania legislature to conduct a lottery." Ibid., p. 117.

51. Catesby, *The Natural History of Carolina, Florida and the Bahama Islands*, quoted in Elsa Allen, "History of American Ornithology before Audubon," p. 474.

52. Wilson, *American Ornithology*, vol. 1, p. 2. William Smellie, translator and editor of a popular edition of Buffon published in Edinburgh, compared Linnaeus to Buffon: the first "has reduced natural history to the austere regularity of scientific method" while the latter "has displayed its [nature's] material in a loose order, and arrayed them in all the pomp of popular eloquence." Linnaean system provided the instruction, and Buffonian description and narrative provided the amusement. "Yet whatever may be their comparative merits, Buffon will long continue one of the most popular writers on scientific subjects: but, it is not probable, that the volumes of Linnaeus will ever be fondly perused by any but men of science." William Smellie, ed., *The System of Natural History, Written by M. de Buffon, Carefully Abridged: And the Natural History of Insects; Compiled, Chiefly from, Swammerdam, Brookes, Goldsmith, &c. Embellished with, Sixty-Four Elegant Copper-Plates* (Edinburgh, 1800), p. iv.

53. Editors of subsequent editions rearranged Wilson's original chapter sequence to put the birds in taxonomic order.

54. Wilson, *American Ornithology*, vol. 1, p. 4. Latham attempted to keep an updated catalog of all known birds, and used more field observation than some other taxonomists. His work incorporated the birds collected on Cook's voyages. He combined Ray's system of classification for the larger divisions with that of Linnaeus for genus and species. On Latham, see Farber, *Emergence of Ornithology*, pp. 70–74.

55. Assessments of Wilson's classification tend

to be based on how many of his species have held up under changing taxonomic criteria. Many of Wilson's names did not stick because the rules of ornithological taxonomy became more strictly codified over time. Moreover, during Wilson's lifetime, the category of family was not well worked out, and, lacking a taxonomic level between order and genus, Wilson tended to ignore the established orders and to focus on genera and species. See Emerson Stingham, *Alexander Wilson, a Founder of Scientific Ornithology* (Kerrville, Texas: Published by the author, 1958), pp. 19–20.

56. Bartram's drawings and engravings placed animals in a landscape, but Bartram did not publish them as illustrations to a text as Wilson did.

57. See Porter, *Eagle's Nest*, p. 45. Porter suggests that the word "nature" misleads, "because the background landscapes and the associations of species illustrated may or may not correspond to actual habitats or the place where the birds are found." To require that the landscape backgrounds represent actual habitat is to judge by a later, indeed a twentieth-century, standard. Wilson based his claims of authenticity—"Drawn from Nature"—on personal eyewitness.

58. Porter, *Eagle's Nest*, pp. 43–45. Porter also points out that Wilson's eagle is visible in the upper left-hand corner of Charles Willson Peale's 1822 painting, *The Artist in His Museum*; see ibid., p. 28, for reproduction and caption.

59. Wilson, "Prospectus," 1807, reprinted in Hunter, *Life and Letters*, pp. 268, 272.

60. Wilson, advertisements in the *Albany Gazette*, 3 November 1808, and *Charleston Currier*, 16 February 1809. Quoted in Hunter, *Life and Letters*, p. 267.

61. Wilson, *American Ornithology*, vol. 1, p. vi.

62. Ibid., vol. 4, p. v.

63. Ibid., vol. 2, p. vi.

64. Ibid., pp. vi–vii.

65. Hunter, *Life and Letters*, p. 111.

66. Wilson, *American Ornithology*, vol. 4, p. v. Surviving copies of the first edition of *American Ornithology* vary considerably in the quality of coloring of the plates.

67. Frank L. Burns, "Miss Lawson's Recollections of Ornithologists," *The Auk* 34, no. 3 (July 1917): 279.

68. Wilson, *American Ornithology*, vol. 4, p. v.

69. Burns, "Miss Lawson's Recollections," p. 276.

70. Hunter, *Life and Letters*, pp. 110–111. No explanation for the discontent and desertion of the colorists has been recorded, although with Wilson in financial difficulties and Bradford withholding reimbursement, the colorists may not have been paid with regularity.

71. Hunter, *Life and Letters*, p. 104.

72. For subscription amounts, see Cantwell, *Wilson*, pp. 237–238; 253–254.

73. Ibid., p. 237.

74. Ibid., pp. 108–109.

75. For a list of subscribers and thumbnail biographical sketches of each, see Cantwell, *Wilson*, appendix 2, pp. 277–305.

76. Wilson, *American Ornithology*, vol. 5, pp. vii–viii.

CHAPTER 2
DIVERGENCE

1. For Lesueur's biography, see Robert William Glenroie Vail, "The American Sketchbooks of Charles Alexandre Lesueur, 1816–1836," *Proceedings of the American Antiquarian Society* (Worcester, Mass.), April 1938. See also Bernard Smith, *European Visions of the South Pacific*, 2d ed. (New Haven, Conn.: Yale University Press, 1985), pp. 197–200, for a discussion of Lesueur's drawings of marine invertebrates on the French Australia expedition, and for reproductions of several of them. Lesueur's engravings of the collections on the expedition were published in François Péron (1775–1810), *Voyage de découvertes aux terres Australes, éxécuté par ordre de Sa Majesté l'empereur et roi, sur les corvettes le Géographe, le Naturaliste, et la goelette la Casuarina, pendant les années 1800, 1801, 1802, 1803 et 1804* (Paris: De l'Imprimerie impériale, 1817).

2. On the early years of the Academy of Natural Sciences, see Edward James Nolan, *A Short History of the Academy of Natural Sciences of Philadelphia* (Philadelphia: Academy of Natural Sciences, 1909); Porter, *Eagle's Nest*, pp. 55–57; and Greene, *American Science*, pp. 57–59.

3. Alberto Gil Novales, *William Maclure: socialismo utópico en Espana (1808–1840)* (Barcelona: Universidad Autónoma de Barcelona, 1979), p. 7.

4. Harry B. Weiss and Grace M. Zeigler, *Thomas Say: Early American Naturalist* (Springfield, Ill., and Baltimore, Md.: Charles C. Thomas, 1931), p. 39.

5. Maclure's essays on political economy, education, publishing, and society were collected in William Maclure, *Opinions on Various Subjects, Dedicated to the Industrious Producers*, 3 vols. (New Harmony, Indiana: School Press, 1831–1838, reprinted by Burt Franklin, New York, 1969).

6. For an account of the invention and early years of lithography, see Michael Twyman, *Lithography*,

*1800–1850: The Techniques of Drawing on Stone in England and France and Their Application in Works of Topography* (London: Oxford University Press, 1970).

7. Philip J. Weimerskirch, "Naturalists and the Beginnings of Lithography in America," in *From Linnaeus to Darwin: Commentaries on the History of Biology and Geology* (London: Society for the History of Natural History, 1985), pp. 167–177. Weimerskirch rightly notes that the contribution by naturalists to the introduction and establishment of lithography in the United States has been overlooked by historians of the medium. In two of the principal histories of American lithography, Harry T. Peter's *America on Stone* (1931; reprint edition, Arno Press, 1976); and Peter C. Marzio, *Chromolithography 1840–1900, The Democratic Art: Pictures for a 19th-Century America* (Boston: David R. Godine, 1979), lithographic scientific illustration is lumped together with other kinds of technical illustration, and the firms that specialized in scientific illustration receive little notice for it. Nicholas B. Wainwright, *Philadelphia in the Romantic Age of Lithography* (Philadelphia: Historical Society of Pennsylvania, 1958), in contrast, credits the major lithography firms of Philadelphia for their scientific work.

8. On the early evidence of relief printing from stone in Boston, see Richard J. Wolfe, *Jacob Bigelow's American Medical Botany, 1817–1821* (Boston: R. Wolfe, 1979).

9. Weimerskirch, "Naturalists," p. 170.

10. Weimerskirch notes that Lesueur did not print the lithographs himself; they were printed in New York by the firm of William Barnet and Isaac Doolittle, established in 1821, the first lithography shop in the country; see "Naturalists," p. 171.

11. Lesueur to Maclure, 29 March 1821, The Workingmen's Institute Library, New Harmony, Indiana. Translated from the original French and quoted by Weimerskirch, "Naturalists," p. 171. According to Lesueur's letter, so unsatisfactory were the first American experiments that the two printers, Barnet and Doolittle, who had studied lithography in Paris and were attempting to establish a shop in New York, offered to sell their business to Lesueur, so little profit did they see in the endeavor. Lesueur mentions two other impediments to the success of lithography. The first illuminates the state of the market for prints: "In order for lithography to succeed here it is necessary for there to be more connoisseurs of fine arts than there are now. You only find a few individuals who have portfolios full of engravings, and those who have them send

them to auction to get rid of them." Second, Lesueur commented on the secrecy that surrounded the technique: "[H]e who draws in this manner is not able to print everything at the same time if he does not want to communicate his art to his apprentices." Quoted in ibid., p. 171.

12. [Benjamin Silliman], "Notice of the Lithographic Art, or the art of multiplying designs, by substituting Stone for Copper Plate, with introductory remarks by the Editor," *American Journal of Science and Arts* 4 (1822): 170.

13. Lesueur to Desmarest, 26 November 1823, Museum d'Histoire Naturelle, Le Havre. Translated and quoted by Weimerskirch, "Naturalists," p. 173.

14. On the difficulties of publishing illustrated works in the 1820s and '30s, see Porter, *Eagle's Nest*, pp. 140–141.

15. The classic life and letters of Thomas Say is Weiss and Zeigler's *Thomas Say*. For more recent discussions of Say's role in American science, see Porter, *Eagle's Nest*, chapters 4, 7, 8, 10; and Greene, *American Science*, pp. 306–311. On the Philadelphia naturalists' economic incentive to go on western expeditions, see Porter, *Eagle's Nest*, pp. 135–136.

16. Thomas Say to John F. Melsheimer, 12 April 1816. Quoted in Weiss and Zeigler, *Thomas Say*, pp. 42–43.

17. Weiss and Zeigler, *Thomas Say*, pp. 88, 111.

18. Thomas Say, *American Entomology*, vol. 1 (Philadelphia: Samuel Augustus Mitchell, 1824), pp. v–vi.

19. On Say's taxonomy, see Porter, *Eagle's Nest*, pp. 59–64. She explains that Say did not consider the Linnaean system "inviolable." Say attempted to designate generic characters, and began each section of the *Entomology* with a description of a "generic type" constructed out of characters shared by species of the genus. The generic type, "being ideal," Porter writes, "cannot be represented by an individual species or contained in a collection." Working as he was from few representative species for each genus, Say focused on characters that helped identify, rather than rigidly define, the lower taxa and postponed final determinations until entomologists had collected and described more species.

20. Weiss and Zeigler, *Thomas Say*, p. 191.

21. Out of the fifty-four plates in the three completed volumes, Peale drew twenty-eight, Lesueur nine. Nine plates were drawn by W. W. Wood, whom Weiss and Zeigler have identified as a Philadelphia stationer and amateur insect collector. H. B. Bridport, an English miniature painter living in Philadelphia, drew two of the remaining plates. Weiss and Zeigler suggest two reasons why publication ceased

with the third volume: (1) the "indifference of the public on account of its high price"; and (2) Say's residence in New Harmony, Indiana, after 1825. See Weiss and Zeigler, *Thomas Say*, pp. 190–191. But also see Porter, *Eagle's Nest*, pp. 140–141, on the rising costs of producing natural history books with original illustrations, the failure of British publishing firms, and the refusal of Say's publisher, Mitchell, to undertake further volumes.

22. Say, *American Entomology*, vol. 1 (1824), p. v.
23. See Porter, *Eagle's Nest*, p. 64.
24. Say, *American Entomology*, vol. 1, p. v.
25. Ibid., vol. 3, n. p.
26. Porter offers an alternative explanation for the appearance of Say's illustrations. She writes: "Because academy authors such as Say and Bonaparte could not make strong [definitive] taxonomic statements, their texts relied heavily on illustrations for scientific credibility. Say admitted that an important function of his later illustrated work [on land shells] was to provide surrogate collections for readers who were isolated from the cabinets of scientific societies. This purpose called for illustrations showing obvious taxonomic characters and little attention to background, shadowing, or animated posture. At the same time, the starkness of the page and the sole emphasis on the specimen's form gave undeserved realism to a method of classification that placed a premium on external structure and none whatsoever on internal anatomy or biological relationships." *Eagle's Nest*, p. 64. Porter's interpretation does not explain, however, why Bonaparte's plates included landscape backgrounds, continuing the pictorial narrative elements initiated by Wilson, and Say's did not, given that both works based their taxonomy on external form. The lack of landscape background in Say's work reflected the lack of narrative in his approach to natural history and his intended specialist audience. The perpetuation of these elements in Bonaparte's continuation of Wilson reflected less Bonaparte's own experience—he, too, worked from preserved specimens—than the influence of Lawson. Furthermore, Bonaparte's draftsman for the first volume, Titian Peale, drew not only from the exhibits in his family's museum, each with its own painted background, but he had had extensive field experience on the Long expedition to the Rocky Mountains, 1819–1820. As the controversy over his illustrations for the later Wilkes expedition would demonstrate, for Peale personal observation and the presence of the naturalist translated into narrative landscape background in natural history illustration.

27. Tiebout was best known for his portraits; he engraved a full-length portrait of Thomas Jefferson in 1801. William Dunlap, author of *History of the Rise and Progress of the Arts of Design in the United States*, first published in 1834, was dismissive of Tiebout. Dunlap considered Tiebout's portrait style "feeble." He continued: "Mr. Tiebout was the first American who went to London for instruction in engraving, and about the same time Alexander Lawson came to America and made himself an excellent engraver without instruction." See vol. 2 (New York: Benjamin Blom, Inc., 1965), p. 156.

28. *The North American Review* 21 (1825): 251. Quoted in Jessie Poesch, "Titian Ramsay Peale, 1799–1885, and His Journal of the Wilkes Expedition," *Memoirs of the American Philosophical Society* 52 (1961): 43. The *North American Review and Miscellaneous Journal*, published in Boston, took a special interest in promoting American imprints, with its principal focus on Boston and New England writers and Boston and Cambridge learned societies. Frank Luther Mott, historian of American magazines, characterized the *Review* as "Bostonian, Harvardian, Unitarian" and noted that it tended to praise rather than criticize. See Mott, *History of American Magazines 1850–1865*, vol. 2 (Cambridge, Mass.: Harvard University Press, 1938), pp. 220–232.

29. Captain Thomas Hamilton (1789–1842), *Men and Manners in America. By the author of "Cyril Thornton"* (Edinburgh: W. Blackwood, 1833). Quoted in Weiss and Zeigler, *Thomas Say*, p. 116.

30. Weiss and Zeigler, *Thomas Say*, pp. 113–114.

31. John Godman, *American Natural History* 3 vols. (Philadelphia: H. C. Carey and I. Lea, 1826–1828). Matthew Carey, founder of Godman's publisher, Carey and Lea, had made his son, Henry Charles Carey, and son-in-law, Isaac Lea, partners in 1824. Matthew established his firm as a leader in Philadelphia publishing. He was known for his success as a bookseller and for improving book distribution. His editions of Guthrie's *Geography* and Goldsmith's *Animated Nature* had proved that a profitable market for imported natural history existed as early as the 1790s. Under the direction of his son the firm would become a major publisher of medical texts, Philadelphia being a center of medical research and education; but under the younger generation, the company became known initially for pirating British novels. Isaac Lea, a largely absent partner, was a member of the Academy and perennial contributor of shell descriptions to its *Journal*. See Tebbel, *History of Book Publishing*, vol. 1, pp. 109, 366–368.

32. The American debate over the definition of species was central to the debate over the unity or diversity of the human species. To adopt a biological definition, that members of the same species produced fertile offspring, meant that blacks and whites were a single species, a conclusion unacceptable for many Americans. On the history of the species question in the United States leading up to the Civil War, see William Stanton, *The Leopard's Spots: Scientific Attitudes Toward Race in America, 1815–1859* (Chicago: University of Chicago Press, 1960). See also chapter 5.

33. Porter, *Eagle's Nest*, p. 141. See ibid., pp. 141–143, on Godman's rivalry with Richard Harlan. Harlan published his book on North American mammals before Godman, but the illustrations in Godman's book insured that it captured public attention and outsold Harlan's.

34. Among the other illustrators for Godman were W. W. Wood, one of Say's illustrators, and Alexander Rider, former colorist for Wilson. Rider also completed the illustrations for Bonaparte's second volume.

35. See Porter, *Eagle's Nest*, p. 142.

36. Say's metonymy to describe Lesueur's skill, in a letter to George Ord, 11 April 1818. Quoted in Weiss and Zeigler, *Thomas Say*, p. 57. Courtesy of the Historical Society of Pennsylvania.

37. Kent Redford pointed out to me that the shape and position of the feet in illustration of mammals indicate whether the drawing has been made from life and direct observation, or from a preserved skin. Wrongly shaped and positioned feet characterize pre-nineteenth-century European illustrations of North and South American mammals.

38. Reynell Coates, "Reminiscences of a Voyage to India," in John Godman, *Rambles of a Naturalist . . . to which are added Reminiscences of a Voyage to India* (Philadelphia: Thomas T. Ash—Key and Biddle, 1833), pp. 130–131.

39. Ibid., p. 133.

40. Thomas Doughty, *Cabinet of Natural History and American Rural Sports*, 3 vols. (Philadelphia, 1831–1834). A friend of the Peale family, Thomas Doughty (1793–1856) is considered the first American artist to have concentrated on landscape painting. His own interests in hunting and fishing informed the subject matter of the magazine's articles. See Poesch, *Titian Peale*, pp. 58–60, and Barbara Novak and Annette Blaugrund, eds., *Next to Nature: Landscape Painting from the National Academy of Design* (New York: Harper and Row, 1980), pp. 35–37. Novak and Blaugrund reproduce one of Doughty's paintings, which was a gift to the National Academy of Design from painter James Smillie, who, with a number of his apprentices, engraved plates for natural history. See chapter 4.

41. On the Long expedition, see Porter, *Eagle's Nest*, especially chapter 2, "The Lessons of Nature," pp. 26–40; Poesch, *Titian Peale*, pp. 23–35; Weiss and Zeigler, *Thomas Say*, pp. 61–83. See also William H. Goetzmann, *Exploration and Empire: The Explorer and the Scientist in the Winning of the American West* (New York: W. W. Norton, 1966), pp. 58–62, 182–184.

42. On George Catlin's observations of Plains Indians in the early 1830s, for comparison to Peale's hunting scenes, see Vincent Crapanzano, "Hermes's Dilemma: The Masking of Subversion in Ethnographic Description," in James Clifford and George E. Marcus, eds., *Writing Culture: The Poetics and Politics of Ethnography* (Berkeley: University of California Press, 1986), pp. 51–76.

43. James, *Account of an Expedition*. The atlas in the books reproduced only one of Peale's sketches, that of the "Moveable Skin Lodge of the Kaskaias," drawn by T. R. Peale, engraved by Young and Delleker. The remaining plates were engraved from drawings by Samuel Seymour, the painter engaged to record landscapes and Indian portraits. See Poesch, *Titian Peale*, p. 24.

44. See for example, James, *Account of an Expedition*, chapter 1, p. 5, for Say's description of *Triton lateralis*, an amphibian. The description is as technical as any he would publish in a scientific journal. Animals discovered and described from the expedition are listed in a systematic appendix to vol. 1, pp. 369–378, including observations of bird migrations.

45. See Porter, *Eagle's Nest*, chapter 10, "The Rise of Peer Review," pp. 125–134; also Charlotte Porter, "'Subsilentio': Discouraged Works of Early Nineteenth-Century American Natural History," *Journal of the Society for the Bibliography of Natural History* 9, no. 2 (1979): 109–119.

46. Porter, *Eagle's Nest*, p. 140.

47. On the relationship between popular natural history texts and school texts and their importance in the establishment of an American culture of science, see Sally Gregory Kohlstedt, "Parlors, Primers, and Public Schooling: Education for Science in Nineteenth-Century America," *Isis* 81, no. 308 (September 1990): 425–445.

48. See W. T. Harris's "Introduction," p. xii, to George B. Lockwood, *The New Harmony Movement* (New York: D. Appleton and Company, 1907 [1905]). See Porter, *Eagle's Nest*, part 3, "New Harmony," for

an excellent account of Maclure's and Say's participation at New Harmony.

49. Quoted in Lockwood, *New Harmony Movement*, p. 221.

50. Ibid., p. 225.

51. Ibid., pp. 217–220.

52. The influence of Owen's curriculum was undermined locally by the effects of child labor—students were pulled from the higher school to go to work at the age of ten—and at the national level by public condemnation of Owen's atheism and therefore rejection of his larger social and educational vision. Keith Taylor, *The Political Ideas of Utopian Socialists* (London: Frank Cass, 1982), pp. 69–99.

53. Ibid., pp. 84–85.

54. Lockwood, *New Harmony Movement*, pp. 272–281, discusses the program of the Schuylkill school based on Neef's treatises on his principles, among them *A Proper System of Education for the Schools of a Free People* (1807), and *Neef's Method of Teaching* (Philadelphia, 1813).

55. Lockwood, *New Harmony Movement*, p. 235.

56. See Novales, *William Maclure*, chapter 6, "La escuela industrial-agraria," pp. 65–78; and Lockwood, *New Harmony Movement*, pp. 232–233.

57. One of the more sympathetic discussions of George Rapp and the Harmonists can be found in William E. Wilson's *The Angel and the Serpent: The Story of New Harmony* (Bloomington: Indiana University Press, 1964).

58. See "Mr. Owen's Address, to the Citizens of the United States," published in *Nile's Register*, 12 November 1825. Reprinted in Karl J. R. Arndt, ed., *Harmony on the Wabash in Transition; From Rapp to Owen, 1824–1826, A Documentary History* (Worcester, Mass.: Harmony Society Press, 1982), pp. 702–705.

59. On Owen's architectural model for New Harmony and its place in the history of the architecture of social control, see Dolores Hayden, *Seven American Utopias: The Architecture of Communitarian Socialism, 1790–1975* (Cambridge, Mass.: MIT Press, 1976).

60. William Maclure, "An Epitome of the Improved Pestalozzian System of Education, as practised by William Phiquepal and Madam Fretageot, formerly of Paris, and now in Philadelphia; communicated at the request of the Editor," *American Journal of Science and Arts* 10, no. 1 (January 1826): 145–151.

61. Ibid., p. 145.

62. Ibid., pp. 145–146.

63. William Maclure, "No. X. Education," 13 February 1828, in *Opinions*, vol. 1, pp. 48–52.

64. Maclure, "Epitome," p. 146.

65. Maclure, "Education," p. 48.

66. Maclure, *Opinions*, vol. 3, 12 September 1835, "The Correct Cultivation of the Senses, the Proper Mode of Acquiring Knowledge," pp. 102–105; and vol. 3, 4 July 1836, "Improving the Senses by Practice—Proper Method of Teaching Children," pp. 204–207. Lockwood notes that the 1824 founding of the Rensselaer Institute in Troy, New York, scooped by one year Maclure's priority in establishing industrial education in the United States. Lockwood, however, considers Rensselaer a technical institute, while Maclure's New Harmony venture he characterizes as a trade school. Lockwood, *New Harmony Movement*, pp. 156–165.

67. Maclure, "Epitome," p. 151.

68. Ibid., p. 150. It is not clear whether the school ever taught lithography.

69. Ibid., pp. 146–147.

70. Lesueur's pencil views of the journey are reproduced in *Dessins de Charles Alexandre Lesueur, exécutés aux Etats Unis de 1816 à 1837.* (Paris: Edité par la Cité Universitaire de Paris, 1933).

71. Diaries of the journey kept by Robert Dale Owen, Robert Owen's son, and by Donald McDonald are reprinted in Arndt, *Harmony*, pp. 785–793. See also the recollections of Victor Colin Duclos reprinted in Harlow Lindley, ed., *New Harmony as Seen by Participants and Travelers* (Philadelphia: Porcupine Press, 1975; Indianapolis: Indiana Historical Society, 1916), n.p.

72. R. Owen, "Address to the New Harmony Community Membership," *New Harmony Gazette* 1, 27 April 1825, p. 1. Reprinted in Arndt, *Harmony*, p. 537.

73. Lockwood, *New Harmony Movement*, pp. 114–115. By this system, New Harmony, at 38.11 N, 87.55 W, became Ipba Veinul.

74. Lockwood, *New Harmony Movement*, p. 83.

75. Lindley, *New Harmony*, n.p. The duke's account was published in English as Karl Bernhardt, Duke of Saxe-Weimar Eisenach, *Travels through North America during the Years 1825 and 1826*, 2 vols. (Philadelphia: Carey, Lea and Carey, 1828).

76. The best account of the internal strife is Wilson's, *Angel and the Serpent*, pp. 177–179.

77. Lockwood, *New Harmony Movement*, p. 252.

78. Maclure's programs underwent a series of changes as the schools became increasingly devoted to teaching trades and skills. See Porter, *Eagle's Nest*, p. 105.

79. *The New Harmony Gazette* moved to New York and continued publication as *The Free Enquirer* under the editorship of Robert Dale Owen

and Frances Wright; Wilson, *Angel and the Serpent*, p. 183.

80. Thomas Say, *American Conchology, or Descriptions of the Shells of North America*, Illustrated with Colored Figures from Original Drawings Executed from Nature (New Harmony, Indiana: School Press, 1830–1834). Quotation is from Say's prospectus for *American Conchology*, reprinted in Weiss and Zeigler, *Thomas Say*, pp. 139–140.

81. Weiss and Zeigler, *Thomas Say*, pp. 192–193. See Porter's account of scientific publishing at New Harmony in *Eagle's Nest*, pp. 105–106, 118–120.

82. Thomas Say to Mme Fretageot, 30 July 1832 and 23 August 1832, Workingmen's Institute Library, New Harmony, Indiana, New Harmony Mss. Say to Fretageot, 21 November 1832, reprinted in Weiss and Zeigler, *Thomas Say*, pp. 150–153.

83. Ian MacPhail, "Natural History in Utopia: The Works of Thomas Say and François-André Michaux at New Harmony, Indiana," in *Contributions to the History of North American Natural History*, papers from the First North American Conference of the Society for the Bibliography of Natural History held at the Academy of Natural Sciences, Philadelphia, 21–23 October 1981. Special Publication no. 2 (London: Society for the Bibliography of Natural History, 1983) pp. 15–33.

84. Weiss and Zeigler, *Thomas Say*, p. 193.

85. Say's prospectus, in Weiss and Zeigler, *Thomas Say*, pp. 139–140.

86. Thomas Say to Mme Fretageot, 23 August 1832; Achille Fretageot to Mme Fretageot, 5 January 1832, Workingmen's Institute Library, New Harmony, Indiana, New Harmony Mss. See also, MacPhail, "Natural History," p. 19.

87. Porter describes Say's espousal of Maclure's principles and the publication of Say's *Conchology* in *Eagle's Nest*, pp. 107–108.

88. Maclure, "Desirable and Probable Diminution of the Cost of Books, for the More General Diffusion of Knowledge," 10 November 1834, in *Opinions*, vol. 2, pp. 374–378.

89. Ibid., p. 374.

90. Ibid., p. 376.

91. Ibid., p. 377; p. 376.

92. Ibid., p. 376.

93. Ibid., p. 375.

94. See Lesueur, *Dessins*. Selected sketches from Lesueur's North American sojourn are reproduced in Vail, *The American Sketchbooks of Charles Alexandre Lesueur*, 1938.

95. Porter, *Eagle's Nest*, p. 105.

96. Richard Owen, letter to David Starr Jordan, 14 December 1886, published in David Starr Jordan, "Sketch of Charles A. Le Sueur," *Popular Science Monthly*, February 1895, pp. 547–550.

97. Ibid., p. 549.

98. "Lesueur refuses to associate Say in his publication and this last refuses the translation of it if he is not put in partnership for the descriptions. So much for author vanity." Mme Fretageot to William Maclure, 28 July 1831, quoted in Weiss and Zeigler, *Thomas Say*, p. 146.

99. Maclure had purchased the plates of Michaux's *Sylva* and issued a reprint from New Harmony. See Porter, " 'Subsilentio,' " pp. 109–119, and MacPhail, "Natural History."

100. Quoted in Porter, " 'Subsilentio,' " p. 116.

101. Lindley, *New Harmony*, n.p.

102. Maximilian, Alexander Philip, Prince of Wied-Neuwied, *Travels in the Interior of North America, 1832–1834*, 2 vols. and atlas, trans. H. Evans Lloyd (London: Ackermann, 1843), reprinted in Reuben Gold Thwaites, ed., *Early Western Travels*, 32 vols. (Cleveland: Arthur H. Clark Co., 1906), vols. 22 and 23.

103. On Lesueur's failure to publish his fish studies, see Porter, *Eagle's Nest*, pp. 116–117. Because of Lesueur's isolation in New Harmony and his subsequent return to Europe, Asa Gray, American botanist of the next generation and professor of natural history at Harvard College, was ready to dismiss his contributions to the taxonomy of American fish. The Swiss zoologist Louis Agassiz, however, agreed with Prince Maximilian in valuing Lesueur's work. See also the opinions of ichthyologists George Brown Goode and David Starr Jordan quoted in Vail, *American Sketchbook*, pp. 18–19.

104. See Mme Fretageot to William Maclure, 24 January 1831, quoted in Weiss and Zeigler, *Thomas Say*, pp. 145–146.

105. Lucy Say, New York, 19 January 1835, letter to William Price, Cincinnati, Ms. Collection 433, Academy of Natural Sciences, Philadelphia.

106. Ibid., n.p. Although at the time of Thomas Say's death some members of the Philadelphia Academy of Natural Sciences published critical notices of his work, his reputation was rehabilitated. For the remainder of her life, Lucy Say cooperated with entomologists to publish her husband's work. In 1841, she was the first woman elected a member of the Academy.

107. Duke of Saxe-Weimar, quoted in Lindley, *Participants and Travelers*, n.p.

108. David Dale Owen's publications and illustrations are discussed and reproduced in chapter 7.

109. See Porter, *Eagle's Nest*, chapter 9, "Maclure's Legacy," pp. 109–121.

CHAPTER 3
FROM NATURALIST-ILLUSTRATOR
TO ARTIST-NATURALIST

1. Audubon biographies abound. His own success as a self-publicist, his deliberate mystification of his past, and the destruction of some of his papers by his family obscured the events of his birth and youth. The adulatory biographies by Lucy Bakewell Audubon and Maria R. Audubon (1843–1925), his granddaughter, contributed to one of the central legends of popular American conceptions of nature. See Maria R. Audubon, *Audubon and His Journals* (New York: Charles Scribner's Sons, 1897). The earliest reliable source on Audubon's life is Francis Hobart Herrick, *Audubon the Naturalist: A History of His Life and Time*, 2 vols. (New York: D. Appleton, 1917). My notes refer to the second edition, 1938.

2. See Alice Ford's edition of *The 1826 Journal of John James Audubon* (Norman: University of Oklahoma Press, 1967), pp. 65–66, n. 15, for refutation of Audubon's claim to have studied with David. The Museum of Comparative Zoology, Harvard University, owns many of Audubon's early drawings. For reproductions of Audubon's original drawings for the *Birds of America*, most of which are now in the collections of the New York Historical Society, see *The Original Water-Color Paintings by John James Audubon for The Birds of America*, Introduction by Marshall B. Davidson, (New York: American Heritage Publishing Co., Inc., 1966). My discussion of Audubon's drawing refers to these sources.

3. George Keats, the brother of the poet John Keats, met Audubon through the Bakewells, and lived with the Audubons in Henderson, Kentucky. George was ruined in one of Audubon's failed ventures. See Hyder Edward Rollins, ed., *The Keats Circle: Letters and Papers, 1816–1878*, 2 vols. (Cambridge, Mass.: Harvard University Press, 1948), vol. 1, pp. cv, cviii.

4. On the meeting between Wilson and Audubon, see Elsa Allen, "History of American Ornithology before Audubon," pp. 561–563. Allen believes that the encounter prompted Audubon's ambition to publish his drawings.

5. Herrick, *Audubon*, vol. 1, pp. 303–306.

6. Poesch, *Titian Peale*, p. 25.

7. During the 1830s, Maria Martin, like Joseph Mason, drew plants and insects for the birds of the Carolinas and Florida. She was the sister-in-law and later wife of John Bachman, Audubon's collaborator on his sequel to the birds, *The Viviparous Quadrupeds of North America*, discussed in chapter 5.

8. Herrick, *Audubon*, vol. 1, pp. 307–326.

9. Charles Lucien Bonaparte, *American Ornithology; or, The Natural History of Birds Inhabiting the United States, Not Given by Wilson*, 4 vols. (Philadelphia: Samuel Augustus Mitchell [vol. 1]; Carey, Lea and Carey [vols. 2–3]; Carey and Lea [vol. 4], 1825–1833).

10. On the collections of Lucien Bonaparte, younger brother of Napoleon, see Francis Haskell, *Rediscoveries in Art: Some Aspects of Taste, Fashion and Collecting in England and France* (Ithaca, N.Y.: Cornell University Press, 1976), pp. 57–60. To compare Audubon's mature style with that of Oudry, see Hal Opperman, *J.-B. Oudry, 1686–1755*.

11. Dunlap, *Rise and Progress*, vol. 3, p. 204.

12. Ibid., p. 208.

13. Ibid., p. 202.

14. Ibid., p. 203.

15. Ibid.

16. Ibid., p. 204.

17. Ibid.

18. Ibid., pp. 204–205.

19. The head and eyes of vertebrate young are larger in proportion to their bodies than those of adults. Audubon's enlargement of his birds' heads and eyes increased the appeal of his drawings. See Stephen Jay Gould, "A Biological Homage to Mickey Mouse," in *The Panda's Thumb: More Reflections on Natural History* (New York: W. W. Norton, 1982 [1980]), pp. 95–107, on neotony and viewer response to Disney cartoon characters. Audubon continued, no doubt unconsciously, to use this trope. The Disney chipmunks, Chip and Dale, bear considerable resemblance to Audubon's later pictures of rodents, reproduced in the lithographic plates of J. J. Audubon and John Bachman, *The Viviparous Quadrupeds of North America* (New York: J. J. Audubon, 1846–1854).

20. Porter, *Eagle's Nest*, p. 64.

21. Bonaparte, *American Ornithology*, vol. 1, p. iv.

22. Ibid., vol. 1, p. iv; vol. 3, p. 21.

23. Ibid., vol. 1, p. iv.

24. Ibid., vol. 4, p. 21.

25. Peale drew only for the first volume of Bonaparte's book. Alexander Rider, the Swiss painter who had worked as a colorist for Wilson and for Bonaparte's first volume, completed the work. Rider also drew from mounted specimens in the Peale museum for Godman's *American Natural History*. See Cantwell, *Wilson*, p. 143, and Porter, *Eagle's Nest*, pp. 45, 56, 141.

26. Wilson used a similar composition in illustrating the cow bunting, *American Ornithology*, vol. 2, plate 18.

27. For examples, see drawings of the swallow-tailed kite, 1821 (Havell 72); red-tailed hawk, 1821 (Havell 51); great blue heron, 1821, and later reworked (Havell 211).

28. Before the Philadelphia visit, Audubon had depicted birds in conflict; for example, Havell plate 51, engraved after a drawing done in Louisiana in 1821, shows two red-tailed hawks fighting over a rabbit.

29. There have been other periodizations of Audubon's work. See Robert Henry Welker, *Birds and Men: American Birds in Science, Art, Literature, and Conservation, 1800–1900* (Cambridge, Mass.: The Belknap Press of Harvard University Press, 1955), chapter 6, "Bird Art and Audubon," pp. 71–90. Welker cites Donald Culross Peattie, who described three phases of Audubon's work: "The first to about Plate 150 [of the *Birds of America*], full of exuberance which may be excessive but is always full of genius; the second, to about Plate 370, with the artist in full command of his mature techniques; and the third, showing a decline signalized by crowding of numerous figures into a single plate, by repetition of backgrounds and artificiality of scenes" (p. 82). Welker agrees with Peattie in the main, and adds an insightful analysis of Audubon's compositional devices.

30. On the distinction between fashionable luxury volumes of ornithological plates and specialized, systematic bird study, see Farber, *Emergence*, pp. 87–90.

31. Porter, *Eagle's Nest*, p. 143.

32. The most interesting—and animated—account is Audubon's own; see Ford, *Journal*. See also, Herrick, *Audubon*, vol. 1, pp. 347–419. For the details of publication, see Waldemar H. Fries, *The Double Elephant Folio: The Story of Audubon's Birds of America* (Chicago: American Library Association, 1973).

33. Audubon, Manchester, 17 September 1826, in Ford, *Journal*, p. 150.

34. Audubon to Sully, 20 December 1826, in ibid., p. 356.

35. Audubon, 30 October 1826, in Ford, *Journal*, p. 244. Although later Audubon was able to strike a collegial tone in print, he cherished an enduring grudge against Ord and Lawson in particular, and in the privacy of his journal he gave free vent to his feelings of injury and his desire to impress them.

36. See Twyman, *Lithography*, chapter 13, pp. 179–200. See also Stafford, *Voyage*.

37. Audubon, 2 November 1826, in Ford, *Journal*, pp. 249–250. Most of the drawings listed in this incident were engraved by Lizars before Audubon moved the project to Havell in London.

38. Fries, *Double Elephant Folio*, pp. 197–198.

39. See Twyman, *Lithography*, pp. 171, 175–176.

40. Audubon observed with excitement that Lizars's colorists worked by gaslight, which suggests that they worked extended hours under dim illumination. See entry for 1 November 1826 in Ford, *Journal*, p. 248. Lizars engraved only the first ten plates. He did not push publication as fast as Audubon wanted, in part because of the strike, but also because he was engraving the illustrations for another ornithological work, Sir William Jardine's and Selby's *Illustrations of Ornithology*. See Fries, *Double Elephant Folio*, p. 24. The Havells reissued the first ten plates. See also Herrick, *Audubon*, vol. 2, pp. 195–199.

41. Fries, *Double Elephant Folio*, p. 19.

42. These paintings are reproduced in Ford, *Journal*.

43. Audubon to Sully, 20 December 1826, in Ford, *Journal*, p. 356.

44. Fries, *Double Elephant Folio*, pp. 360–364.

45. Maria R. Audubon, ed., *Journals*, vol. 1, p. 289. Quoted in Fries, *Double Elephant Folio*, p. 38.

46. Lehman painted the landscape backgrounds, and possibly some of the birds, for the plates of the Louisiana heron (Havell 217), the roseate spoonbill (Havell 321), the great white heron (Havell 281), and the long-billed curlew (Havell 231). Lehman also painted plants for compositions similar to those by Joseph Mason. Audubon's son, John Woodhouse, was responsible for drawing some of the birds that Audubon worked on in Charleston, S.C., and his son Victor Gifford may have painted the background of the distant landscape for the brown pelican (Havell 421). See *Watercolors*, plate 396.

47. See Fries, *Double Elephant Folio*, pp. 440–451. Audubon lost many subscribers along the way and had difficulty collecting payment from others. Some American subscribers dropped out during the economic depression of the late 1830s following the collapse of the banking system. In the lean years of the late '30s and '40s, American reviews of Audubon's work referred to the economically inauspicious times for expensive publication projects.

48. William Swainson, "Some Account of the Work now publishing by M. Audubon, entitled The Birds of America," *The Magazine of Natural History and Journal of Zoology, Botany, Mineralogy, Geology and Meteorology* 1, part 2 (May 1828): 43–52.

49. Ibid., p. 45.

50. Swainson was a proponent of the quinary system, a method of classification based on determining the affinities within and between five orders, arranged in a circle, so that the first connected with the last. See Farber, *Emergence*, pp. 111–113; and Robert O'Hara, "Diagrammatic Classifications of Birds, 1819–1901: Views of the Natural System in 19th-Century British Ornithology," *Proceedings of the Nineteenth International Ornithological Congress*, Ottawa, 1986.

51. Swainson, "Account," p. 45.

52. Ibid., p. 45.

53. Ibid., pp. 46–47.

54. Ibid., p. 49.

55. Ibid., p. 48.

56. Ibid., p. 51.

57. "Introduction," *The Magazine of Natural History* 1, no. 1 (May 1828): 3.

58. Ibid.

59. Fries notes that the separate publication of the plates and letterpress freed Audubon from the requirement under British copyright of depositing complete copies of the work in nine libraries in the United Kingdom. Nine copies of the *Birds of America* would have cost more than Audubon could afford. See Fries, *Double Elephant Folio*, p. 47.

60. Audubon to William Swainson, 22 August 1830. Reprinted in Herrick, *Audubon*, vol. 2, pp. 101–103.

61. Ibid.

62. Swainson to Audubon, 2 October 1830, reprinted in ibid., pp. 106–108.

63. See Farber, *Emergence*, p. 96.

64. See ibid., especially chapter 6, "Focus on Classification: Ornithology, 1800–1820," and chapter 7, "The Emergence of a Discipline: Ornithology, 1820–1850," for discussion of the growing specialization of the field and the increasingly sharp distinction between natural history and the scientific and museum-based study of birds.

65. Herrick, *Audubon*, vol. 2, p. 108.

66. John James Audubon, *Ornithological Biography, or an Account of the Habits of the Birds of the United States of America; Accompanied by Descriptions of the Objects Represented in the Work Entitled, The Birds of America, and Interspersed with Delineations of American Scenery and Manners*, 5 vols. (Edinburgh: Adam Black, 1831–1838).

67. Alexander Wilson and Charles Lucien Bonaparte, in Robert Jameson, ed., *American Ornithology, or the Natural History of Birds of the United States*, 4 vols. (Edinburgh and London: Constable and Co., 1831); Alexander Wilson, *American Ornithology; or the Natural History of Birds of the United States*, with a Continuation by Charles Lucien Bonaparte . . . The Illustrative Notes and Life of Wilson by Sir William Jardine, 3 vols. (London and Edinburgh, 1832); Thomas Brown, *Illustrations of the American Ornithology of Alexander Wilson and Charles Lucien Bonaparte* . . . (Edinburgh and London: Frazer and Co., 1835). See Herrick, *Audubon*, vol. 1, pp. 438–444, especially p. 442, nn. 5 and 6.

68. Audubon, *Ornithological Biography*, vol. 3 (1835), p. ix.

69. Ibid., pp. 629–632.

70. [W.B.O. Peabody], "*History of British Birds*, by William MacGillivray, vol. 1; *Ornithological Biography*, by John James Audubon, vol. 5," *North American Review* 50, no. 107 (April 1840): 381–404.

71. See Herrick's bibliography of reviews and criticism of Audubon's work, *Audubon*, vol. 2, pp. 424–440, for a discussion of Waterton's role in the rattlesnake debate, and its sequel, the controversy over sense of smell versus sense of sight in vultures.

72. Georges Cuvier, "Rapport verbal fait à L'Académie des Sciences, sur L'histoire naturelle des oiseaux de L'Amerique septentrionale, de M. Audubon," *Le Moniteur Universel* (Paris) 275 (1 October 1828): 1528.

73. Frédéric Cuvier, "Ornithological Biography, or an account of the habits of birds, . . . by John James Audubon," *Journal des Savants* (Paris), December 1833, p. 706. My translation. Quoted in translation by Farber, *Emergence*, p. 106.

74. The folios of Daniel Giraud Elliot (1835–1915) aimed at a luxury market and followed the Audubon model of bird-landscape compositions, but were realized with a heavy hand. See Daniel Giraud Elliot, *A Monograph of the Phasianidae, or Family of Pheasants* (New York, 1872); *A Monograph of the Paradiseidae, or Birds of Paradise* (London, 1873); *A Monograph of the Felidae, or Family of Cats* (London: For the subscribers, by the author, 1883); *A Monograph of the Bucerotidae, or Family of Hornbills* (London: For the subscribers, by the author, 1882).

75. Farber, *Emergence*, pp. 107–108. See also Gordon C. Sauer, *John Gould the Bird Man: A Chronology and Bibliography* (London: Henry Sotheran Ltd., 1982).

76. Quoted in Herrick, *Audubon*, vol. 2, pp. 128–129.

77. The illustrations to the geological survey of Maine depict the interrelationship between public funding and amateur science: gentlemen in tailcoats

and top hats, their wagons and surveying equipment, fill the foregrounds of the geological landscapes. Charles T. Jackson, *Atlas of plates, illustrating the geology of the state of Maine, accompanying the first report of the geology of that state* (Augusta, Maine: Smith and Robinson, 1837–1839).

78. See Porter, *Eagle's Nest*, pp. 129–130. Thomas Nuttall, *A Manual of Ornithology of the United States and Canada, Land Birds* (Cambridge, Mass.: Hilliard and Brown, 1832); John Kirk Townsend, *Ornithology of the United States, or; Descriptions of the Birds Inhabiting the States and Territories of the Union*, vol. 1 (Philadelphia: J. B. Chevalier, 1839).

79. See Barbara Novak, *Nature and Culture: American Landscape and Painting, 1825–1875* (New York: Oxford University Press, 1980).

80. [W.B.O. Peabody], "Audubon's Biography of Birds. Ornithological Biography of Birds . . . Philadelphia, 1831," *North American Review* 34 (April 1832): 364–405.

81. Ibid., p. 379.

82. Anon., "A Synopsis of the Birds of North America; by John James Audubon . . . The Birds of America, from drawings made in the United States and their Territories, . . . Vol. I. New York, 1840," *American Journal of Science and Arts* 39, no. 2 (July–September 1840): 343–357.

83. Ibid., p. 348.

84. Ibid., pp. 349–350.

85. Ibid. Novak notes that American artists inspired by European history painting translated ambition into size, as had Audubon, "engaging the great European masters in rivalry through an appropriately aggressive scale, as well as an aggressively announced 'nobility.' Perhaps for these artists large size was automatically endowed with monumentality and grandeur." Novak also discusses how the American "public had to be trained to respect Ambition"; *Nature and Culture*, p. 19. Audubon's reviewer had learned this lesson well.

86. By the late nineteenth century Audubon was popularly considered the founder of American ornithology, and his name could be marshalled to unite the divergent strands of professional ornithology and popular bird lore in the common cause of bird protection under the Audubon societies of the 1890s. See Welker, *Birds and Men*, pp. 204–212.

87. See Goetzmann, *Exploration and Empire*, chapter 5, "Something More than Beaver," pp. 146–180.

88. Thomas Cole, "Essay on American Scenery" (1835), in John McCoubrey, ed., *American Art, 1700–1960*, Sources and Documents in the History of Art Series (Englewood Cliffs, N.J.: Prentice-Hall,

1965), p. 102. Quoted in Novak, *Nature and Culture*, pp. 4–5.

89. Audubon, *Ornithological Biography*, vol. 5, p. vii.

90. Ibid., p. viii.

91. Ibid., p. x.

## CHAPTER 4
## SCIENTIFIC PRESTIGE, NATIONAL HONOR

1. The initial proliferation of natural history societies began to slow around 1825, according to George H. Daniels, *Science in American Society: A Social History* (New York: Alfred A. Knopf, 1971), p. 148. Before 1840 almost twice as many agricultural societies as geological and other scientific societies had been established, speaking to the practical orientation of memberships recruited from the general public. See Ralph S. Bates, *Scientific Societies in the United States*, 3d ed. (Cambridge, Mass.: MIT Press, 1965), tables on pp. 51 and 61. Nevertheless, twenty-two states instituted geological surveys between 1822 and the close of the 1840s; Howard S. Miller, "The Political Economy of Science," in George C. Daniels, ed., *Nineteenth-Century American Science: A Reappraisal* (Evanston, Ill.: Northwestern University Press, 1972), pp. 95–112. Despite the number of new agricultural societies compared to geological societies, the state geological surveys paid less attention to agriculture than one might expect. The surveys' success seems to have depended largely on the strong personalities of governors and survey leaders, their lobbying skills, and the profuse acknowledgments of local participation and cooperation in survey work. See Michele Aldrich, "American state geological surveys, 1820–1845" in Cecil Schneer, ed., *Two Hundred Years of Geology in America* (Hanover, N.H.: University Press of New England, 1979), pp. 133–144.

2. Samuel George Morton's *Crania Americana*, for example, was illustrated with drawings of skulls by Philadelphia lithographer John Collins. John Edward Holbrook's *North American Herpetology* contained plates drawn by the Italian drawing master J. Sera and the scientific illustrator John H. Richard, and printed in part by the experimental color medium of the lithotint: Samuel George Morton, *Crania Americana* (Philadelphia, 1839); John Edwards Holbrook, *North American Herpetology*, 4 vols. (Philadelphia, 1836–1840). Richard's origins are unclear. Some sources cite his German origins: George C. Groce and David W. Wallace, *The New*

The tensions inherent in the distinction be-

York Historical Society's Dictionary of American Artists, 1564–1860 (New Haven, Conn.: Yale University Press, 1957), p. 534. Goetzmann states that Richard worked in Paris; Goetzmann, Exploration and Empire, p. 345. Theodore Gill, for whom Richard drew fish, called him Alsacian French; "The first edition of Holbrook's North American Herpetology," Science 17, no. 440 (5 June 1903): 910–912. Joseph Drayton, artist on the United States Exploring Expedition, 1838–1840, called him "my Dutchman"; Drayton to Charles Wilkes, 11 December 1852, Wilkes Letter Books, vol. 4, quoted in Daniel C. Haskell, The United States Exploring Expedition, 1838–1842, and Its Publications, 1844–1874 (New York: New York Public Library, 1942), p. 77. The illustrators and illustrations of Morton's and Holbrook's works are treated at greater length in the following chapter.

3. William Stanton finds the history of debate on federal funding for Pacific exploration rife with arguments about prestige. One senator resented that civilian scientists would benefit from the prestige of a successful expedition; any "glory" ought to go to the public body which provided the funding. An editor of the New York Mirror, on the other hand, castigated the federal government's "narrow policy and contemptible economy" for failing to support exploration, and praised private supporters, the "active and enlightened merchants" whose capital would "promote the cause of science, and add to the reputation of the country." William Stanton, The Great United States Exploring Expedition of 1838–1842 (Berkeley: University of California Press, 1975), pp. 24–26.

4. A congressional committee offered the following justification for the 1836 appropriations for the United States Exploring Expedition to the Pacific: "In the seas which it was proposed to explore, the whale fishery alone gave employment to more than one-tenth of our tonnage, manned by twelve thousand men, and requiring capital then estimated at twelve millions of dollars" (29th Cong., 1st sess. Sen. Doc. 405, p. 1), quoted in Haskell, Exploring Expedition and Publications, p. 1, n. 3.

5. A. Hunter Dupree, Science in the Federal Government: A History of Policies and Activities to 1940 (Cambridge, Mass.: The Belknap Press of Harvard University Press, 1957), p. 57.

6. Charles Wilkes to the Secretary of the Navy, 16 July 1842, Wilkes Papers, Library of Congress, reprinted in Nathan Reingold, Science in Nineteenth-Century America: A Documentary History (New York: Hill and Wang, 1964), p. 125.

7. Porter, Eagle's Nest, p. 141.

8. The tensions inherent in the distinction between specialist and amateur are an important theme in the literature on American science of the 1830s through '40s. See Daniels, American Science in the Age of Jackson, chapter 2, "The Scientific Profession"; Dupree, Science in the Federal Government, chapter 3, "Practical Achievements in the Age of the Common Man." For an alternative discussion of professionalization in nineteenth-century science, see Susan Faye Cannon, Science in Culture: The Early Victorian Period (New York: Dawson and Science History Publications, 1978), chapter 5, "Professionalization."

9. See Porter on Constantine Rafinesque (1783–1840) in Eagle's Nest, pp. 147–149.

10. Ibid., p. 11. Porter provides the best account of the American shift from the values and practices of field natural history to institutional science. See Eagle's Nest, chapters 10–12, pp. 125–158.

11. Although one strand of American portrait and landscape painting developed from craft origins, similar distinctions were emerging between painters in oils and book illustrators and printmakers. The establishment of painting as a profession, like that of science, brought about new distinctions among practitioners.

12. Helen E. Lawson practiced engraving and drew for S. S. Haldemann, entomologist, conchologist, and member of the Academy of Natural Sciences. She also exhibited her portraits and decorative paintings in Philadelphia from 1830 to 1842. Oscar A. Lawson (1813–1854) also had public exhibitions of his work, in addition to drawing and engraving natural history subjects. After an eleven-year career with the U.S. Coast Survey he returned to Philadelphia and engraving in 1851. See Groce and Wallace, Dictionary of American Artists, p. 388. For examples of work from the Maverick shop, see Annals of the Lyceum of Natural History of New-York, vols. 1 and 2. Emily Maverick engraved John LeConte's drawings of beetles to illustrate LeConte's "Description of some new species of North American Insects," vol. 1 (1824), pp. 169–173, plate 11. Emily Maverick and Maria A. Maverick engraved Titian Peale's drawings for Richard Harlan, "Observations on the Genus Salamandra, with the anatomy of the Salamandra Gigantea (Barton) or S. Alleghaniensis (Michaux) and two new genera proposed," vol. 1 (1824), pp. 222–234, plates 16 and 17. Peter Maverick lithographed a plate published in vol. 2 of the Annals (1828). His figure of a segment of an Ammonite looks strikingly like a small fish with a human face. See plate 5, figure 1, illustrating James De Kay, "Report on several fossil multilocular shells from the State of

364

Delaware: With observations on a second specimen of the new fossil Genus Eurypterus," vol. 2 (1828), pp. 273–279.

13. These statistics refer to printing in general, not necessarily to engraving and lithography establishments. Moreoever, many small firms persisted; the median number of employees per Philadelphia printing firm in 1850 was 9.50. See Bruce Laurie, Theodore Hershberg, and George Alter, "Immigrants and Industry: The Philadelphia Experience, 1850–1880," *Journal of Social History* 9, no. 2 (Winter 1975): 219–248.

14. The federal government had already embarked on a literature of overland exploration of the trans-Mississippian west. See Goetzmann, *Exploration and Empire*, chapters 5 and 6, pp. 146–198.

15. My account draws primarily on the histories of the expedition and its results by Stanton and Haskell and the volume edited by Herman J. Viola and Carolyn Margolis, *Magnificent Voyagers: The U.S. Exploring Expedition, 1838–1842* (Washington, D.C.: Smithsonian Institution, 1985). Stanton incorporates the full breadth of primary sources pertaining to the expedition, its preparation, the voyage, and the aftermath. Haskell provides meticulous bibliographic detail and quotes a rich fund of primary documents on the publication program. Jessie Poesch's biography of Titian Peale offers valuable perspectives on the voyage and the publication of its results: Poesch, *Titian Peale*, chapters 6–10, pp. 63–103. See also David B. Tyler, *The Wilkes Expedition: The First United States Exploring Expedition (1838–1842)* (Philadelphia: The American Philosophical Society, 1968). These works document the expedition thoroughly; the Viola and Margolis volume contains some splendid reproductions of the report illustrations. Rather than reiterating the well-told story of the expedition and the politics surrounding it, I use these sources to interpret changes in illustration practices and meanings.

16. Some reports never saw publication, notably that on the fish collections, contracted to Louis Agassiz. Decades later, Henry Weed Fowler described the expedition fish; his unpublished manuscript is at the Smithsonian Institution Archives, RU7180. The botanical volumes were the most troublesome, the last of them, edited by Asa Gray, finally appearing in 1874. For the complete catalog and collation of the expedition publications, see Haskell, *Publications*, pp. 31–110.

17. Reynolds began his involvement in the project of mounting a polar exploring expedition as a proponent of John Cleves Symmes's theory of a hollow earth. He made a strategic move away from the theoretical margin to argue for the expedition based on the need for a geographic and nautical scientific survey and for the commercial advantages of exploration; these projects won him the ear of President John Quincy Adams. See Stanton, *Exploring Expedition*, pp. 13–19.

18. Jeremiah Reynolds to Andrew Jackson, 16 November 1836, Wilkes Expedition Letters, National Archives, reprinted in Reingold, *Documentary History*, pp. 112–116.

19. Ibid., p. 116.

20. Ibid.

21. Ibid. Cole seems to have reciprocated the respect given him by scientists and taken an interest in scientific developments; see Novak, *Nature and Culture*, p. 57. The production of the expedition atlases and the developing field of American landscape painting had many connecting threads. John Casilear (1811–1893), among the ranks of engravers to work on the massive atlases, would later gain recognition for his landscapes in oils. Casilear's brother and the engraver James Smillie, both engravers for the expedition reports, also produced illustrations for the popular fiction of Washington Irving. See also Roger B. Stein, *John Ruskin and Aesthetic Thought in America, 1840–1900* (Cambridge, Mass.: Harvard University Press, 1967), p. 3.

22. American Philosophical Society to Wilkes, October 1836, Wilkes Papers, Library of Congress, quoted in Poesch, *Titian Peale*, p. 64.

23. Stanton, *Exploring Expedition*, pp. 19–20.

24. Charles Wilkes, Memorandum, 1838, Wilkes Expedition Letters, National Archives, reprinted in Reingold, *Documentary History*, p. 119.

25. Stanton, *Exploring Expedition*, p. 63. Poesch, who provides an excellent account of the expedition from Peale's point of view, does not mention his role in Wilkes's controversial decision.

26. Wilkes, Memorandum, in Reingold, *Documentary History*, p. 119.

27. Joel R. Poinsett, endorsement of Wilkes's Memorandum, Wilkes Expedition Letters, National Archives, in Reingold, *Documentary History*, p. 120.

28. In Charles Wilkes, *Narrative*, vol. 1, p. 353, quoted in Dupree, *Federal Government*, p. 59.

29. Dunlap described Agate as "a good and rapidly improving miniature painter" with an "amiable character." Dunlap, *Rise and Progress*, vol. 3, p. 243. Agate exhibited his work in New York from about 1831. See Peggy and Harold Samuels, *The Illustrated Biographical Encyclopedia of Artists of the American West* (Garden City, N.Y.: Doubleday and Co., 1970), p. 5.

30. See plates illustrating articles by Isaac Lea,

vol. 3, no. 4 (1830) through vol. 6 (1839), *Transactions of the American Philosophical Society*, Philadelphia. Drayton was engraving in Philadelphia by 1819, and his work included aquatint landscapes. He signed some of his plates for the *Transactions* "Drayton Aqf."—abreviation for aquafortis. See Samuels, *Artists of the American West*, p. 144, and Groce and Wallace, *Dictionary of Artists in America*, p. 15.

31. Stanton provides a lively and detailed account of the voyage, *Exploring Expedition*, chapters 4–16.

32. On the courts martial, see Stanton, *Exploring Expedition*, chapter 17, pp. 281–289.

33. Charles Wilkes to the Secretary of the Navy, 16 July 1842, Wilkes Papers, Library of Congress, reprinted in Reingold, *Documentary History*, pp. 123–126. In addition to the difficulties caused by the expedition's internal disputes, lack of interest on the part of the Tyler administration created further administrative obstacles; see Stanton, *Exploring Expedition*, pp. 282–283.

34. Haskell, *Publications*, p. 9.

35. John Torrey to Brackenridge, 5 June 1848, U.S. National Museum, quoted in Haskell, *Exploring Expedition and Publications*, pp. 10–11. See also Stanton, *Exploring Expedition*, p. 331.

36. See Douglas E. Evelyn, "The National Gallery at the Patent Office" in Viola and Margolis, eds., *Magnificent Voyagers*, chapter 11, pp. 226–241.

37. Titian Peale, "The South Sea Surveying and Exploring Expedition," *American Historical Record* 3 (1874): 250, quoted in Haskell, *Exploring Expedition and Publications*, p. 17.

38. Stanton, *Exploring Expedition*, p. 302.

39. Peale to John Frazer, 15 May 1844, John Fries Frazer Papers, American Philosophical Society, quoted in Poesch, *Titian Peale*, p. 96.

40. George Brown Goode, "The Beginnings of American Science. The Third Century," *Proceedings of the Biological Society of Washington* 4 (1887): 109–110, quoted in Haskell, *Exploring Expedition and Publications*, p. 73.

41. James Dwight Dana, preface, *Crustacea* (Philadelphia: C. Sherman, 1852), quoted in Haskell, *Exploring Expedition and Publications*, p. 7.

42. Dana to Asa Gray, 16 March 1846, Historic Letter File, Gray Herbarium, Harvard University, quoted in ibid., p. 11.

43. Wilkes to Tappan, 23 July 1842, Benjamin Tappan Papers, vol. 17, f. 2618–2619, Manuscript Division, Library of Congress, quoted in ibid., p. 35.

44. Report of the Library Committee, 29th Cong., 1st. sess. Sen. Doc. 405, p. 7, quoted in ibid., p. 9.

45. *An Act to provide for publishing an account of the discoveries made by the Exploring Expedition, under the command of Lieutenant Wilkes, of the United States Navy*, 26 August 1842, 27th Cong., 2d sess. Ch. 204, reprinted in ibid., p. 150.

Dickerson had in hand the report for the voyage from 1826–1829: Dumont d'Urville, Jules-Sebastien-Cesar, 1790–1842, *Voyage de la corvette l'Astrolabe . . . 1826–29* (Paris, 1830–1835). The first volume of the later report had only recently been published at the time of the 1842 statement: *Voyage au pole sud et dans l'Océanie sur les corvettes l'Astrolabe et la Zelée* (Paris, 1841–1846). Although it seems unlikely that the second publication could have reached Washington by 1842, the phrase "lately received" suggests that it may have. However, the plates of the zoological atlas of the earlier voyage clearly had a strong influence on the style of the American publication, which suggests that it was to those volumes that the congressional act refers. Both the earlier and the later *Astrolabe* reports would have a strong influence on the illustrations of publications of the Smithsonian Institution. See chapter 5 on Spencer Fullerton Baird's superintendence of federal zoological collections.

46. Stanton, *Exploring Expedition*, p. 35.

47. Ibid., p. 56.

48. Wilkes's claim to have discovered Antarctica was one of the key points in his court martial. See Stanton, *Exploring Expedition*, pp. 286–288.

49. A Resolution for distributing the work on the Exploring Expedition. 20 February 1845, 28th Cong., 2d sess. Joint res. 5, reprinted in Haskell, *Exploring Expedition and Publications*, p. 151.

50. Wilkes to Senator Benjamin Tappan, Chair of the Library Committee, 30 November 1848, Tappan Papers, vol. 21, f. 3398, Manuscript Division, Library of Congress, quoted in ibid., p. 10, n. 26.

51. Torrey to Brackenridge, 25 March 1851, William Dunlop Brackenridge Papers, RU7189, Smithsonian Institution Archives, quoted in ibid., p. 22.

52. The lithographic plates for Dana's geology atlas were executed by the New York firm of Sarony and Major, which became known for its publication of popular prints. On the history of the firm, see Marzio, *Chromolithography. The Democratic Art*, pp. 49–51.

53. There are drawings from the expedition in the Boston Museum of Science, and the United States Exploring Expedition Collection, Boxes 3 and 4, RU7186, Smithsonian Institution Archives.

54. Charles Wilkes, *Narrative of the United States Exploring Expedition, during the Years 1838, 1839, 1840, 1841, 1842*, 5 vols. and atlas (Philadelphia: Printed by C. Sherman, 1844).

55. For an early description of the invention of steel engraving for bank notes by Jacob Perkins,

American inventor, see Dunlap, *Rise and Progress*, vol. 3, pp. 326–337, and the appendix transcribing Jacob Perkins's Memorial to the Massachusetts Senate and House of Representatives, 1806, in ibid., pp. 344–345.

56. Wilkes to Tappan, 14 April 1851, Tappan Papers, vol. 21, f. 3450, Manuscript Division, Library of Congress, quoted in Haskell, *Exploring Expedition and Publications*, p. 78. Dougal, born in New Haven, Connecticut, apprenticed to the New York map engravers Sherman and Smith. He opened his own studio in Washington around 1845, where he would engrave plates for later federal expedition reports as well as for the Exploring Expedition. Sketches of San Francisco and California were found among his possessions when he died. See Samuels, *Artists of the American West*, p. 142.

57. Haskell, *Exploring Expedition and Publications*, pp. 16–17.

58. The scientists made $2,500 per year, the artists, $2,000; see ibid., p. 4.

59. Over the course of the project, Drayton recieved $28,147 at $166.66 per month; Dana, $16,200 for four volumes; Peale, $6,840 for one, suppressed on publication; Rich, $4,540 for an unsatisfactory manuscript; Hale, $2,158 for one volume; John Cassin, the author of the volume replacing Peale's, received $2,999.93; Augustus A. Gould, $3,210 for his volume on Couthouy's collections; Asa Gray, $5,400 for two volumes, one never published; and Louis Agassiz, $5,916.66 for the two contracted volumes that he never delivered. John H. Richard, who drew fish for Agassiz's unfinished volumes, was paid 75 cents per hour; he noted the numbers of hours for each drawing in the upper right-hand corner of the sheet. See ibid., p. 17.

60. Drayton to Wilkes, 6 August 1850, Charles Wilkes Papers, Letters and Papers, Exploring Expedition, Library of Congress, quoted in ibid., p. 23.

61. Augustus Addison Gould, *Report on the Invertebrates of Massachusetts* (Cambridge, Mass., 1841). Gould drew the illustrations for his report.

62. Amos Binney to Tappan, 2 December 1843, Benjamin Tappan Papers, Library of Congress, quoted in Stanton, *Exploring Expedition*, p. 322.

63. Gould to Tappan, 16 October 1843, Tappan Papers, vol. 18, f. 2844, quoted in Haskell, *Exploring Expedition and Publications*, p. 73.

64. Ibid., p. 74.

65. Stanton, *Exploring Expedition*, p. 366.

66. Peale to George Ord, 26 November 1844, Misc. Manuscript Collection, American Philosophical Society, quoted in Poesch, *Titian Peale*, p. 97.

67. Haskell, *Exploring Expedition and Publications*, p. 55.

68. See Peale to George Ord, 26 November 1844, Misc. Manuscript Collection, American Philosophical Society, quoted in Poesch, *Titian Peale*, p. 97.

69. Wilkes to Senator Pearce, the new chair of the Library Committee, 23 June 1847, Wilkes Papers, Letter Book, 1841–1847, Manuscript Division, Library of Congress, quoted in Haskell, *Exploring Expedition and Publications*, p. 55.

70. Titian Peale, Introduction, Manuscript Collection, American Museum of Natural History, reprinted in ibid., p. 60.

71. Wilkes to Pearce, 2 June 1847, Wilkes Papers, Letterpress copies of letters sent, f. 134–136, Library of Congress, quoted in ibid., p. 56.

72. Ibid.

73. Ibid.

74. See Poesch, *Titian Peale*, pp. 100–101.

75. Peale, *American Historical Record*, vol. 3, pp. 307–308, quoted in Haskell, *Exploring Expedition and Publications*, pp. 56–57; Poesch, *Titian Peale*, pp. 98–99.

76. Poesch, *Titian Peale*, pp. 105–117. In 1833 Peale had produced lithographic plates for a prospectus of *Lepidoptera Americana*. The proofs are in the Academy of Natural Sciences, Philadelphia. Peale's later manuscript and drawings on Lepidoptera are preserved in the American Museum of Natural History: *Butterflies of North America. Diurnal Lepidoptera. Whence They Come; Where They Go; and What They Do. Illustrated and Described*, ms. copy, unbound, 5 vols.

77. Gombrich has noted: "Even in scientific illustrations it is the caption which determines the truth of the picture." *Art and Illusion*, p. 68.

78. Wilkes to Pearce, 1 May 1850, Wilkes Papers, Letterpress copies of letters sent, f. 134–136, Library of Congress, quoted in Haskell, *Exploring Expedition and Publications*, p. 57.

79. Drayton to Wilkes, 26 November 1850, Wilkes Papers, Library of Congress, quoted in Poesch, *Titian Peale*, p. 101.

80. Plate 42, John Cassin, *Atlas. Mammalogy and Ornithology* (Philadelphia: C. Sherman and Son, 1858). See Haskell, *Exploring Expedition and Publications*, p. 62.

81. See George E. Watson, "Vertebrate Collections: Lost Opportunities" in Viola and Margolis, eds., *Magnificent Voyagers*, chapter 3, pp. 42–69. On the plate of the Hawaiian io, *Buteo solitaris*, whose prey Cassin had changed from a honeycreeper to a fish, *Mammalogy and Ornithology* (1858), plate 4, atlas, see Viola and Margolis, eds., *Magnificent Voyagers*, p. 57.

82. Richard D. Worthington and Patricia H. Worthington, "John Edwards Holbrook, Father of

American Herpetology," in Kraig Adler, ed., *North American Herpetology*, reprint edition (Society for the Study of Amphibians and Reptiles, 1976), pp. xiii–xxvii.

83. Holbrook, *North American Herpetology*, vol. 1 (1842), p. xi.

84. Rutledge, "Artists in the Life of Charleston" *Transactions of the American Philosophical Society* 39, part 2 (1949): 217.

85. Theodore Gill, "John Edwards Holbrook, 1794–1871," in *Biographical Memoirs, National Academy of Sciences*, vol. 5, p. 54.

86. William Goetzmann, pers. comm., 1980.

87. While the lithotint was not yet a full color process with each color printed from a separate stone, it was a step toward color printing and away from hand coloring, the only step of the process employing primarily women. Two lithotints appeared in the *Herpetology*; they were printed from one stone in black ink with additional neutral tones, then hand colored and finished with gum-arabic. The remaining plates were printed only in black ink with the color applied by hand. On the lithotint process, see Charles B. Wood III, "Prints and Scientific Illustration in America" in John D. Morse, ed., *Prints in and of America*, Winterthur Conference Report (Charlottesville: University Press of Virginia, 1970), pp. 182–184. The fullest account of the career of Peter Duval is in Nicholas B. Wainwright, *Philadelphia in the Romantic Age of Lithography* (Philadelphia: The Historical Society of Pennsylvania, 1958), pp. 30–45, 61–76.

88. Wainwright, *Philadelphia Lithography*, pp. 38, 62.

89. Theodore Gill, Holbrook's biographer of a later generation, considered the illustrations comparable "to most of those published contemporaneously in Europe" but criticized the plates for their lack of background: "The illustration was confined ... to the bare animals, and no background (or ground to stand on) was ever represented, nor were any accessory figures illustrating details of structure furnished." In asking for landscape context and anatomical detail, Gill conflated the conventions of two distinct generations of taxonomic method and illustration. Gill, "Holbrook," p. 58.

90. Many of the drawings for the *Herpetology* are preserved in the Archives, Museum of Comparative Zoology, Harvard University.

91. Porter, *Eagle's Nest*, p. 155.

92. Dana to Tappan, 18 September 1843, Tappan Papers, vol. 18, f. 2808, Library of Congress, quoted in Haskell, *Exploring Expedition and Publications*, p. 50.

93. Gray to Tappan, 11 March 1846, Tappan Papers, vol. 21, f. 3285, Library of Congress, quoted in ibid., pp. 50–51.

94. James Dwight Dana, *Atlas. Crustacea* (Philadelphia: C. Sherman, Printer, 1855), quoted in ibid., p. 82.

95. Ibid.

96. Using his observations made during the voyage, Pickering wrote a volume on the geographic distribution of plants and animals. Pearce, however, believed that the termination of the publication program was long overdue, and the work never appeared in the official series, although Pickering published a portion of the work himself. See Haskell, *Exploring Expedition and Publications*, pp. 97–99.

97. Appropriations for the publication program exceeded $350,000 over a thirty-year period; ibid., p. 16.

98. Dana, *Atlas. Crustacea*, quoted in ibid., p. 82.

CHAPTER 5
"A BETTER STYLE OF ART"

1. *Congressional Globe*, 35th Cong., 2d sess., pp. 1616–1618, quoted in Haskell, *Exploring Expedition and Publications*, p. 15.

2. The appropriation passed 29 to 6. 21 February 1861, debate reprinted in William Jones Rhees, *The Smithsonian Institution: Documents relative to its Origin and History*, Smithsonian Miscellaneous Collections, vol. 17 (Washington, D.C., 1880), pp. 654–672. A breakdown of appropriations and costs is available in Haskell, *Exploring Expedition and Publications*, pp. 16–17.

3. Smithson's will is reprinted in Rhees, *Documents*, pp. 1–2.

4. Dupree, *Science in the Federal Government*, chapter 4, "The Fulfillment of Smithson's Will, 1829–1861," pp. 66–90.

5. Evelyn, "The National Gallery at the Patent Office," pp. 226–240, and Nathan Reingold and Marc Rothenberg, "The Exploring Expedition and the Smithsonian Institution," chapter 12, in ibid., pp. 242–253.

6. Dupree, *Science in the Federal Government*, pp. 85–86.

7. See Jefferson Davis's 1851 comments on federal support of science in Rhees, *Documents*, pp. 471–473.

8. Haskell, *Exploring Expedition and Publications*, p. 151.

9. For a full discussion of federal expeditionary science, see William H. Goetzmann, *Army Explora-*

*tion in the American West, 1803–1863* (New Haven, Conn.: Yale University Press, 1959).

10. Rhees, *Documents*, p. 622.

11. Ibid., p. 624.

12. For a history of social and biological evolutionary thought, from the late eighteenth through the nineteenth centuries, see Robert M. Young, *Darwin's Metaphor: Nature's Place in Victorian Culture* (Cambridge, U.K.: Cambridge University Press, 1985).

13. See George H. Daniels, "The Process of Professionalization in American Science: The Emergent Period, 1820–1860," *Isis* 58, no. 2 (1967): 151–166.

14. Hall's paleontological survey grew out of the natural history survey of New York, led by James De Kay. See Michele Aldrich, "New York Natural History Survey, 1836–1845," Ph.D. diss., University of Texas, Austin, 1974.

15. William Healey Dall, *Spencer Fullerton Baird: A Biography* (Philadelphia: J. B. Lippincott Company, 1915), chapter 2, "Childhood and Youth," pp. 19–109.

16. Baird to his brother, William M. Baird, 20 December 1841, in ibid., pp. 57–59.

17. Ibid., p. 58.

18. Ibid.

19. Ibid., p. 59.

20. Ibid., p. 58.

21. Baird to W. M. Baird, 7 January 1842, in ibid., pp. 60–62.

22. Ibid., pp. 70–80.

23. "Plan of the National Gallery Containing the Collections of the Exploring Expedition," lithograph, 1844–1849, Smithsonian Archives, reproduced in Viola and Margolis, eds., *Magnificent Voyagers*, p. 239.

24. Roger B. Stein analyzes the standards and criteria of American art criticism in the 1830s and 40s in the first chapter of *John Ruskin*, pp. 1–31.

25. *The Iconographic Encyclopedia*, published in New York by Charles Rudolph Garrigue in 1852, was a translation of Brockhaus's *Bilder Atlas zum Konversations Lexikon*, published in Leipzig. Dall, *Baird*, pp. 184–185.

26. Baird to Marsh, 9 February 1851, in Dall, *Baird*, p. 256.

27. Baird to Marsh, 5 June 1851, in ibid.

28. On Baird's "amiable" wish to avoid controversy, see Henry Bryant to Baird, 5 February 1860, in ibid., pp. 335–336.

29. This brief sketch of the debate over Morton's work understates its importance in American antebellum science and society. William Stanton's *The Leopard's Spots* provides a detailed account of the controversy and its social context. Stephen J. Gould's *The Mismeasure of Man* (New York: W. W. Norton, 1981) puts Morton's anthropometry within the historical context of biological determinism and analyzes its social implications for social theories of race, class, and gender. Morton's emphasis on exact measurement should inform our evaluation of the lithographic illustrations by John Collins for *Crania Americana* (1839).

30. Baird to Richard, 18 May 1855, Smithsonian Archives, RU53, Official Outgoing Correspondence, Box 2, vol. 11.

31. Richard to Baird, 17 December 1858, Smithsonian Archives, RU52, Incoming Correspondence, Box 9, folder 22.

32. For a history of American color lithography, see Marzio, *Democratic Art*.

33. Henry R. Schoolcraft, *Historical and Statistical Information Respecting the History, Condition and Prospects of the Indian Tribes of the United States*, 6 vols. (Philadelphia: J. B. Lippincott and Co., 1851–1857). See Marzio, *Democratic Art*, pp. 28–31.

34. The record is contradictory about Sonrel's first name. He may have changed his name on immigration. Sources tracing him to Europe refer to him as Auguste; see Maryse Surdez, *Catalogue des archives de Louis Agassiz (1807–1873)*, Le Fonds appartient à l'Institut de Géologie (Neuchâtel, Switz.: Secrétariat de l'Université, 1973), and Edward Lurie, *Louis Agassiz: A Life in Science* (Chicago: University of Chicago Press, 1960). Sources based on American documentation record him as Antoine Sonrel. The notice of his death gives Antoine as his name. See Pamela Hoyle, *The Development of Photography in Boston, 1840–1875*, exhibition catalog (Boston: Boston Atheneum, 1979).

35. Lurie, *Louis Agassiz*, chapter 4, "The American Welcome, 1846–1850," pp. 122–165.

36. Agassiz to Baird, 10 April 1847, in Elmer Charles Herber, *Correspondence between Spencer Fullerton Baird and Louis Agassiz: Two Pioneer Naturalists* (Washington, D.C.: Smithsonian Institution, 1963), pp. 23–25.

37. Lurie, *Louis Agassiz*, chapter 3, "From Switzerland to Boston, 1832–1846," pp. 72–121.

38. Fries, *Double Elephant Folio*, p. 355. Baird did not consider Bien's work adequate for scientific illustration, but sent him transfer work, discussed below, and maps for illustrating the western surveys: Baird to Bien, 22 December 1852, Smithsonian Archives, RU53, Official Outgoing Correspondence, Box 2, vol. 5. On Prang as a forty-eighter, see Marzio, *Democratic Art*, p. 51.

39. See especially Wainwright, *Philadelphia Li-*

*thography,* and Peter Marzio, *Democratic Art,* for discussions of major trends in commercial lithography.

40. John M. Clarke, *James Hall of Albany: Geologist and Paleontologist, 1811–1898,* second printing (Albany, 1923), p. 180.

41. Silliman, Review: "Paleontology of New York," vol. 1, *American Journal of Science and Arts,* series 2, 5 (May 1848): 149–150.

42. See Ann S. Blum, "'A Better Style of Art': The Illustrations of the *Paleontology of New York,*" *Earth Science History* 6, no. 1, (1987): 72–85.

43. Herber, *Correspondence,* pp. 27–29.

44. Baird to Sonrel, 16 October 1850, Smithsonian Archives, RU53, Official Outgoing Correspondence, vol. 1, folder 13.

45. Baird to Sonrel, 28 November 1851, Smithsonian Archives, RU53, Official Outgoing Correspondence, vol. 2, p. 125.

46. On Sonrel's work for Tappan and Bradford, see Bettina A. Norton, "Tappan and Bradford: Boston Lithographers with Essex County Associations," *Essex Institute Historical Collections* 109 (July 1978): 149–160.

47. For another example of negotiations around drawing and printing scientific illustration, see the correspondence between Fielding Bradford Meek, paleontologist, and Thomas Sinclair, Smithsonian Archives, RU7062.

48. Tappan and Bradford to Baird, 11 September 1851, Smithsonian Archives, RU52, Assistant Secretary, 1850–1877, Incoming Correspondence, Box 1, folder 15.

49. Dall, *Baird,* p. 389.

50. Ibid.

51. "Lithography," *Encyclopaedia Britannica,* vol. 14 (1861), pp. 212–213.

52. Baird to Sonrel, 21 February 1852, Smithsonian Archives, RU53, Official Outgoing Correspondence, Box 1, vol. 2, p. 371.

53. Tappan and Bradford to Baird, 11 September 1851, Smithsonian Archives, RU52, Assistant Secretary, 1850–1877, Incoming Correspondence, Box 1, folder 15.

54. Sonrel to Baird, 28 June 1852, Smithsonian Archives, RU52 Official Incoming Correspondence, vol. 4, p. 9.

55. Joseph Leidy to Baird, 23 June 1852, Smithsonian Archives, RU52, Official Incoming Correspondence, vol. 4, p. 10.

56. Leidy to Baird, 29 June 1852, Smithsonian Archives, RU52, Official Incoming Correspondence, vol. 4, p. 7.

57. See, for example, the somewhat crude plates drawn by French at Sinclair's, ca. 1847, *Journ. Acad. Nat. Sci. Phil.,* n.s., vol. 1, part 1 (1847).

58. Wainwright, *Philadelphia Lithography,* p. 82.

59. Sonrel to Baird, 28 June 1852, Smithsonian Archives, RU52, Official Incoming Correspondence, vol. 4, p. 5. Sonrel's prices also forced Baird to make special arrangements to pay Sonrel. Baird had to draw from separate funds for drawing and printing to pay Sonrel for the Stansbury plates: Baird to Sonrel, 6 October 1852, Smithsonian Archives, RU53, Official Outgoing Correspondence, Box 1, vol. 3.

60. Joseph Leidy, *Fresh-Water Rhizopods of North America* (Washington, D.C.: U.S. Geological Survey of the Territories, 1879), "References to the Plates," opposite plate 1.

61. Sonrel to Baird, 6 July 1852, Smithsonian Archives, RU52, Official Incoming Correspondence, vol. 4, p. 45.

62. Baird to Sonrel, [22] December 1853, Smithsonian Archives, RU53, Official Outgoing Correspondence, Box 2, vol. 7, p. 350.

63. Baird to Sonrel, 12 April 1854, Box 2, vol. 8, p. 326; 3 November 1854, vol. 10, p. 9; 16 January 1855, vol. 10, p. 233, Smithsonian Archives, RU53, Official Outgoing Correspondence.

64. Sonrel to Baird, 31 March 1853, Smithsonian Archives, RU52, Official Incoming Correspondence, vol. 6, p. 349.

65. Wainwright, *Philadelphia Lithography,* p. 89.

66. For a history of technological change in the New England paper industry, see Judith A. McGaw, *Most Wonderful Machine: Mechanization and Social Change in Berkshire Paper Making, 1801–1885* (Princeton, N.J.: Princeton University Press, 1987).

67. Baird to Thomas Sinclair, 29 December 1851, Smithsonian Archives, RU53, Official Outgoing Correspondence, Box 1, vol. 2, p. 218. Occasionally uncut lithographic proof sheets have survived; see Othneil Charles Marsh, Lithographs, Manuscript Collection 275, Academy of Natural Sciences, Philadelphia.

68. Baird to Sinclair, 6 January 1852, Smithsonian Archives, RU53, Box 1, vol. 2, p. 237.

69. Baird to Sinclair, 18 April 1852, Smithsonian Archives, RU53, Box 1, vol. 3, p. 79.

70. For discussions of the art:science relationships in American painting and criticism, see Stein, *Ruskin,* and Novak, *Nature and Culture.*

71. Baird to Sinclair, 30 May 1853, Smithsonian Archives, RU53, Assistant Secretary, Official Outgoing Correspondence, Box 1, vol. 6, p. 267.

72. Baird to Sinclair, April 1857[?], Smithsonian Archives, RU53, Official Outgoing Correspondence, Box 5, vol. 16, p. 141.

73. Baird to Sinclair, 17 December 1853, Smithsonian Archives, RU53, Official Outgoing Correspondence, Box 2, vol. 7, p. 328; 24 May 1854, Box 3, vol. 9, p. 84.

74. Baird to Sinclair, [22] October 1855, Smithsonian Archives, RU53, Official Outgoing Correspondence, Box 4, vol. 13, p. 38.

75. Baird to Sinclair, [2] February 1857, Smithsonian Archives, RU53, Official Outgoing Correspondence, Box 4, vol. 15, p. 541.

76. Philadelphians, including Oscar Lawson, had made an early shift from engraving to lithography, even as the city lost primacy in publishing to New York. Joseph Drayton had produced lithographic illustration there before sailing with the Wilkes Expedition. John Collins, the lithographer of skull illustrations for Morton's *Crania Americana* had run a small lithographic establishment, although he had a reputation as a poor businessman. Sinclair, not long after arriving from Scotland, got his start buying Collins's equipment. John Cassin became the business partner of Lavinia Bowen, widow of the founder of one of Philadelphia's leading lithography firms. Wainwright, *Philadelphia Lithography*, pp. 58, 89.

77. Richard to Baird, 11 January 1865, Smithsonian Archives, RU52, Assistant Secretary, 1850–1877, Incoming Correspondence, Box 23, folder 10.

78. Weaver, d. 1847, Diary, 6 January 1842, quoted in Wainwright, *Philadelphia Lithography*, p. 82.

79. M. Toumey and F. S. Holmes, *Pleiocene Fossils of South-Carolina: Containing Descriptions and Figures of the Polyparia, Echinodermata and Mollusca* (Charleston, S.C.: Russell and Jones, 1857). The plates for Toumey's report had originally been engraved, but were ruined, and the report appeared without illustrations in 1848. At a meeting of the American Association for the Advancement of Science, Louis Agassiz and A. A. Gould, among others, urged raising the money to complete the illustrations. Holmes oversaw the work, commissioning lithographs supported mainly by private subscriptions. Additional aid from the South Carolina legislature saved the project from taking a loss. See Preface, ibid.

80. Ibid., p. vi.

81. Emphasis added. James Hall, *Paleontology of New York: Organic Remains of the Lower Helderberg Group and the Oriskany Sandstone*, vol. 3 (Albany: C. van Benthuysen, 1859), pp. xi–xii.

82. Hall, *Geology of New York*, part 4 (Albany, 1843), p. viii.

83. Hall, *Paleontology of New York: Organic Remains of the Lower Division of the New-York System*, vol. 1 (Albany: C. van Benthuysen, 1847), p. xii.

84. Clifford M. Nelson, "Meek at Albany, 1852–58," *Earth Science History* 6, no. 1 (1987): 40–46. On Meek's post-Albany career, see Clifford M. Nelson and Fritiof M. Fryxell, "The Ante-Bellum Collaboration of Meek and Hayden in Stratigraphy," in Schneer, ed., *Two Hundred Years of Geology in America*, pp. 187–200.

85. Roger L. Batten, "Robert Parr Whitfield: Hall's Assistant Who Stayed too Long," *Earth Science History* 6, no. 1 (1987): 61–71; and Blum, "'A Better Style of Art,'" ibid.: 72–85.

86. On Baird's collecting policies and his corps of collectors, see William A. Deiss, "Spencer F. Baird and His Collectors," *Journal of the Society for the Bibliography of Natural History* 9, no. 4 (1980): 635–645.

87. Baird to John H. Clark, 14 May 1852, in Dall, *Baird*, p. 276–277.

88. Agassiz to Baird, 10 April 1847, in Herber, *Correspondence*, pp. 23–25.

89. Baird to Agassiz, 23 January 1861, reprinted in Herber, *Correspondence*, pp. 162–164.

90. Baird to George Gibbs, 3 March 1861, in Dall, *Baird*, pp. 338–339.

CHAPTER 6
ILLUSTRATIONS OF THEORY, ILLUSTRATIONS
OF PRACTICE

1. Louis Agassiz, *Contributions to the Natural History of the United States*, 4 vols. (Boston: Little, Brown and Company, 1857–1862).

2. Edward Lurie, "Editor's Introduction," in Louis Agassiz, *Essay on Classification* (Cambridge, Mass.: Harvard University Press, 1962), p. xviii.

3. The illustrators who drew for Louis's son, Alexander Agassiz, a marine biologist, emulated Sonrel's plates. See also the chromolithographs from drawings by Alfred Goldsborough Mayer in his *Medusae of the World*, 3 vols. (Washington, D.C.: Carnegie Institution, 1910). Mayer, who was a student of Alexander Agassiz's, would have become familiar with Sonrel's work. Mayer's illustrations will be discussed in chapter 8. Ernst Haeckel (1834–1919), one of Germany's most enthusiastic Darwinians, whose own illustrations of Radiolaria emphasized—even dramatized—structure, borrowed from Sonrel's plates in the illustrations for his popular *Kunstformen der Natur* (Leipzig and Vienna, 1899–1904). The plates of this popularly oriented work touched the imagination of the young Paul Klee, who incor-

porated versions of the organic forms into his personal iconography. Peter Walsh of the Fogg Art Museum called my attention to Klee's imagery and its source.

4. Agassiz had established his scientific reputation with his monumental *Recherches sur les poissons fossiles* (1833–1843). In his work on glaciation, *Etudes sur les glaciers* (1840) and *Système glaciares* (1847), he believed he had established definitively that each ice age and its glaciation caused the extinction of existing life, lending support to his insistence on successive creations of entire floras and faunas. Lurie, "Introduction," p. xiii.

5. Ibid., pp. xx–xxii; Agassiz, "Essay on Classification," p. 21–23. See also Ernst Mayr, "Agassiz, Darwin, and Evolution," *Harvard Library Bulletin* 13 (1959): 165–194.

6. Agassiz, "Essay on Classification," chapter 2, pp. 137–196. I have taken my synopsis of Agassiz's views on organic structure largely from the *Contributions*, vol. 3, *Acalephs*, in which Agassiz recapitulated his argument in the "Essay on Classification" in specific reference to Darwin; see "Acalephs in General," sec. 5: "Individuality and Specific Differences Among Acalephs," pp. 88–99.

7. Emmanuel Bénézit, *Dictionnaire critique et documentaire des peintres, sculpteurs, dessinateurs et graveurs de tous les temps et de tous les pays*, new ed., vol. 2 (Paris: Grund, 1976), p. 403.

8. Agassiz, *Contributions*, vol. 1 (1857), p. xvi. Clark and Agassiz later engaged in a dispute over the scientific property of Clark's embryological investigations. See Lurie, *A Life in Science*, pp. 318–323.

9. Agassiz, *Contributions*, vol. 1, p. xvi. On Agassiz's illustrators in Neuchatel and at Harvard, see Ann Blum and Sarah Landry, "In Loving Detail," *Harvard Magazine* 79, no. 7 (May–June 1977): 38–51.

10. Mary P. Winsor, "Louis Agassiz and the Species Question," *Studies in the History of Biology*, (1979): 89–117.

11. Agassiz, *Contributions*, vol. 3 (1860), p. 89.

12. Ibid.

13. Ibid.

14. Ibid., p. 88.

15. Ibid.

16. Ibid., p. 92, n. 1.

17. Agassiz, "Essay," p. 136. Quoted in Mayr, *The Growth of Biological Thought*, p. 865.

18. Agassiz, "Essay," p. 94.

19. Agassiz, *Contributions*, vol. 4 (1862), pp. 88–89. The description refers to plate 3, vol. 3.

20. Ibid., vol. 3, p. vi.

21. Ibid., p. 92.

22. Ibid., p. 88.

23. Lurie, *A Life in Science*, pp. 307–308.

24. Stein, *Ruskin*, pp. 159–162.

25. James Jackson Jarves, *The Art-Idea* (1864), quoted in Novak, *Nature and Culture*, pp. 27, 47–48.

26. Ibid., p. 69. Novak in *Nature and Culture* and Stein in *John Ruskin and Aesthetic Thought in America* provide rich analyses of the influences of Ruskin and Humboldt and American science on landscape painting and criticism. A study like Peter Marzio's history of American chromolithography indicates the directions and tastes of the commercial print market, including reproductions of landscape paintings. On Church, see also Barbara Novak, *The Thyssen-Bornemisza Collection: Nineteenth-Century American Painting* (New York: The Vendome Press, 1986), pp. 90–101.

27. Charles Darwin, *On the Origin of Species*, facsimile edition, introduction by Ernst Mayr (Cambridge, Mass.: Harvard University Press, 1964), p. 16.

28. Clarke, *Hall*, p. 184.

29. Allen announced his theory at a regular meeting of the Boston Society of Natural History. See Joel Asaph Allen, "[S]ome remarks concerning geographic variation in mammals and birds," 17 April 1872, *Proceedings of the Boston Society of Natural History* 15 (1872–1873): 156–159.

30. Lane Cooper, *Louis Agassiz as a Teacher: Illustrative Abstracts on His Method of Instruction* (Ithaca, N.Y.: Comstock Publishing Co., 1945). Scudder's description was originally published as "In the Laboratory with Agassiz," *Every Saturday* 16 (4 April 1874): 369–370.

31. The article on geological illustration to 1840 by Martin J.S. Rudwick provides a model and a comparable example of the relationship between illustration and the internal and external status of the field. Rudwick argues "that the diverse forms of visual expression in the new science were derived from extremely varied social and cognitive sources, and were inter-related conceptually and historically. Furthermore, each developed in the course of time toward greater abstraction and formalization, and thereby became able to bear an increasing load of theoretical meaning." See Rudwick, "The emergence of a visual language for geological science, 1760–1840," *History of Science* 14 (1976): 149–195. In nineteenth-century zoological illustration, especially taxonomic illustration, the link between abstraction and theory is problematic. As I argue, I see a stronger link between representation and practice. For studies in contemporary illustration, laboratory practice, and publishing genres, see Michael Lynch and Steve Woolgar, eds., *Representation in Scientific Practice*, special issue of *Human Studies* 11, nos. 2–3 (April–July 1988): 99–359.

32. John Strong Newberry, *Modern Scientific Investigation: Its Methods and Tendencies. An Address Delivered before the American Association for the Advancement of Science, August, 1867* (Salem, Mass.: Essex Institute Press, 1867), p. 3.

33. Ibid.

34. Ibid., pp. 15–16.

35. Ibid., p. 17. The defense of Darwinian theory by Asa Gray, America's first proponent of evolution, was characterized by his desire to make it accommodate religion. See A. Hunter Dupree, *Asa Gray, 1810–1888* (Cambridge, Mass.: The Belknap Press of Harvard University Press, 1959).

36. Newberry, *Modern Scientific Investigation*, p. 17.

37. Edward Sylvester Morse, "What American Zoologists have done for Evolution: An Address by Vice-President E. S. Morse, before the American Association for the Advancement of Science, at Buffalo, N.Y., August, 1876," reprinted from vol. 25 of *Proceedings of the American Association for the Advancement of Science* (Salem, Mass.: Naturalist's Agency, 1876), p. 8.

38. Ibid., p. 4.

39. Ibid.

40. Ibid., p. 8.

41. Othneil C. Marsh, "Introduction and Succession of Vertebrate Life in America," an address delivered before the American Association for the Advancement of Science, at Nashville, Tenn., 30 August 1877, p. 4.

42. Marsh's finds were popularized as the all-important "missing links" of evolution. The explorations for and excavations of Marsh's fossils were also closely associated with the political and military process of appropriating land from western Indians. See William C. Wyckoff, "A Perilous Fossil Hunt: Professor Marsh's Last Trip to the Bad Lands," *The New York Tribune*, 22 December 1874, pp. 3–4.

43. Charles Schuchert and Clara Mae LeVene, *O. C. Marsh, Pioneer in Paleontology* (New Haven, Conn.: Yale University Press, 1940), p. 275.

44. Edward Drinker Cope, letter of 16 March 1873, Washington, D.C., quoted in Henry Fairfield Osborn, *Cope: Master Naturalist* (Princeton, N.J.: Princeton University Press, 1931), p. 192.

45. On the history of federal paleontology, see Clifford M. Nelson and Ellis L. Yochelson, "Organizing Federal Paleontology in the U.S.," *Journal of the Society for the Bibliography of Natural History* 9, no. 4 (1980): 607–618.

46. Schuchert and LeVene, *Marsh*, p. 474. Othneil Charles Marsh, *Dinocerata: A Monograph of an Extinct Order of Gigantic Mammals*, Monographs of the United States Geological Survey, no. 10 (Washington, D.C.: Government Printing Office, 1886).

47. Osborn, *Cope*, pp. 360, 394, 401–402.

48. Othneil Charles Marsh, *Odontornithes: A Monograph on the Extinct Toothed Birds of North America* (Washington, D.C.: Government Printing Office, 1880), vol. 7 of U.S. Geological Exploration of the Fortieth Parallel. The volume's thirty-four plates cost $8,000. Marsh had an author's edition of 250 copies printed at his own expense (New Haven, 1880). See Schuchert and LeVene, *Marsh*, p. 428.

49. Archibald Geikie, "The Toothed Birds of Kansas," *Nature*, 16 September 1880, pp. 457–458, quoted in Schuchert and LeVene, *Marsh*, p. 430.

50. Ibid., p. 431. While studying in Germany in 1864, Marsh had written in his lecture notebook: "The figures of a paleontological work are the main thing and should always be good, and they may thus be of value after the text has become antiquated." Schuchert and LeVene, *Marsh*, p. 62.

51. George Russell Agassiz, *Letters and Recollections of Alexander Agassiz* (Boston: Houghton Mifflin, 1913), provides the classic "life and letters" treatment of his father's career.

52. Roetter illustrated Alexander Agassiz, *Report on the Echinoidea*, Challenger Zoological Reports, vol. 3, part 4, of Great Britain Challenger Office, *Report on the Scientific Results of the Voyage of the H.M.S. Challenger during the years 1873–1876* (Edinburgh: 1880–1895). Earlier, Roetter had drawn for George Engelmann's "Cactaceae of the Boundary," in *United States and Mexican Boundary Survey*, vol. 2, part 1, (Washington, D.C.: A.O.P. Nicholson, Printer, 1859). Paulus Roetter, born in Nuremberg in 1806, immigrated to the United Sates in 1845 as part of a group intent on founding a religious community. He was the first instructor of drawing at Washington University in St. Louis. After serving in the Civil War, he became associated with the Museum of Comparative Zoology until he returned to St. Louis in 1884. Throughout his career as a botanical and zoological illustrator, he continued to paint in oils. See Mary M. Powell, "Three Artists of the Frontier," *Bulletin of the Missouri Historical Society* 5, no. 1 (October 1948): 34–43.

53. Walter Faxon, *The Stalk-Eyed Crustacea*, vol. 18 of *Reports on an Exploration off the West Coasts of Mexico, Central and South America, and off the Galapagos Islands, in charge of Alexander Agassiz, by the U.S. Fish Commission Steamer "Albatross," during 1891*, Memoirs of the Museum of Comparative Zoology, at Harvard College, 1895.

54. Samuel Garman, *The Fishes*, vol. 24 of *Reports on an Exploration off the West Coasts of Mexico, Central and South America, and off the Galapa-*

gos Islands, in charge of Alexander Agassiz, by the U.S. Fish Commission Steamer "Albatross," during 1891, Memoirs of the Museum of Comparative Zoology, at Harvard College, 1899.

55. Dupree, "Alexander Agassiz on Government Science and the Theory of Laissez Faire," in Science in the Federal Government, pp. 220–227.

56. Ibid., p. 225. John Wesley Powell couched his successful defense of the Geological Survey in terms of the public benefit from democratic access to knowledge.

57. Mary P. Winsor's history of the Museum of Comparative Zoology, Reading the Shape of Nature: Comparative Zoology at the Agassiz Museum (University of Chicago Press, 1991) treats at length Alexander Agassiz's administrative decisions.

58. C. O. Whitman, "Introduction," Journal of Morphology 1 (1887): p. ii.

59. Similarly, viewed in sequence, the individual exposures in Eadweard Muybridge's Animal Locomotion of 1887 depicted movement.

CHAPTER 7
THE LENS AND THE LINE

1. Louis Agassiz to Lodowick H. Bradford, 22 December 1862, Letter Books, vol. 2, p. 99, Agassiz Papers, Museum of Comparative Zoology, Harvard University.

2. See Alpers, The Art of Describing, chapter 2, "'Ut pictura, ita visio': Kepler's Model of the Eye and the Nature of Picturing in the North," pp. 26–71, for an extended discussion of the relationship between seventeenth-century optics and painting.

3. Gail Buckland's Fox Talbot and the Invention of Photography (Boston: David R. Godine, 1980), gives a full account of the invention of the early versions of the process, including the role of the camera obscura and camera lucida in Fox Talbot's initial concept.

4. Ibid., pp. 97–98.

5. David Dale Owen, Report of the Geological Survey of Wisconsin, Iowa, and Minnesota; and Incidentally of a Portion of Nebraska Territory. Atlas. Made under instructions from the United States Treasury Department (Philadelphia: Lippincott, Grambo and Co., 1852).

6. Rolf H. Kraus cites Josef Berres, The Phototyp, According to the Invention of Professor Berres of Vienna (1840), containing five prints from etched daguerreotypes, a rare use of the technique. Kraus continues, "Even the famous Excursions Daguerriennes, which were published between 1840 and

1843 by Lerebours, in co-operation with Rittner, Goupil and Bossange, contained (besides 109 lithographs, copper engravings and copper etchings based on daguerreotypes) only 3 images which were printed directly from etched daguerreotypes": Rolf H. Kraus, "Photographs as Early Scientific Illustrations," History of Photography 2, no. 4 (October 1978): 292.

7. After the appearance of Owen's report, George Mathiot of the United States Coast Survey used engraved daguerreotypes for printing maps and described the process in the Survey's annual report for 1854. Mathiot's experiments with engraved daguerreotypes led to further experiments with electrotyping. See Ripley and Dana, "Engraving," in The New American Cyclopedia, vol. 7 (New York: D. Appleton, 1871), pp. 211–212.

8. "Letter to Robert Owen, from his Son, David Dale Owen, 24 July 1852," Robert Owen's Journal (London), part 25 (25 September 1852): 167–168. Courtesy of the New Harmony Workingmen's Institute, New Harmony, Indiana.

9. Ibid.

10. James Deane, Ichnographs from the Sandstone of Connecticut River (Boston: Little, Brown and Company, 1861), illustrated with lithographs by Thomas Sinclair, from drawings by J. Deane, and with photographs.

11. Elizabeth Cock writes: "[T]he first use of a photographic print in an American magazine occurred in the Photographic Art Journal, 1853." She adds that it was a photograph taken from a daguerreotype, like most of the photographic illustration until 1857, and tipped into the magazine pages. See Elizabeth M. Cock, "The Influence of Photography on American Landscape Painting, 1839–1880," Ph.D. diss., New York University, 1967. Deane's fossil footprint illustrations were also tipped or pasted into the text. The photographs have none of the clarity or tonal range of a photograph taken of a daguerreotype, suggesting that they were taken directly from the specimen.

12. Quoted in Sylvester Rosa Koehler, Exhibition Illustrating the Technical Methods of the Reproductive Arts, From the XV. Century to the Present Time, with Special Reference to Photo-Mechanical Methods (Boston: Museum of Fine Arts, 1892), p. iii.

13. On the quest for satisfactory photomechanical reproduction, see Estelle Jussim, Visual Communication and the Graphic Arts (New York: R. R. Bowker Company, 1974), especially chapter 8, "The New Technologies and the History of Art History," pp. 237–278.

14. Norton, "Tappan and Bradford," p. 159; also

Josef Maria Eder, *History of Photography*, trans. Edward Epstean (New York: Columbia University Press, 1945), p. 611.

15. For an account of Brewer's career, see George E. Gifford, Jr., "Thomas Mayo Brewer, M.D., . . . A Blackbird and Duck, Sparrow and Mole," *Harvard Medical Alumni Bulletin* 37, no. 1 (Fall 1962): 32–35.

16. Thomas Mayo Brewer, "Introduction," *North American Oölogy*, part 1, *Raptores and Fissirostres*, in *Smithsonian Contributions to Knowledge* (Washington, D.C.: Smithsonian Institution, and New York: D. Appleton and Co., 1857), p. iv.

17. Like Baird, Brewer had cultivated his interest in birds while still studying medicine, and had also conducted an extended correspondence with Audubon. In the introduction to his study, Brewer begged not to be interpreted as ungrateful where his conclusions differed from those of his former mentor. Excusing Audubon for conclusions now considered erroneous, Brewer wrote: "It is hardly possible even for the most exact and cautious to avoid falling into mistakes" (ibid., p. iv).

18. Ibid., p. v. A bound collection of Trudeau's drawings of eggs is preserved in the Library, Museum of Comparative Zoology, Harvard University.

19. Ibid., p. v.

20. Norton, "Tappan and Bradford," p. 160. Despite the short-lived application of Bradford's technique, Appleton's *New American Cyclopedia* continued to describe it as late as 1875; see "Lithography," vol. 10 (1875), p. 529. See also Eder, *History of Photography*, p. 611.

21. For unknown reasons, the plates were published in two versions. Bradford's photolithographs were printed in color in Brewer, *North American Oölogy*. They were also published as monochrome lithographs, apparently redrawn by Otto Knirsch and printed by Bowen and Co., Philadelphia, under the same title, in the *Smithsonian Contributions to Knowledge*, vol. 2, article 2 (Washington, D.C.: Smithsonian Institution, 1859).

22. Charles Emil Bendire, *Life Histories of American Birds, with Special Reference to the Breeding Habits and Eggs* (Washington, D.C.: Government Printing Office, 1892), Special bulletin of the Smithsonian Institution, United States National Museum, no. 1. Issued also as *Smithsonian Contributions to Knowledge*, vol. 28; a supplement is *Life Histories of American Birds, from the Parrots to the Grackles, with Special Reference to the Breeding Habits and Eggs*, in *Smithsonian Contributions to Knowledge*, vol. 32. (Washington, D.C.: Smithsonian Institution, 1895).

23. James Hall, *Illustrations of Devonian Fossils: Gasteropoda, Pteropoda, Cephalopoda, Crustacea and Corals from the Upper Helderberg, Hamilton and Chemung Groups* (Albany: Weed, Parsons and Company, and New York: E. Bierstadt, 1876).

24. See William Welling, *Collector's Guide to Nineteenth-Century Photographs* (New York: Macmillan, 1976), p. 85.

25. Ripley and Dana, "Photography," in *American Cyclopedia*, vol. 13 (1875), pp. 468–473.

26. Ibid., p. 472.

27. James Hall, *Paleontology of New York: Gasteropoda, Pteropoda and Cephalopoda of the Upper Helderberg, Hamilton, Portage and Chemung Groups*, vol. 5, part 2 (Albany: C. van Benthuysen, 1879).

28. Alexander Agassiz, "Application of Photography to Illustrations of Natural History. With two figures printed by the Albert and Woodbury Processes," *Bulletin of the Museum of Comparative Zoology* 3, 2 (1871): 47.

29. Alexander Agassiz to Elizabeth Cary Agassiz, 6 June 1870, quoted in George Russell Agassiz, *Letters and Recollections*, p. 104.

30. A. Agassiz, "Application of Photography to Illustrations," p. 47.

31. Ibid., pp. 47–48.

32. Ibid., p. 47.

33. Examples of Sonrel's cartes-de-visite and larger-format portraits, including stereo views, can be found in the Portrait File, Library, Museum of Comparative Zoology, and in the photographic collections of the Massachusetts Society for the Preservation of New England Antiquities. See also Hoyle, *Photography in Boston*, pp. 17, 30.

34. Theodore Lyman, Diary, 1879, Massachusetts Historical Society, Boston.

35. Alexander Agassiz, *Revision of the Echini*, Illustrated Catalogue of the Museum of Comparative Zoology, no. 7, in four parts (Cambridge, Mass.: Harvard University Press, 1872–1874).

36. Charles Darwin, *The Expression of the Emotions in Man and Animals* (London: John Murray, 1872).

37. Alexander Agassiz to Charles Darwin, 9 December 1872, quoted in G. R. Agassiz, *Letters and Recollections*, pp. 120–121.

38. To make a woodburytype, a kind of relief print, a photographic negative was first projected onto a light-sensitive gelatin surface that was then washed to remove the unexposed parts. The gelatin relief, subjected to hydraulic pressure against a soft metal surface, usually lead, molded the surface of the metal. The lowest part of the metal mold corre-

sponded to the darkest areas of the picture, the highest parts to the lightest. A solution of warm gelatin with suspended ink was poured into the metal mold and the printing paper laid on the gelatin. As the gelatin set, it adhered to the paper and was finally fixed chemically to make it durable. Ripley and Dana, "Photography," in *American Cyclopedia*, vol. 13 (1875), p. 473. The principal drawback of woodburytypes was that over time the gelatin became brittle and cracked or chipped off the paper support. S. R. Koehler adopted a latter-day, quasi-Buffonian interpretation of the fragility of the woodburytype: "The Woodburytype gives very beautiful results, but unfortunately the prints are apt to become brittle, from drying out, and to chip off. This seems to be more especially the case in America, owing, no doubt, to climatic differences" (*The Reproductive Arts*, p. 63).

39. Agassiz to Darwin, 9 December 1872, quoted in G. R. Agassiz, *Letters and Recollections*, p. 121.

40. Agassiz published his studies of the development of young fish in several parts; see Alexander Agassiz, "On the Young Stages of Some Osseous Fishes. I. Development of the Tail," *Proceedings of the American Academy of Arts and Sciences* 13 (1877): 117–124; "Development of the Flounders," ibid., 14 (1878–1879): 1–25; "On the Young Stages of Some Osseous Fishes. IV," ibid., 17 (1881–1882): 271–298. The reproduction of Agassiz's fish drawings constituted a kind of informal publication; the plate numbers and captions were handwritten.

41. Koehler, *Reproductive Arts*, p. iv.

42. Ibid.

43. A description of a technique for photography on the block appeared as early as 1857. See Ripley and Dana, "Engraving," in *American Cyclopedia*, vol. 7 (1871), pp. 212–213.

44. Dard Hunter, *Papermaking: The History and Technique of an Ancient Craft*, 2d ed. (New York: Alfred A. Knopf, 1947). See also McGaw, *Most Wonderful Machine*.

45. James Shirley Hodson, *An Historical Guide to Art Illustration, in Connection with Books, Periodicals, and General Decoration* (London: Sampson Low, Marston, Searle and Rivington, 1884), p. 99.

46. Harris's work was reprinted by the Commonwealth of Massachusetts for its value to lay as well as scientific readers. The edition of 1862, edited by C. L. Flint, included new illustrations provided through Louis Agassiz's volunteering his illustrators to the project, and lending the work increased scientific weight. Thaddeus William Harris, *A Treatise on Some Insects Injurious to Vegetation*, 2d ed. (Boston: W. White, printer to the state, 1862).

47. Anna Botsford Comstock, *The Comstocks of Cornell: John Henry Comstock and Anna Botsford Comstock, an Autobiography of Anna Botsford Comstock*, edited by Glenn W. Herrick and Ruby Green Smith (Ithaca, N.Y.: Comstock Publishing Associates, Cornell University Press, 1953); Jussim, *Visual Communication*, p. 159. For a study in the social context of wood engraving, see Diana Korzenik, *Drawn to Art: A Nineteenth-Century American Dream* (Hanover, N.H.: University Press of New England, 1985).

48. Comstock, *The Comstocks*, pp. 152–153.

49. John P. Davis, "A Symposium of Wood-Engravers," *Harper's New Monthly Magazine* (February 1880), quoted in Jussim, *Visual Communication*, pp. 157–158.

50. Ibid., quoted in Jussim, *Visual Communication*, p. 158.

51. Ibid.

52. W. J. Linton, "Art in Engraving on Wood," *The Atlantic Monthly* 43 (June 1879): 706–707.

53. Ibid., p. 707.

54. Ibid.

55. Ibid.

56. Ibid., pp. 709–713.

57. Davis, "Symposium," quoted in Jussim, *Visual Communication*, p. 158.

58. William J. Linton, *The History of Wood-Engraving in America* (London: Bell, 1882), p. 34, quoted in Jussim, *Visual Communication*, p. 159.

59. Comstock, *The Comstocks*, p. 181.

60. Ibid., pp. 182, 192, 244, 264.

61. Sanborn Tenney became a professor of natural history at Williams College. Abby Tenney was his co-author for some publications. His *Natural History: A Manual of Zoology for Schools, Colleges, and the General Reader* (New York: Charles Scribner and Co., 1865), had reached its fifth edition by 1867. See also Sanborn Tenney, *Elements of Zoology: A Text-Book* (New York: Charles Scribner's Sons, 1875), which contained 750 wood engravings. See also wood engravings illustrating the textbook by Vassar College professor James Orton, *Comparative Zoology, Structural and Systematic. For use in Schools and Colleges* (New York: Harper and Brothers, 1876).

62. Alfred Edmund Brehm (1829–1884) produced a nineteenth-century equivalent of Buffon's *Histoire Naturelle*. His illustrated *Thierleben* appeared in many editions, translated into several languages, and has been kept in print almost to the present. See John Sterling Kingsley, ed., *The Standard Natural History*, 6 vols. (Boston: S. E. Cassino and Company, 1884–1885), also in subsequent editions as *The Riv-*

*erside Natural History.* On the bird illustrations of Brehm's *Thierleben,* see Jean Anker, *Bird Books and Bird Art: An Outline of the Literary History and Iconography of Descriptive Ornithology* (Copenhagen: Levin and Munksgaard, 1938), p. 49.

63. See, for example, Samuel G. Goodrich, *Illustrated Natural History of the Animal Kingdom, being a Systematic and Popular Description of the Habits, Structure, and Classification of Animals, from the Highest to the Lowest Forms, with their Relations to Agriculture, Commerce, Manufactures, and the Arts,* 2 vols. (New York, 1859). The two volumes contain 1,400 wood engravings. Samuel Goodrich was a prolific author of popular children's books. Ever alert to the market, he dedicated his natural history to Louis Agassiz. Goodrich's illustrations of the "types of mankind" mark an extreme of racial stereotyping: Genghis Kahn represents the Mongolian, and the Apollo Belvedere represents the Caucasian. An illustration of a marble statue of an angel, thoroughly white, drives the point home even without a caption. Goodrich devotes his first volume of 680 pages to mammals. Volume 2 begins with 352 pages on birds, the illustrations of which are copied almost entirely from Audubon, and gives a diminishing number of pages to the remaining animal kingdom, down to a mere 8 pages for protozoa.

64. On the response in popular illustation drawing techniques to the introduction of photomechanical processes, see Jussim, *Visual Communication,* for a general discussion, and Henry C. Pitz, *Howard Pyle: Writer, Illustrator, Founder of the Brandywine School* (New York: Clarkson N. Potter, 1975), for an account of one illustrator's work.

65. See Keith R. Benson, "From Museum Research to Laboratory Research: The Transformation of Natural History into Academic Biology," in Ronald Rainger, Keith R. Benson, and Jane Maienschein, eds., *The American Development of Biology* (Philadelphia: University of Pennsylvania Press, 1988), pp. 49–83.

66. Edward Sylvester Morse Papers, Box 58, Natural History Manuscripts Collection, Phillips Library, Peabody Museum of Salem, Salem, Mass.

67. Samuel Hubbard Scudder, *The Butterflies of the Eastern United States and Canada, with Special Reference to New England,* 3 vols. (Cambridge, Mass.: Published by the author, 1889).

68. Scudder, "Preface," in *Butterflies of New England,* vol. 1, pp. ix–x.

69. Scudder to L. Trouvelot and Smith, 1869–1870, Scudder Papers, Boston Museum of Science.

70. Scudder kept a record book for his illustrations. It includes a list of the Abbot drawings he studied at the British Museum, Natural History, and ranks the drawings from "good" to "undesirable." Scudder Papers, Boston Society of Natural History, Boston Museum of Science.

71. With the support of advance subscriptions, Scudder paid nearly $6,000 for printing the plates to illustrate his monograph; payment to his illustrators amounted to only a third of what he paid for printing. Scudder's accounts for the publication are in his papers, in the manuscript collections of the Boston Society of Natural History, Boston Museum of Science.

72. Sinclair charged $175 each for plates 6, illustrating Lycaenids, and 8, illustrating swallowtails. Each plate was printed from fifteen stones. About plate 6, Scudder noted: "Returned for improvements, the entire edition being faulty; in November. Received new edition and returned old Dec 31." Of plate 8, he recorded: "Edition finished but rejected and now undergoing retreatment." See "Illustrations and References to Plates," Scudder Papers, Boston Museum of Science.

73. In *America on Stone,* Harry T. Peters dismisses the firm of Augustus and Berthold Meisel: "The Meisels seem to have done nothing but technical prints of various kinds," p. 279.

74. Scudder, "Illustrations and References to Plates," Scudder Papers, Boston Museum of Science.

75. Addison E. Verrill, "The Cephalopods of the North-eastern Coast of America: Part I, The Gigantic Squids (*Architeuthis*) and Their Allies; With Observations on Similar Large Species from Foreign Localities," *Transactions of the Connecticut Academy of Arts and Sciences* 5 (December 1879): 177–257; "Part II, The Smaller Cephalopods, Including the 'Squids' and the Octopi, with Other Allied Forms," ibid. (June 1880): 259–446.

76. Verrill, "Part I" (1879), note, pp. 184–185.

77. Ibid., p. 185.

78. For examples of nineteenth-century applications of both traditional conventions and innovations in engineering illustration, see Ken Baynes and Francis Pugh, *The Art of the Engineer* (Woodstock, N.Y.: The Overlook Press, 1981).

79. Ripley and Dana, "Drawing," in *American Cyclopedia,* vol. 6 (1867), p. 612.

80. On the twentieth-century convention of the photograph-diagram pair, see Michael Lynch, "Science in the Age of Mechanical Reproduction," *Biology and Philosophy* 6, no. 2 (April 1991): 205–226.

81. See Brian Bracegirdle, *A History of Microtechique* (Ithaca, N.Y.: Cornell University Press, 1978).

82. See, for example, S. Bradbury, "The Quality of

the Image Produced by the Compound Microscope, 1700–1840," in S. Bradbury and G.L'E. Turner, eds., *Historical Aspects of Microscopy*, papers read at a one-day conference held by the Royal Microscopical Society at Oxford, 18 March 1966 (Cambridge, U.K.: Heffer for the Royal Microscopical Society, 1967). Bradbury examined historical instruments and compared their chromatic and spherical aberrations; he describes the contribution of optical phenomena to theories of biological structure in the early nineteenth century.

83. For examples of popular manuals, see Philip Henry Gosse, *Evenings at the Microscope; or, Researches among the Minuter Organs and Forms of Animal Life* (New York: D. Appleton and Company, 1879); the elementary introduction by Samuel Wells, Mary Treat, and Frederick Leroy Sargent, *Through a Microscope: Something of the Science Together with Many Curious Observations Indoor and Out, and Directions for a Home-Made Microscope* (Chicago: The Interstate Publishing Company, 1886); and a slightly more advanced work by Alfred C. Stokes, M.D., *Microscopy for Beginners; or, Common Objects from the Ponds and Ditches* (New York: Harper and Brothers, 1887), which attempts to provide an American equivalent of the imported popular English texts on microscopy, for example, that of William B. Carpenter, *The Microscope and Its Revelations*, 5th ed. (London: J. and A. Churchill, 1875).

84. Simon Gage, "Preface to Part I," in *The Microscope and Histology*, 3d ed. (Ithaca, N.Y., 1891), n.p.

85. Ibid.

86. Beale, *How to Work with a Microscope*, 4th ed. (London: Harrison, 1865), pp. 32–33.

87. Ibid., p. 33.

88. Albert McCalla, "The Verification of Microscopic Observation," in *Proceedings of the American Society of Microscopists*, Sixth Annual Meeting (Buffalo, N.Y.: Steam Printing House of Haas and Klein, 1883).

89. Ibid., p. 5.

90. Ibid., p. 9.

91. Ibid., p. 10.

92. Ibid.

93. Ibid., p. 11.

94. "Photography," in Ripley and Dana, *American Cyclopedia*, vol. 13 (1870), p. 291.

95. Ibid.

96. McCalla, "Verification of Microscopic Observation," p. 11.

97. Ibid.

98. Ibid.

99. Ibid., p. 10.

CHAPTER 8
THE ZOOLOGIST'S PROVINCE

1. Mayer, *Medusae of the World*.

2. Ibid., vol. 1, p. 3.

3. For a discussion of Gibson's career, see Jussim, *Visual Communication*, chapter 6, "William Hamilton Gibson (1850–1896): The Artist Using the Camera's Eye," pp. 149–194.

4. The decorative quality of Gibson's work resembled that of M. H. Giacomelli, illustrator of the popular natural histories by the French author Jules Michelet. Compare, for example, Giacomelli's illustrations for the English edition of Jules Michelet, *The Insect* (London: T. Nelson and Sons, 1875), with those of Gibson's *Highways and Byways: Or Saunterings in New England* (New York: Harper and Brothers, 1882).

5. Gibson, *Highways and Byways*, list of illustrations, pp. 9–11.

6. Henry Christopher McCook, *Tenants of an Old Farm: Leaves from the Note-Book of a Naturalist* (New York: Fords, Howard and Hulbert, 1885); *American Spiders and their Spinning Work. A Natural History of the Orbweaving Spiders of the United States, with Special Regard to their Industry and Habits*, 3 vols. (Philadelphia: The Author, Academy of Natural Science of Philadelphia, 1889).

7. George Brown Goode, *The Fisheries and Fishery Industries of the United States. Prepared through the Co-operation of the Commissioner of Fisheries and the Superintendent of the Tenth Census*, 5 vols. (Washington, D.C.: Government Printing Office, 1884–1887).

8. Among the best-known women writers on birds, and their books, were Florence Merriam Bailey, *Birds through an Opera Glass* (Boston and New York: Houghton, Mifflin and Company, 1889), *A-Birding on a Bronco* (Boston: Houghton, Mifflin and Company, 1896), *Birds of Village and Field: A Bird Book for Beginners* (Boston and New York: Houghton, Mifflin and Company, 1898); Olive Thorne Miller [Harriet Mann Miller], *Bird Ways* (Boston: Houghton, Mifflin and Company, 1885), *In Nesting Time* (Boston: Houghton, Mifflin and Company, 1888), *Little Brothers of the Air* (Boston: Houghton, Mifflin and Company, 1892), *The First Book of Birds* (Boston and New York: Houghton, Mifflin and Company, 1899); Mabel Osgood Wright, *Birdcraft: A Field Book of Two Hundred Song, Game, and Water Birds* (New York: Macmillan and Co., 1895), *Citizen Bird: Scenes from Bird-Life in Plain English for Beginners*, with Elliott Coues (New

York: The Macmillan Company, 1897), *Gray Lady and the Birds: Stories of the Bird Year for Home and School* (New York: The Macmillan Company, 1907); Neltje Blanchan Doubleday, *Bird Neighbors: An Introductory Acquaintance with One Hundred and Fifty Birds Commonly Found in the Gardens, Meadows, and Woods about Our Homes* (New York: Doubleday and MacClure, 1897) (in its 14th edition by 1907), *Birds That Hunt and Are Hunted: Life Histories of One Hundred and Seventy Birds of Prey, Game Birds and Water Fowls* (New York: Doubleday and MacClure, 1898), *How to Attract the Birds: And Other Talks about Bird Neighbors* (New York: Doubleday Page, 1902), *Birds That Every Child Should Know: The East* (New York: Grosset and Dunlap, 1907).

9. Joel A. Allen, "Correspondence," *The Auk* 1, no. 3 (July 1884): 302.

10. William Brewster, *The Auk* (October 1893), quoted in Harriet Kofalk, *No Woman Tenderfoot: Florence Merriam Bailey, Pioneer Naturalist* (College Station: Texas A & M University Press, 1989), p. 59.

11. Elliott Coues, *A Key to North American Birds; Containing a Concise Account of Every Species of Living and Fossil Bird at Present Known from the Continent North of the Mexican and United States Boundary* (Salem, Mass.: Naturalists' Agency, and New York: Dodd and Mead, 1872).

12. Erwin Stresemann, *Ornithology from Aristotle to the Present*, trans. Hans J. Epstein and Cathleen Epstein (Cambridge, Mass.: Harvard University Press, 1975), pp. 244–246.

13. Robert Ridgway, correspondence with Spencer F. Baird, 1864–1866, Box 1; and examples of Ridgway's drawings and illustrations edited for publication, Boxes 2–6, in Robert Ridgway Papers, RU7167, Smithsonian Archives.

14. Spencer F. Baird, Thomas M. Brewer, and Robert Ridgway, *A History of North American Birds: Land Birds*, illustrated with 64 plates and 593 woodcuts (Boston: Little, Brown and Company, 1874).

15. See especially Robert Ridgway, *A Manual of North American Birds* (Philadelphia: J. B. Lippincott, 1887); the volume contains 464 outline illustrations of generic characters.

16. Robert Ridgway, *A Nomenclature of Colors for Naturalists, and Compendium of Useful Knowledge for Ornithologists* (Boston: Little, Brown and Company, 1886).

17. Wilhelm von Bezold, *The Theory of Color in Its Relation to Art and Art-Industry*, trans. S. R. Koehler (Boston: L. Prang and Company, 1876).

18. Ridgway, *Nomenclature of Colors*, p. 15.

19. Ibid., pp. 19–20.

20. Ibid., p. 26.

21. Robert Ridgway's brother, John Livesy Ridgway, was an illustrator and prepared a brochure intended to standardize geological illustration: *The Preparation of Illustrations for Reports of the U.S. Geological Survey with Brief Descriptions of the Processes of Reproduction* (Washington, D.C.: United States Geological Survey, 1920).

22. Welker, *Birds and Men*.

23. The organizations of the 1890s were second-generation Audubon societies. The first had been sponsored by the New York weekly journal of hunting and fishing, *Field and Stream*, in 1886. By the summer of 1887, the membership had swelled to 38,000, but by 1889 the society virtually disappeared. See Welker, *Birds and Men*, pp. 204–205.

24. See Robert McCracken Peck, *A Celebration of Birds: The Life and Art of Louis Agassiz Fuertes*, published for the Academy of Natural Sciences of Philadelphia (New York: Walker and Company, 1982).

25. Wright and Coues, *Citizen Bird*.

26. On Fuertes's debut at the A.O.U., see Frank M. Chapman, *Autobiography of a Bird-Lover* (New York: D. Appleton-Century, 1933), pp. 76–77.

27. Sharon Kingsland, "Abbott Thayer and the Protective Coloration Debate," *Journal for the History of Biology* 11, no. 2 (Fall 1978): 223–244. Abbott H. Thayer, "The Law Which Underlies Protective Coloration," *The Auk* 13 (April 1896): 124–129; part 2, pp. 318–320. Reprinted in *Annual Report of the Board of Regents of the Smithsonian Institution. 1897* (Washington, D.C.: 1898), pp. 477–482. See also Norelli, *American Wildlife Painting*.

28. Chapman, *Autobiography*, p. 78.

29. Ibid.

30. Ibid.

31. Ibid., p. 79.

32. Kingsland, "Abbott Thayer," p. 223.

33. Alfred Goldsborough Mayer, "On the Color and Color-Pattern of Moths and Butterflies," *Proceedings of the Boston Society of Natural History* 27, no. 14 (1897): 243–330. Also published in the *Bulletin of the Museum of Comparative Zoology* 30, no. 4 (1897): 243–330.

34. Kingsland describes Beddard's work on coloration in "Abbott Thayer," pp. 223–225: Beddard, *Animal Coloration: An Account of the Principal Facts and Theories* (London: Swan, Sonnenschein, 1892).

35. Gerald H. Thayer, *Concealing Coloration in the Animal Kingdom: An Exposition of the Laws of Disguise through Color and Pattern: Being a Summary of Abbott H. Thayer's Discoveries* (New

York: The Macmillan Company, 1909), legend, plate 1, n.p.

36. Abbott H. Thayer, "Introduction," in Gerald Thayer, *Concealing Coloration*, p. 3.

37. Ibid.

38. Ibid.

39. Chapman, *Autobiography*, p. 80.

40. Thayer, "Introduction," in Gerald Thayer, *Concealing Coloration*, p. 12.

41. Ibid.

42. Kingsland, "Abbott Thayer," pp. 231–240.

43. Ibid., p. 243.

44. A study of reef fishes confirmed Thayer's conclusions, "especially in relation to the low conspicuousness of brightly colored forms." During World War II, the successful use of camouflage confirmed Thayer's theory of "ruptive" patterning. Kingsland, "Abbott Thayer," pp. 242–243.

45. Ibid.

46. Ibid., p. 237.

# Bibliography

———◉———

THIS list does not include all references. It includes primary and secondary sources from which I have gathered quotations and works referred to in the notes. Books and journals from which illustrations have been reproduced are given in the List of Figures and List of Plates. There is some duplication: where I have both quoted from a work and reproduced an illustration from it, the work appears in this Bibliography as well as in the List of Figures and List of Plates.

## MANUSCRIPTS

Academy of Natural Sciences of Philadelphia
  Chinese Fish in Original Watercolors, 1794?, Collection 624
  Alexander Lawson, Scrapbooks of Engravings, 1808–1831?, Collection 79
  Alexander Lawson, 1772–1846, Biographical Papers, n.d., Collection 420
  Charles Alexander Lesueur, Papers and Drawings, Collection 136
  Othneil Charles Marsh, Lithographs, Collection 275
  Lucy Way Say, 1880–1886. Papers, 1822–1885, Collection 433
  Thomas Say, 1787–1834, Letters to the Melsheimers, 1813–1825, Collection 13
American Museum of Natural History
  Edward Drinker Cope, Correspondence, 1846–1897
  Titian Ramsay Peale, Introduction, to Zoology
  Titian Ramsay Peale, *The Butterflies of North America: Diurnal Lepidoptera. Whence They Come; Where They Go; and What They Do. Illustrated and Described*, ms. copy, unbound, 5 vols.
American Philosophical Society
  APS Archives
  John Fries Frazer, Papers, 1834–1871 (B/F865)
  Miscellaneous Manuscript Collection
  Titian Ramsay Peale, Sketches, 1817–1875 (B/P31.15d)
  Thomas Say, Papers, 1819–1883 (B/Sa95.g)

Boston Museum of Science
  Samuel Hubbard Scudder Papers
Gray Herbarium, Harvard University
  Historic Letter File
Historical Society of Pennsylvania
  Dreer Collection
Library of Congress
  Benjamin Tappan Papers
  Charles Wilkes Papers, Wilkes Letter Books
Massachusetts Historical Society
  Theodore Lyman, Diary
Museum of Comparative Zoology, Harvard University
  Louis Agassiz, Proof Plates, *Contributions to the Natural History of the United States*, n.d.
  Agassiz Papers
  Alexander Wilson, Drawings
Museum d'Histoire Naturelle, Le Havre
  Charles Alexandre Lesueur, Correspondence
National Archives
  Wilkes Expedition Letters
New-York Historical Society
  John James Audubon, Drawings
Peabody Museum, Salem, Massachusetts
  Edward Sylvester Morse Papers, Natural History Manuscripts Collection, Phillips Library
The Pierpont Morgan Library
  J. J. Audubon, Drawings
Private Collections
  Mrs. Willa Funck
Smithsonian Institution Archives
  Assistant Secretary (Spencer F. Baird), 1850–1877, Incoming Correspondence, RU52
  Assistant Secretary (Spencer F. Baird), 1850–1877, Outgoing Correspondence, RU53
  Spencer F. Baird Papers, 1833–1889, Incoming Correspondence, 1833–1849, RU7002
  William Dunlop Brackenridge Papers, circa 1838–1875, RU7189
  Collected Notes, Lists, Catalogs, Illustrations, and Records on Fishes, circa 1835–1974, RU7220
  Fielding B. Meek Papers, 1843–1877, RU7062
  Prints and Drawings, 1840– , RU92
  Robert Ridgway Papers, circa 1850s-1919, RU7167
  United States Exploring Expedition Collection, 1838–1885, RU7186

United States National Museum of Natural History Division of Fishes, Illustrations and Photographs
Workingmens' Institute, New Harmony, Indiana
New Harmony Manuscripts

PUBLISHED WORKS AND
DISSERTATIONS

Ackley, Clifford S. *Printmaking in the Age of Rembrandt.* Boston: Boston Museum of Fine Arts, 1981.

Agassiz, Alexander. "Application of Photography to Illustrations of Natural History. With two figures printed by the Albert and Woodbury Processes." *Bulletin of the Museum of Comparative Zoology* 3, no. 2 (1871): 47–48.

Agassiz, George Russell. *Letters and Recollections of Alexander Agassiz.* Boston: Houghton Mifflin, 1913.

Agassiz, Louis. *Recherches sur les poissons fossiles.* 5 vols. in 2 and atlas. Neuchâtel, Switz.: L'Auteur, 1833–1843.

Agassiz, Louis. *Études sur les glaciers.* Neuchâtel, Switz.: Jent et Gassmann, 1840.

Agassiz, Louis. *Nouvelle études et expériences sur les glaciers actuels, leur structure, leur progression et leur action physique sur le sol.* Paris: V. Masson, 1847.

Agassiz, Louis. *Contributions to the Natural History of the United States.* 4 vols. Boston: Little, Brown and Company, 1857–1862.

Agassiz, Louis. *Essay on Classification.* Edited by Edward Lurie, Cambridge, Mass.: Harvard University Press, 1962.

Albinus, Bernhard Siegfried. *Tabulae sceleti et musculorum corporis humani.* Lugduni Batavorum, Prostant apud Joannen & Hermannum Verbeek, 1747–1753.

Aldrich, Michele. "The New York Natural History Survey, 1836–1845." Ph.D. diss., University of Texas, Austin, 1974.

Allen, David E. *The Naturalist in Britain: A Social History.* London: Allen Lane, 1976.

Allen, Elsa G. "The History of American Ornithology before Audubon." *Transactions of the American Philosophical Society,* n.s., 41, part 3 (1951): 387–591.

Allen, Joel A. "[S]ome Remarks Concerning Geographic Variation in Mammals and Birds." *Proceedings of the Boston Society of Natural History* 15 (1872–1873): 156–159.

Allen, Joel A. "Correspondence." *The Auk* 1, no. 3 (July 1884): 302.

Alpers, Svetlana. *The Art of Describing: Dutch Art in the Seventeenth Century.* Chicago: University of Chicago Press, 1983.

Anker, Jean. *Bird Books and Bird Art: An Outline of the Literary History and Iconography of Descriptive Ornithology.* Copenhagen: Levin & Munksgaard, 1938.

Arndt, Karl J. R., ed. *Harmony on the Wabash in Transition: From Rapp to Owen, 1824–1826. A Documentary History.* Worcester, Mass.: Harmony Society Press, 1982.

Audubon, John James. *Ornithological Biography; or an Account of the Habits of the Birds of the United States of America; Accompanied by Descriptions of the Objects Represented in the Work Entitled, The Birds of America, and Interspersed with Delineations of American Scenery and Manners.* 5 vols. Edinburgh: Adam Black, 1831–1838.

Audubon, John James. *The Original Water-Color Paintings by John James Audubon for The Birds of America.* Introduction by Marshall B. Davidson. New York: American Heritage Publishing Co., 1966.

Audubon, Maria R. *Audubon and His Journals.* 2 vols. New York: Charles Scribner's Sons, 1897.

Bailey, Florence Merriam. *Birds through an Opera Glass.* Boston and New York: Houghton, Mifflin and Company, 1889.

Bailey, Florence Merriam. *A-Birding on a Bronco.* Boston: Houghton, Mifflin and Company, 1896.

Bailey, Florence Merriam. *Birds of Village and Field: A Bird Book for Beginners.* New York: Houghton, Mifflin and Company, 1898.

Bartram, William. *Travels through North & South Carolina, Georgia, East & West Florida, the Cherokee country, the extensive territories of the Muscogulges, or Creek Confederacy, and the country of the Chactaws; containing an account of the soil and natural productions of those regions, together with observations on the manners of the Indians.* Philadelphia: James and Johnson, 1791.

Bastide, Françoise. "Iconograpie des textes scientifiques: Principes d'analyse." *Culture Technique* 14 (June 1985): 133–151.

Bates, Ralph S. *Scientific Societies in the United States.* 3d. ed. Cambridge, Mass.: MIT Press, 1965.

Batten, Roger L. "Robert Parr Whitfield: Hall's Assistant Who Stayed Too Long." *Earth Science History* 6, no. 1 (1987): 40–46.

Baynes, Ken, and Francis Pugh. *The Art of the Engineer.* Woodstock, N.Y.: The Overlook Press, 1981.

Beale, Lionel S. *How to Work with a Microscope.* 4th ed. London: Harrison, 1865.

Beddard, F. E. *Animal Coloration: An Account of*

the Principal Facts and Theories. London: Swan, Sonnenschein, 1892.

Bénézit, Emmanuel, Dictionnaire critique et documentaire des peintres, sculpteurs, dessinateurs et graveurs de tous les temps et de tous les pays. New ed. 10 vols. Paris: Grund, 1976.

Berger, John. Ways of Seeing. London: British Broadcasting Corporation and Penguin Books, 1972.

Berger, John. About Looking. New York: Pantheon Books, 1979.

Bernhardt, Karl, Duke of Saxe-Weimar Eisenach. Travels through North America during the Years 1825 and 1826. Philadelphia: Carey, Lea and Carey, 1828.

Bewick, Thomas. A General History of Quadrupeds. Newcastle, U.K.: R. Beilby and T. Bewick, 1791.

Bewick, Thomas. A History of British Birds. 2 vols. Newcastle, U.K.: Beilby and Bewick, 1797–1804.

Bewick, Thomas. A General History of Quadrupeds. Engravings by Alexander Anderson. New York: Printed by G. and R. Waite, 1804.

Bezold, Wilhelm von. The Theory of Color in Its Relation to Art and Art-Industry. Trans. S. R. Koehler. Boston: L. Prang and Company, 1876.

Blum, Ann S. "'A Better Style of Art': The Illustrations of the Paleontology of New York." Earth Science History 6, no. 1 (1987): 72–85.

Blum, Ann, and Sarah Landry. "In Loving Detail." Harvard Magazine 79 (May–June 1977): 38–51.

Blunt, Wilfred, and William T. Stearn. The Art of Botanical Illustration. New York: Charles Scribner's Sons, 1951.

Bonaparte, Charles Lucien. American Ornithology; or, The Natural History of Birds Inhabiting the United States, Not Given by Wilson. 4 vols. Philadelphia: Samuel Augustus Mitchell; Carey, Lea and Carey, 1825–1834.

Bracegirdle, Brian. A History of Microtechique. Ithaca, N.Y.: Cornell University Press, 1978.

Bradbury, S. "The Quality of the Image Produced by the Compound Microscope, 1700–1840." In Historical Aspects of Microscopy Papers in the Royal Microscopical Society at Oxford. Edited by S. Bradbury and G.L'E. Turner. Cambridge, U.K.: Heffer, for the Royal Microscopical Society, 1967.

Brehm, Alfred Edmund. Brehm's Thierleben; allgemeine Kunde des Thierreichs. 10 vols. Leipzig: Bibliographischen Instituts, 1876–1879.

Brewer, Thomas M. North American Oölogy, part 1: Raptores and Fissirostres. Washington: Smithsonian Institution, and New York: D. Appleton and Co., 1857.

Brown, Thomas. Illustrations of the American Ornithology of Alexander Wilson and Charles Lucien Bonaparte Prince of Musignano. With the addition of numerous Recently Discovered Species, and Representations of the Whole Sylvae of North America. Edinburgh and London: Frazer and Co., 1835.

Buckland, Gail. Fox Talbot and the Invention of Photography. Boston: David R. Godine, 1980.

Burns, Frank L. "Miss Lawson's Recollections of Ornithologists." The Auk 34, no. 3 (July 1917): 275–282.

Cabinet of Natural History and American Rural Sports. 3 vols. Philadelphia: J. and T. Doughty, 1830–1834.

Cain, A. J. Animal Species and Their Evolution. London: Hutchinson's University Library, 1954.

Cannon, Susan Faye. Science in Culture: The Early Victorian Period. New York: Dawson and Science History Publications, 1978.

Cantwell, Robert. Alexander Wilson: Naturalist and Pioneer. Philadelphia: J. B. Lippincott Company, 1961.

Carpenter, William B. The Microscope and Its Revelations. 5th ed. London: J. and A. Churchill, 1875.

Catesby, Mark. The Natural History of Carolina, Florida and the Bahama Islands: Containing the Figures of Birds, Beasts, Fishes, Serpents, Insects, and Plants: Particularly the Forest-Trees, Shrubs, and other Plants, not hitherto Described, or Very Incorrectly Figured by Authors. Together with their Descriptions in English and French. To which, are added Observations on the Air, Soil, and Waters: With Remarks upon Agriculture, Grain, Pulse, Roots, &c. To the Whole, is Prefixed a New and Correct Map of the Countries Treated Of. London: Printed at the expence of the author, 1731–1743.

Chapman, Frank M. Autobiography of a Bird-Lover. New York: D. Appleton-Century Company, 1933.

Choulant, Ludwig. History and Bibliography of Anatomic Illustration. Translated and annotated by Mortimer Frank. New York: Hafner Publishing Company, 1962.

Clarke, John M. James Hall of Albany: Geologist and Paleontologist, 1811–1898. Albany, 1923.

Clifford, James, and George E. Marcus, eds. Writing Culture: The Poetics and Politics of Ethnography. Berkeley: University of California Press, 1986.

Coates, Reynell. "Reminiscences of a Voyage to India." In John Godman, Rambles of a Naturalist . . . to which are added Reminiscences of a Voyage to India. Philadelphia: Thomas T. Ash—Key and Biddle, 1833.

Cock, Elizabeth M. "The Influence of Photography

on American Landscape Painting, 1839–1880." Ph.D. diss., New York University, 1967.

Comstock, Anna Botsford. *The Comstocks of Cornell: John Henry Comstock and Anna Botsford Comstock, an Autobiography of Anna Botsford Comstock.* Edited by Glenn W. Herrick and Ruby Green Smith. Ithaca, N.Y.: Comstock Publishing Associates, Cornell University Press, 1953.

Cooper, Lane. *Louis Agassiz as a Teacher: Illustrative Abstracts on His Method of Instruction.* Ithaca, N.Y.: Comstock Publishing Co., 1945.

Cuvier, Frédéric. "Ornithological Biography, or an account of the habits of birds, . . . by John James Audubon." *Journal des Savants* (Paris), December 1833, pp. 705–719.

Cuvier, Georges. "Rapport verbal fait à l'Académie Royale des Sciences, sur l'histoire naturelle des oiseaux de l'Amerique s[e]ptentrionale, de M. Audubon." *Le Moniteur Universel* (Paris) 275 (1 October 1828): 1528.

Dall, William Healey. *Spencer Fullerton Baird: A Biography.* Philadelphia: J. B. Lippincott Company, 1915.

Dana, James Dwight. *Atlas. Crustacea.* Philadelphia: C. Sherman, Printer, 1855.

Dance, S. Peter. *The Art of Natural History: Animal Illustrators and Their Work.* Woodstock, N.Y.: Overlook Press, 1978.

Daniels, George H. *American Science in the Age of Jackson.* New York: Columbia University Press, 1968.

Daniels, George H. *Science in American Society: A Social History.* New York: Alfred A. Knopf, 1971.

Daniels, George H. "The Process of Professionalization in American Science: The Emergent Period, 1820–1860." *Isis* 58, no. 2 (1967): 151–166.

Daniels, George H., ed. *Nineteenth-Century American Science: A Reappraisal,* Evanston, Ill.: Northwestern University Press, 1972.

Darwin, Charles. *The Expression of the Emotions in Man and Animals.* London: John Murray, 1872.

Darwin, Charles. *On the Origin of Species.* Facsimile edition. Introduction by Ernst Mayr. Cambridge, Mass.: Harvard University Press, 1964.

Deane, James. *Ichnographs from the Sandstone of Connecticut River.* Boston: Little, Brown and Company, 1861.

Dear, Peter. "*Totius in verba*: Rhetoric and Authority in the Early Royal Society." *Isis* 76, no. 282 (June 1985): 145–161.

Deiss, William A. "Spencer F. Baird and His Collectors." *Journal of the Society for the Bibliography of Natural History* 9, no. 4 (1980): 635–645.

Doubleday, Neltje Blanchan. *Bird Neighbors: An Introductory Acquaintance with One Hundred and Fifty Birds Commonly Found in the Gardens, Meadows, and Woods about Our Homes.* New York: Doubleday and MacClure, 1897.

Doubleday, Neltje Blanchan. *Birds That Hunt and Are Hunted: Life Histories of One Hundred and Seventy Birds of Prey, Game Birds and Water Fowls.* New York: Doubleday and MacClure, 1898.

Doubleday, Neltje Blanchan. *How to Attract the Birds; and Other Talks about Bird Neighbors.* New York: Doubleday Page, 1902.

Doubleday, Neltje Blanchan. *Birds That Every Child Should Know: The East.* New York: Grosset and Dunlap, 1907.

Dumont d'Urville, Jules-Sébastien-César. *Voyage de la corvette l'Astrolabe: Exécuté par ordre du roi, pendant les années 1826–1827–1828–1829, sous le commandement de J. Dumont d'Urville.* 14 vols. in 13, and 7 vols. of atlas. Paris: J. Tastu, 1830–1835.

Dumont d'Urville, Jules-Sébastien-César. *Voyage au pole sud et dans l'Océanie sur les corvettes l'Astrolabe et la Zelée, éxécuté par ordre du roi pendant les années 1837–1838–1839–1840, sous le commandement de M. J. Dumont d'Urville.* 10 vols in 8. Paris: Gide, 1841–1846.

Dunlap, William. *History of the Rise and Progress of the Arts of Design in the United States.* Edited by Frank W. Bayley and Charles E. Goodspeed. 3 vols. New York: Benjamin Blom, 1965.

Dupree, A. Hunter. *Science in the Federal Government: A History of Policies and Activities to 1940.* Cambridge, Mass.: The Belknap Press of Harvard University Press, 1957.

Dupree, A. Hunter. *Asa Gray, 1810–1888.* Cambridge, Mass.: The Belknap Press of Harvard University Press, 1959.

Eder, Josef Maria. *History of Photography.* Translated by Edward Epstean. New York: Columbia University Press, 1945.

Edgerton, Samuel Y., Jr. *The Renaissance Rediscovery of Linear Perspective.* New York: Harper and Row, 1975.

Eisenstein, Elizabeth L. *The Printing Press as an Agent of Change: Communications and Cultural Transformations in Early-Modern Europe.* 2 vols. New York: Cambridge University Press, 1980.

Elliot, Daniel Giraud. *A Monograph of the Phasianidae, or Family of Pheasants.* New York, 1872.

Elliot, Daniel Giraud. *A Monograph of the Paradiseidae, or Birds of Paradise.* London, 1873.

Elliot, Daniel Giraud. *A Monograph of the Bucerotidae, or Family of Hornbills.* London: For the subscribers, by the author, 1882.

Elliot, Daniel Giraud. *A Monograph of the Felidae, or Family of Cats.* London: For the subscribers, by the author, 1883.

*Encyclopaedia Britannica; or, Dictionary of Arts, Sciences, and General Literature.* 8th ed. 21 vols. Boston: Little, Brown, and Company, 1852–1860.

Ewan, Joseph, ed. *William Bartram: Botanical and Zoological Drawings, 1756–1788.* Philadephia: American Philosophical Society, 1968.

Ewan, Joseph. "The Natural History of John Abbot: Influences and Some Questions." *Bartonia* 51 (1985): 37–45.

Farber, Paul Lawrence. *The Emergence of Ornithology as a Scientific Discipline, 1760–1850.* Studies in the History of Modern Science, vol. 12. Dordrecht, Holland: D. Reidel, 1982.

Feduccia, Alan. *Catesby's Birds of Colonial America.* Chapel Hill: University of North Carolina Press, 1985.

Ford, Alice. *The 1826 Journal of John James Audubon.* Norman: University of Oklahoma Press, 1967.

Foucault, Michel. *The Order of Things: An Archaeology of the Human Sciences.* New York: Pantheon Books, 1971.

Fries, Waldemar H. *The Double Elephant Folio: The Story of Audubon's Birds of America.* Chicago: American Library Association, 1973.

Fyfe, Gordon, and John Law. "Introduction: On the Invisibility of the Visual." In *Picturing Power: Visual Depictions of Social Relations,* pp. 1–14. Edited by Gordon Fyfe and John Law. London: Routledge, 1988.

Gage, Simon Henry. *The Microscope and Histology for the Use of Laboratory Students in the Anatomical Department of Cornell University.* Part 1, *The Microscope and Microscopical Methods.* 3d ed. Ithaca, N.Y., 1891.

Geikie, Archibald. "The Toothed Birds of Kansas." *Nature,* 16 September 1880, pp. 457–458.

Gibson, William Hamilton. *Highways and Byways; or, Saunterings in New England.* New York: Harper and Brothers, 1882.

Gifford, George E., Jr. "Thomas Mayo Brewer, M.D., ... A Blackbird and Duck, Sparrow and Mole." *Harvard Medical Alumni Bulletin* 37, no. 1 (Fall 1962): 32–35.

Gill, Theodore. "The First Edition of Holbrook's North American Herpetology." *Science* 17, no. 440 (5 June 1903): 910–912.

Gill, Theodore. "John Edwards Holbrook, 1794–1871." *Biographical Memoirs of the National Academy of Sciences* 5 (1905): 47–77.

Gillispie, Charles Coulston, ed. in chief, *Dictionary of Scientific Biography.* 18 vols. New York: Charles Scribner's Sons, 1970–1990.

Godman, John M. *American Natural History.* 3 vols. Philadelphia: H. C. Carey and I. Lea, 1826–1828.

Goetzmann, William H. *Army Exploration in the American West, 1803–1863.* New Haven, Conn.: Yale University Press, 1959.

Goetzmann, William H. *Exploration and Empire: The Explorer and the Scientist in the Winning of the American West.* New York: W. W. Norton, 1966.

Goldsmith, Oliver. *An History of the Earth and Animated Nature.* 4 vols. Philadelphia: Mathew Carey, 1795.

Gombrich, E. H. *Art and Illusion: A Study in the Psychology of Pictorial Representation.* Princeton, N.J.: Princeton University Press, 1969.

Goode, George Brown. "The Beginnings of American Science. The Third Century." *Proceedings of the Biological Society of Washington* 4 (1887): 9–134.

Goodrich, Samuel G. *Illustrated Natural History of the Animal Kingdom, being a Systematic and Popular Description of the Habits, Structure, and Classification of Animals, from the Highest to the Lowest Forms, with their Relations to Agriculture, Commerce, Manufactures, and the Arts.* 2 vols. New York, 1859.

Gosse, Philip Henry. *Evenings at the Microscope; or, Researches among the Minuter Organs and Forms of Animal Life.* New York: D. Appleton and Company, 1879.

Gould, Augustus Addison. *Report on the Invertebrates of Massachusetts. . . .* Cambridge, Mass.: Commissioners on the Zoological and Botanical Survey; Folsom, Wells and Thurston, printers, 1841.

Gould, Stephen J. *The Mismeasure of Man.* New York: W. W. Norton, 1981.

Gould, Stephen J. *The Panda's Thumb: More Reflections on Natural History.* New York: W. W. Norton, 1982.

Greene, John C. *American Science in the Age of Jefferson.* Ames: University of Iowa Press, 1984.

Groce, George C., and David W. Wallace. *The New York Historical Society's Dictionary of American Artists, 1564–1860.* New Haven, Conn.: Yale University Press, 1957.

Haeckel, Ernst Heinrich Philipp August. *Kunst-*

*formen der Natur.* Leipzig and Vienna: Verlag des Bibliographischen Instituts, 1899–1904.

Hall, James. *Geology of New York. Part 4. Survey of the Fourth Geological District.* Albany: Carrol and Cook, 1843.

Hall, James. *Paleontology of New York: Organic Remains of the Lower Division of the New-York System.* Vol. 1. Albany: C. van Benthuysen, 1847.

Hall, James. *Paleontology of New York: Organic Remains of the Lower Helderberg Group and the Oriskany Sandstone.* Vol. 3. Albany: C. van Benthuysen, 1855–1859.

Hall, James. *Illustrations of Devonian Fossils: Gasteropoda, Pteropoda, Cephalopoda, Crustacea and Corals from the Upper Helderberg, Hamilton and Chemung Groups.* New York: E. Bierstadt, and Albany: Weed, Parsons and Company, 1876.

Hall, James. *Paleontology of New York: Gasteropoda, Pteropoda and Cephalopoda of the Upper Helderberg, Hamilton, Portage and Chemung Groups.* Vol. 5, part 2. Albany: C. van Benthuysen, 1879.

Hamilton, Thomas. *Men and Manners in America.* Edinburgh: W. Blackwood, 1833.

Haskell, Daniel C. *The United States Exploring Expedition, 1838–1842, and Its Publications, 1844–1874.* New York: New York Public Library, 1942.

Haskell, Francis. *Rediscoveries in Art: Some Aspects of Taste, Fashion and Collecting in England and France.* Ithaca, N.Y.: Cornell University Press, 1976.

Hayden, Dolores. *Seven American Utopias: The Architecture of Communitarian Socialism, 1790–1975.* Cambridge, Mass.: MIT Press, 1976.

Heck, Johann Georg. *Iconographic Encyclopedia of Science, Literature, and Art.* Translated and edited by Spencer Fullerton Baird. New York: R. Garrigue, 1852.

Herber, Elmer Charles. *Correspondence between Spencer Fullerton Baird and Louis Agassiz—Two Pioneer Naturalists.* Washington, D.C.: Smithsonian Institution, 1963.

Herrick, Francis Hobart. *Audubon the Naturalist: A History of His Life and Time.* 2 vols. New York: D. Appleton, 1917. Second ed., 1938.

Hodson, James Shirley. *An Historical Guide to Art Illustration, in Connection with Books, Periodicals, and General Decoration.* London: Sampson Low, Marston, Searle and Rivington, 1884.

Honour, Hugh. *The New Golden Land: European Images of America from the Discoveries to the Present Time.* New York: Pantheon Books, 1975.

Hooke, Robert. *Micrographia, or Some Physiological Descriptions of Minute Bodies Made by Magnifying Glasses, with Observations and Inquiries Thereupon.* New York: Dover Publications, 1961.

Hoyle, Pamela. *The Development of Photography in Boston, 1840–1875.* Boston: Boston Atheneum, 1979.

Hunter, Clark, ed. *The Life and Letters of Alexander Wilson.* Philadelphia: American Philosophical Society, 1983.

Hunter, Dard. *Papermaking: The History and Technique of an Ancient Craft.* 2d. ed. New York: Alfred A. Knopf, 1947.

"Introduction." *The Magazine of Natural History, and Journal of Zoology, Botany, Mineralogy, Geology, and Meteorology* 1, no. 1 (May 1828): 1–9.

Ivins, William M. *Prints and Visual Communication.* New York: Da Capo Press, 1969.

Ivins, William M. *The Rationalization of Sight.* New York: Plenum Press, 1973.

James, Edwin. *Account of an Expedition from Pittsburgh to the Rocky Mountains, Performed in the Years 1819 and '20, by Order of the Hon. J. C. Calhoun, Sec'y of War: Under the Command of Major Stephen H. Long. From the Notes of Major Long, Mr. T. Say, and Other Gentlemen of the Exploring Party.* Philadelphia: H. C. Carey and I. Lea, 1822–1823.

Johnson, Allen, et al., eds. *Dictionary of American Biography.* 23 vols. and supplements. New York: Charles Scribner's Sons, 1928–1955.

Jordan, David Starr. "Sketch of Charles A. Le Sueur." *Popular Science Monthly,* February 1895, pp. 547–550.

Jussim, Estelle. *Visual Communication and the Graphic Arts: Photographic Technologies in the Nineteenth Century.* New York: R. R. Bowker, 1974.

Keller, Evelyn Fox. *Reflections on Gender and Science.* New Haven: Yale University Press, 1985.

Kingsland, Sharon. "Abbott Thayer and the Protective Coloration Debate." *Journal for the History of Biology* 11, no. 2 (Fall 1978): 223–244.

Koehler, Sylvester Rosa. *Exhibition Illustrating the Technical Methods of the Reproductive Arts, from the XV. Century to the Present Time, with Special Reference to Photo-Mechanical Methods.* Boston: Museum of Fine Arts, 1892.

Kofalk, Harriet. *No Woman Tenderfoot: Florence Merriam Bailey, Pioneer Naturalist.* College Station: Texas A and M University Press, 1989.

Kohlstedt, Sally Gregory. "Parlors, Primers, and Public Schooling: Education for Science in Nineteenth-Century America." *Isis* 81, no. 308 (1990): 425–445.

Koreny, Fritz. *Albrecht Dürer, and the Animal and Plant Studies of the Renaissance*. Boston: A New York Graphic Society Book; Little, Brown and Company, 1985.

Korzenik, Diana. *Drawn to Art: A Nineteenth-Century American Dream*. Hanover, N.H.: University Press of New England, 1985.

Kraus, Rolf H. "Photographs as Early Scientific Illustrations." *History of Photography* 2, no. 4 (October 1978): 291–314.

Lapage, Geoffrey. *Art and the Scientist*. Bristol, U.K.: John Wright and Sons, 1961.

Laqueur, Thomas. *Making Sex: Body and Gender from the Greeks to Freud*. Cambridge, Mass.: Harvard University Press, 1990.

Latour, Bruno. "Visualization and Cognition: Thinking with Eyes and Hands." *Knowledge and Society: Studies in the Sociology of Culture Past and Present* 6 (1986): 1–40.

Latour, Bruno. *Science in Action: How to Follow Scientists and Engineers through Society*. Cambridge, Mass.: Harvard University Press, 1987.

Laurie, Bruce, Theodore Hershberg, and George Alter. "Immigrants and Industry: The Philadelphia Experience, 1850–1880." *Journal of Social History* 9, no. 2 (Winter 1975): 219–248.

Leidy, Joseph. *Fresh-Water Rhizopods of North America*. Washington, D.C.: U.S. Geological Survey of the Territories, 1879.

Lesueur, Charles Alexandre. *Dessins de Charles Alexandre Lesueur. Exécutés aus Etats Unis de 1816 à 1837*. Paris: Edité par la Cité Universitaire de Paris, 1933.

Lindley, Harlow, ed. *New Harmony as Seen by Participants and Travelers*. Philadelphia: Porcupine Press, 1975.

Lindroth, Sten. "Carl Linnaeus (or von Linné)." In Gillispie, ed., *Dictionary of Scientific Biography*, vol. 8 (1973), pp. 374–381.

Linton, William J. "Art in Engraving on Wood." *The Atlantic Monthly* 43 (June 1879): 705–715.

Linton, William J. *The History of Wood-Engraving in America*. London: Bell, 1882.

Lockwood, George B. *The New Harmony Movement*. New York: D. Appleton and Company, 1907.

Lurie, Edward. *Louis Agassiz: A Life in Science*. Chicago: University of Chicago Press, 1960.

Lurie, Edward. "Editor's Introduction." In *Essay on Classification*, by Louis Agassiz. Cambridge, Mass.: Harvard University Press, 1962.

Lynch, Michael. "Science in the Age of Mechanical Reproduction: Moral and Epistemic Relations between Diagrams and Photographs." *Biology and Philosophy* 6, no. 2 (April 1991): 205–226.

Lynch, Michael, and Steve Woolgar, eds. *Representation in Scientific Practice*. Special issue of *Human Studies* 11, nos. 2–3 (April–July 1988): 99–359.

Lyon, John, and Phillip R. Sloan, eds. *From Natural History to the History of Nature: Readings from Buffon and His Critics*. Notre Dame, Indiana: University of Notre Dame Press, 1981.

McCalla, Albert. "The Verification of Microscopic Observation." *Proceedings of the American Society of Microscopists*, Sixth Annual Meeting. Buffalo, N.Y.: Steam Printing House of Haas and Klein, 1883.

McCook, Henry Christopher. *Tenants of an Old Farm: Leaves from the Note-Book of a Naturalist*. New York: Fords, Howard and Hulbert, 1885.

McCoubrey, John, ed. *American Art, 1700–1960*. Sources and Documents in the History of Art Series. Englewood Cliffs, N.J.: Prentice-Hall, 1965.

McGaw, Judith A. *Most Wonderful Machine: Mechanization and Social Change in Berkshire Paper Making, 1801–1885*. Princeton, N.J.: Princeton University Press, 1987.

Maclure, William. "An Epitome of the Improved Pestalozzian System of Education, as practised by William Phiquepal and Madam Fretageot, formerly of Paris, and now in Philadelphia; communicated at the request of the Editor." *American Journal of Science and Arts* 10, no. 1 (January 1826): 145–151.

Maclure, William. *Opinions on Various Subjects, Dedicated to the Industrious Producers*. 3 vols. New Harmony, Indiana: School Press, 1831–1838, reprinted by Burt Franklin, New York, 1969.

MacPhail, Ian. "Natural History in Utopia: The Works of Thomas Say and François-André Michaux at New Harmony, Indiana." In *Contributions to the History of North American Natural History Papers from the First North American Conference of the Society for the Bibliography of Natural History Held at the Academy of Natural Sciences in Philadelphia, 21–23 October, 1981*, Special Publication no. 2, pp. 15–33. London: Society for the Bibliography of Natural History, 1983.

Marsh, Othneil C. "Introduction and Succession of Vertebrate Life in America." An address delivered before the American Association for the Advancement of Science, at Nashville, Tenn., 30 August 1877.

Marzio, Peter C. *Chromolithography 1840–1900. The Democratic Art: Pictures for a 19th-Century America*. Boston: David R. Godine, in association with the Amon Carter Museum of Western Art, Fort Worth, Texas, 1979.

Maximilian, Alexander Philip, Prince of Wied-Neuwied. *Travels in the Interior of North America, 1832–1834*. Vols. 22 and 23. Translated by H. Evans Lloyd. Reprinted in Reuben Gold Thwaites, ed., *Early Western Travels*. 32 vols. Cleveland: Arthur Clark Co., 1906.

Mayer, Alfred Goldsborough. "On the Color and Color-Pattern of Moths and Butterflies." *Proceedings of the Boston Society of Natural History* 27, no. 14 (1897): 243–330.

Mayer, Alfred Goldsborough. *Medusae of the World*. 3 vols. Washington, D.C.: Carnegie Institution, 1910.

Mayr, Ernst. "Agassiz, Darwin, and Evolution." *Harvard Library Bulletin* 13 (1959): 165–194.

Mayr, Ernst. *The Growth of Biological Thought: Diversity, Evolution, and Inheritance*. Cambridge, Mass.: The Belknap Press of Harvard University Press, 1982.

Michelet, Jules. *The Insect*. London: T. Nelson and Sons, 1875.

Miller, Olive Thorne [Harriet Mann]. *Bird Ways*. Boston: Houghton, Mifflin and Company, 1885.

Miller, Olive Thorne [Harriet Mann]. *In Nesting Time*. Boston: Houghton, Mifflin and Company, 1888.

Miller, Olive Thorne [Harriet Mann]. *Little Brothers of the Air*. Boston: Houghton, Mifflin and Company, 1892.

Miller, Olive Thorne [Harriet Mann]. *The First Book of Birds*. Boston and New York: Houghton, Mifflin and Company, 1899.

Mitchell, W.J.T. *Iconology: Image, Text, Ideology*. Chicago: University of Chicago Press, 1986.

Mongan, Elizabeth. *Wenceslaus Hollar, 1607–1677: Birds, Beasts and Grotesques*. Gloucester, Mass.: The Hammond Museum, 1979.

Morse, Edward Sylvester. "What American Zoologists have done for Evolution: An Address by Vice-President E. S. Morse, before the American Association for the Advancement of Science, at Buffalo, N.Y., August, 1876." Reprinted from the *Proceedings of the American Association for the Advancement of Science* 25 (1876). Salem, Mass.: Naturalist's Agency, 1876.

Mott, Frank Luther. *History of American Magazines, 1850–1865*. 4 vols. Cambridge, Mass.: Harvard University Press, 1938.

Muybridge, Eadweard. *Animal Locomotion: An Electro-Photographic Investigation of Consecutive Phases of Animal Movements*. 16 vols. Philadelphia: Published under the auspices of the University of Pennsylvania, 1887.

Neef, Joseph. *Sketch of a Plan and Method of Education, Founded on an Analysis of Human Faculties, and Natural Reason, Suitable for the Offspring of a Free People, and for all Rational Beings*. Philadelphia: Printed by the author, 1808.

Neef, Joseph. *The Method of Instructing Children Rationally, in the Arts of Writing and Reading*. Philadelphia: Printed by the author, 1813.

Nelson, Clifford M. "Meek at Albany, 1852–58." *Earth Science History* 6, no. 1 (1987): 40–46.

Nelson, Clifford M., and Ellis L. Yochelson. "Organizing Federal Paleontology in the U.S." *Journal of the Society for the Bibliography of Natural History* 9, no. 4 (1980): 607–618.

Newberry, John Strong. *Modern Scientific Investigation: Its Methods and Tendencies. An Address Delivered before the American Association for the Advancement of Science, August, 1867*. Salem, Mass.: Essex Institute Press, 1867.

Nolan, James. *A Short History of the Academy of Natural Sciences of Philadelphia*. Philadelphia: Academy of Natural Sciences, 1909.

Norelli, Martina R. *American Wildlife Painting*. New York: Watson-Guptill, in association with the National Collection of Fine Arts, Smithsonian Institution, Washington, D.C., 1975.

Norton, Bettina A. "Tappan and Bradford: Boston Lithographers with Essex County Associations." *Essex Institute Historical Collections* 109 (July 1978): 149–160.

Novak, Barbara. *Nature and Culture: American Landscape and Painting, 1825–1875*. New York: Oxford University Press, 1980.

Novak, Barbara. *The Thyssen-Bornemisza Collection: Nineteenth-Century American Painting*. New York: The Vendome Press, 1986.

Novak, Barbara, and Annette Blaugrund, eds. *Next to Nature: Landscape Painting from the National Academy of Design*. New York: Harper and Row, 1980.

Novales, Alberto Gil. *William Maclure: Socialismo utópico en España (1808–1840)*. Barcelona: Universidad Autonoma de Barcelona, 1979.

Nuttall, Thomas. *A Manual of Ornithology of the United States and Canada: Land Birds*. Cambridge, Mass.: Hilliard and Brown, 1832.

O'Hara, Robert. "Diagrammatic Classifications of Birds, 1819–1901: Views of the Natural System in 19th-Century British Ornithology." In *Proceed-*

BIBLIOGRAPHY

ings of the Nineteenth International Ornithological Congress, Ottawa, 1986.

Opperman, Hal. J.-B. Oudry, 1686–1755. Fort Worth, Texas: Kimbell Art Museum, 1983.

Orton, James. Comparative Zoology, Stuctural and Systematic. For Use in Schools and Colleges. New York: Harper and Brothers, 1876.

Osborn, Henry Fairfield. Cope: Master Naturalist. Princeton, N.J.: Princeton University Press, 1931.

Owen, David Dale. Report of the Geological Survey of Wisconsin, Iowa, and Minnesota; and Incidentally of a Portion of Nebraska Territory. Atlas. Philadelphia: Lippincott, Grambo and Co., 1852.

Owen, David Dale. "Letter to Robert Owen, from his son, David Dale Owen, July 24, 1852." Robert Owen's Journal, part 25 (25 September 1852): 167–168.

Owen, Robert. "Address to the New Harmony Community Membership." New Harmony Gazette 1 (27 April 1825): 1.

Peabody, W.B.O. "Audubon's Biography of Birds. Ornithological Biography of Birds . . . Philadelphia, 1831." North American Review 34 (April 1832): 364–405.

[Peabody, W.B.O.] "1. History of British Birds, by William MacGillivray. 2. Ornithological Biography, by John James Audubon." North American Review 50, no. 107 (April 1840): 381–404.

Peale, Rembrandt. Historical Disquisition on the Mammoth, or Great American Incognitum, an Extinct, Immense, Carnivorous Animal, whose Fossil Remains Have Been Found in North America. London: E. Lawrence, 1803.

Peale, Titian. "The South Sea Surveying and Exploring Expedition." American Historical Record 3 (1874): 244–251, 305–311.

Peck, Robert McCracken. A Celebration of Birds: The Life and Art of Louis Agassiz Fuertes. Published for the Academy of Natural Sciences of Philadelphia. New York: Walker and Company, 1982.

Péron, François. Voyage de découvertes aux terres Australes, éxecuté par ordre de Sa Majesté l'empereur et roi, sur les corvettes le Géographe, le Naturaliste, et la goélette la Casuarina, pendant les années 1800, 1801, 1802, 1803 et 1804. Paris: De l'Imprimerie impériale, 1817.

Peters, Harry T. America on Stone. Reprint edition. Arno Press, 1976.

Pitz, Henry C. Howard Pyle: Writer, Illustrator, Founder of the Brandywine School. New York: Clarkson N. Potter, 1975.

Poesch, Jessie. "Titian Ramsay Peale, 1799–1885, and His Journal of the Wilkes Expedition." In Memoirs of the American Philosophical Society 52 (Philadelphia, 1961).

Porter, Charlotte. " 'Subsilentio': Discouraged Works of Early Nineteenth-Century American Natural History." Journal of the Society for the Bibliography of Natural History 9, no. 2 (1979): 109–119.

Porter, Charlotte. The Eagle's Nest: Natural History and American Ideas, 1812–1842. Tuscaloosa: University of Alabama Press, 1986.

Powell, Mary M. "Three Artists of the Frontier." Bulletin of the Missouri Historical Society 5, no. 1 (October 1948): 34–43.

Rainger, Ronald, Keith R. Benson, and Jane Maienschein, eds. The American Development of Biology. Philadelphia: University of Pennsylvania Press, 1988.

Reeds, Karen. "Publishing Scholarly Books in the Sixteenth Century." Scholarly Publishing 14, no. 3 (April 1983): 258–274.

Reingold, Nathan. Science in Nineteenth-Century America: A Documentary History. New York: Hill and Wang, 1964.

Reingold, Nathan, and Marc Rothenberg. "The Exploring Expedition and the Smithsonian Institution." In Magnificent Voyagers. Edited by H. Viola and C. Margolis. Washington, D.C.: Smithsonian Institution, 1985.

Rhees, William Jones. The Smithsonian Institution: Documents Relative to Its Origin and History. Smithsonian Miscellaneous Collections, vol. 17. Washington, D.C.: Smithsonian Institution, 1880.

Richardson, Edgar P., Brooke Hindle, and Lillian B. Miller. Charles Willson Peale and His World. New York: Harry N. Abrams, 1983.

Ridgway, John Livesy. The Preparation of Illustrations for Reports of the U.S. Geological Survey with Brief Descriptions of the Processes of Reproduction. Washington, D.C.: United States Geological Survey, 1920.

Ridgway, Robert. A Nomenclature of Colors for Naturalists, and Compendium of Useful Knowledge for Ornithologists. Boston: Little, Brown and Company, 1886.

Ripley, George, and Charles A. Dana, eds. The New American Cyclopedia: A Popular Dictionary of General Knowledge. 16 vols. New York: D. Appleton and Company, 1869–1870.

Roger, Jacques. "Georges-Louis Leclerc, Comte de Buffon." In Dictionary of Scientific Biography, vol. 2, pp. 576–582. Edited by Charles Coulston Gillispie. New York: Charles Scribner's Sons, 1970.

Roger, Jacques. Buffon, un philosophe au jardin du roi. Paris: Fayard, 1989.

Rollins, Hyder Edward, ed. *The Keats Circle: Letters and Papers, 1816–1878*. Cambridge, Mass.: Harvard University Press, 1948.

Ronsil, René. "L'art français dans le livre d'oiseaux: Eléments d'une iconographie ornithologique française." *Mémoires de la Société Ornithologique de France et de l'Union Française*, no. 6, suppl. to *L'Oiseau et la Revue Française d'Ornithologie* 27 (4). Paris, 1957.

Rudwick, Martin J.S. "The Emergence of a Visual Language for Geological Science 1760–1840." *History of Science* 14 (1976): 149–195.

Rudwick, Martin J.S. *The Meaning of Fossil: Episodes in the History of Paleontology*. 2d ed. Chicago: University of Chicago Press, 1985.

Rutledge, Anna Wells. "Artists in the Life of Charleston through Colony and State from Restoration to Reconstruction." *Transactions of the American Philosophical Society* 39, no. 2 (1949): 101–260.

Samuels, Peggy, and Harold Samuels. *The Illustrated Biographical Encyclopedia of Artists of the American West*. Garden City, N.Y.: Doubleday and Co., 1970.

Sauer, Gordon C. *John Gould the Bird Man: A Chronology and Bibliography*. London: Henry Sotheran Ltd., 1982.

Saunders, J. B. de C.M., and Charles D. O'Malley. *The Anatomical Drawings of Andreas Vesalius*. New York: Bonanza Books, 1982.

Say, Thomas. *American Entomology, or Descriptions of the Insects of North America*. 3 vols. Philadelphia: Samuel Augustus Mitchell, 1824–1828.

Say, Thomas. *American Conchology, or Descriptions of the Shells of North America*. New Harmony, Indiana: School Press, 1834.

"Say's American Entomology." *North American Review and Miscellaneous Journal* 21, no. 48 (July 1825): 251–252.

Schiebinger, Londa. "Skeletons in the Closet: The First Illustrations of the Female Skeleton in Eighteenth-Century Anatomy." *Representations* 14 (Spring 1986): 42–82.

Schneer, Cecil J., ed. *Two Hundred Years of Geology in America: Proceedings of the New Hampshire Bicentennial Conference on the History of Geology, October 15–19, 1976*. Hanover, N.H.: University Press of New England, 1979.

Schoolcraft, Henry R. *Historical and Statistical Information Respecting the History, Condition and Prospects of the Indian Tribes of the United States*. Philadelphia: J. B. Lippincott and Co., 1851–1857.

Schuchert, Charles, and Clara Mae LeVene. *O. C. Marsh, Pioneer in Paleontology*. New Haven, Conn.: Yale University Press, 1940.

Scudder, Samuel Hubbard. *The Butterflies of the Eastern United States and Canada, with Special Reference to New England*. 3 vols. Cambridge, Mass.: Published by the author, 1889.

Sellars, Charles Coleman. *Mr. Peale's Museum: Charles Willson Peale and the First Popular Museum of Natural Science and Art*. New York: W. W. Norton, 1980.

Silliman, Benjamin. "Notice of the Lithographic Art, or the art of multiplying designs, by substituting Stone for Copper Plate, with introductory remarks by the Editor." *American Journal of Science and Arts* 4 (1822): 170.

Silliman, Benjamin. "Paleontology of New York, vol. I." *American Journal of Science and Arts*, series 2, vol. 5 (May 1848): 149–150.

Smellie, William, ed. *The System of Natural History, Written by M. de Buffon, Carefully Abridged: And the Natural History of Insects; Compiled, Chiefly from, Swammerdam, Brookes, Goldsmith, &c*. 2 vols. Edinburgh: J. Ruthven and Sons, 1800.

Smith, Bernard. *European Visions of the South Pacific*. 2d. ed. New Haven, Conn.: Yale University Press, 1985.

Sprat, Thomas. *History of the Royal Society*. Reproduction of the 1667 edition. Edited by Jackson I. Cope and Harold Whitmore Jones. St. Louis, Mo.: Washington University, 1958.

Stafford, Barbara M. "Characters in Stones, Marks on Paper: Enlightenment Discourse on Natural and Artificial Taches." *Art Journal*, Fall 1984, pp. 233–240.

Stafford, Barbara M. *Voyage into Substance: Art, Science, Nature and the Illustrated Travel Account, 1760–1840*. Cambridge, Mass.: MIT Press, 1984.

Stanton, William. *The Leopard's Spots: Scientific Attitudes toward Race in America, 1815–1859*. Chicago: University of Chicago Press, 1960.

Stanton, William. *The Great United States Exploring Expedition of 1838–1842*. Berkeley: University of California Press, 1975.

Stearn, Wilfred Blunt, and William T. Stearn. *The Art of Botanical Illustration*. New York: Charles Scribner's Sons, 1951.

Stein, Roger B. *John Ruskin and Aesthetic Thought in America, 1840–1900*. Cambridge, Mass.: Harvard University Press, 1967.

Stingham, Emerson. *Alexander Wilson, a Founder of Scientific Ornithology*. Kerrville, Texas: Published by the author, 1958.

Stokes, Alfred C. *Microscopy for Beginners; or, Common Objects from the Ponds and Ditches.* New York: Harper and Brothers, 1887.

Stresemann, Erwin. *Ornithology from Aristotle to the Present.* Translated by Hans J. Epstein and Cathleen Epstein. Cambridge, Mass.: Harvard University Press, 1975.

Surdez, Maryse. *Catalogue des archives de Louis Agassiz (1807–1873).* Neuchâtel, Switz.: Secrétariat de l'Université, 1973.

Swainson, William. "Some Account of the Work now publishing by M. Audubon, entitled The Birds of America." *The Magazine of Natural History and Journal of Zoology, Botany, Mineralogy, Geology and Meteorology* 1, no. 1 (May 1828): 43–52.

"A Symposium of Wood-Engravers." *Harper's New Monthly Magazine* 50 (February 1880): 442–453.

"A Synopsis of the Birds of North America; by John James Audubon . . . The Birds of America, from drawings made in the United States and their Territories, . . . Vol. I. New York, 1840." *American Journal of Science and Arts* 39, no. 2 (July–September 1840): 343–357.

Taylor, Basil. *Stubbs.* London: Phaidon Press Limited, 1971.

Taylor, Keith. *The Political Ideas of Utopian Socialists.* London: Frank Cass, 1982.

Taylor, Peter J., and Ann S. Blum. "Pictorial Representation in Biology." *Biology and Philosophy* 6, no. 2 (April 1991): 125–134.

Tebbel, John William. *A History of Book Publishing in the United States.* 4 vols. New York: R. R. Bowker Co., 1972.

Tenney, Sanborn. *Elements of Zoology: A Text-Book.* New York: Charles Scribner's Sons, 1875.

Tenney, Sanborn, and Abby Tenney. *Natural History: A Manual of Zoology for Schools, Colleges, and the General Reader.* New York: Charles Scribner and Co., 1865.

Thayer, Abbott H. "The Law Which Underlies Protective Coloration." *The Auk* 13 (April 1896): 124–129; part 2, pp. 318–320.

Thayer, Gerald H. *Concealing Coloration in the Animal Kingdom: An Exposition of the Laws of Disguise through Color and Pattern: Being a Summary of Abbott H. Thayer's Discoveries.* New York: The Macmillan Company, 1909.

Thomas, Keith. *Man and the Natural World: Changing Attitudes in England, 1500–1800.* London: Allen Lane, 1983.

Toumey, Michael, and Francis S. Holmes. *Pleiocene Fossils of South-Carolina: Containing Descriptions and Figures of the Polyparia, Echinodermata and Mollusca.* Charleston, S.C.: Russell and Jones, 1857.

Townsend, John Kirk. *Ornithology of the United States; or, Descriptions of the Birds Inhabiting the States and Territories of the Union.* Vol. 1. Philadelphia: J. B. Chevalier, 1839.

Tufte, Edward R. *Visual Display of Quantitative Information.* Cheshire, Conn.: Graphics Press, 1983.

Turner, James. *The Politics of Landscape: Rural Scenery and Society in English Poetry, 1630–1660.* Cambridge, Mass.: Harvard University Press, 1979.

Twyman, Michael. *Lithography, 1800–1850: The Techniques of Drawing on Stone in England and France and Their Application in Works of Topography.* London: Oxford University Press, 1970.

Tyler, David B. *The Wilkes Expedition: The First United States Exploring Expedition (1838–1842).* Philadelphia: The American Philosophical Society, 1968.

United States Congress. *The Public Statutes at Large of the United States of America.* Vol. 5, ed. George Minot and George P. Sanger. Boston: Charles C. Little and James Brown, 1850.

United States Congress. *The Statutes at Large and Treaties, of the United States of America.* Vol. 11, ed. Richard Peters. Boston: Little, Brown and Company, 1867.

Vail, Robert W.G. "The American Sketchbooks of Charles Alexandre Lesueur, 1816–1836." *Proceedings of the American Antiquarian Society,* Worcester, Mass., April 1938.

Verrill, Addison E. "The Cephalopods of the Northeastern Coast of America: Part I, The Gigantic Squids (Architeuthis) and Their Allies; with Observations on Similar Large Species from Foreign Localities." *Transactions of the Connecticut Academy of Arts and Sciences* 5 (December 1879): 177–257.

Verrill, Addison E. "Part II, The Smaller Cephalopods, Including the 'Squids' and the Octopi, with Other Allied Forms." *Transactions of the Connecticut Academy of Arts and Sciences* 5 (June 1880): 259–446.

Viola, Herman J., and Carolyn Margolis. *Magnificent Voyagers: The U.S. Exploring Expedition, 1838–1842.* Washington, D.C.: Smithsonian Institution, 1985.

Wainwright, Nicholas B. *Philadelphia in the Romantic Age of Lithography.* Philadelphia: Historical Society of Pennsylvania, 1958.

Webster, Charles. "John Ray." In Gillispie, ed., *Dictionary of Scientific Biography,* vol. 11 (1975), pp. 313–318.

Weimerskirch, Philip J. "Naturalists and the Beginnings of Lithography in America." In *From Linnaeus to Darwin: Commentaries on the History of Biology and Geology*, pp. 167–177. London: Society for the History of Natural History, 1985.

Weiss, Harry B., and Grace M. Zeigler. *Thomas Say: Early American Naturalist*. Springfield, Ill., and Baltimore, Md.: Charles C. Thomas, 1931.

Welker, Robert Henry. *Birds and Men: American Birds in Science, Art, Literature, and Conservation, 1800–1900*. Cambridge, Mass.: The Belknap Press of Harvard University Press, 1955.

Welling, William. *Collector's Guide to Nineteenth-Century Photographs*. New York: Macmillan, 1976.

Wells, Samuel, Mary Treat, and Frederick Leroy Sargent. *Through a Microscope: Something of the Science together with Many Curious Observations Indoor and Out, and Directions for a Home-Made Microscope*. Chicago: The Interstate Publishing Company, 1886.

Whitman, C. O. "Introduction." *Journal of Morphology* 1, no. 1 (1887): i–iii.

Wilkes, Charles. *Narrative of the United States Exploring Expedition, during the Years 1838, 1839, 1840, 1841, 1842*. Philadelphia: C. Sherman, 1844.

Williams, Raymond. *Problems in Materialism and Culture*. London: Verso, 1980.

Wilson, Alexander. *American Ornithology; or, The Natural History of the Birds of the United States*. 9 vols. Philadelphia: Bradford and Inskeep, 1808–1814.

Wilson, Alexander. *American Ornithology; or the Natural History of Birds of the United States, with a Continuation by Charles Lucien Bonaparte . . . The Illustrative Notes and Life of Wilson by Sir William Jardine*. 3 vols. London: Whittaker, Treacher and Arnot; Edinburgh: Stirling and Kenner, 1832.

Wilson, Alexander, and Charles Lucien Bonaparte. *American Ornithology; or, The Natural History of Birds of the United States*. 4 vols. Edited by Robert Jameson. Edinburgh and London: Constable and Co., 1831.

Wilson, William E. *The Angel and the Serpent: The Story of New Harmony*. Bloomington: Indiana University Press, 1964.

Winsor, Mary P. "Louis Agassiz and the Species Question." *Studies in the History of Biology*, vol. 3 (1979): 89–117.

Winsor, Mary P. *Reading the Shape of Nature: Comparative Zoology at the Agassiz Museum*. Chicago: University of Chicago Press, 1991.

Wolfe, Richard J. *Jacob Bigelow's American Medical Botany, 1817–1821*. Boston: R. Wolfe, 1979.

Wood, Charles B. III. "Prints and Scientific Illustration in America." In *Prints in and of America*, Winterthur Conference Report, 1970, pp. 161–191. Edited by John D. Morse. Charlottesville: University Press of Virginia, 1970.

Worster, Donald. *Nature's Economy: The Roots of Ecology*. Garden City, N.Y.: Anchor Press, Doubleday, 1979.

Worthington, Richard D., and Patricia H. Worthington. "John Edwards Holbrook, Father of American Herpetology." In *North American Herpetology*, reprint edition, pp. xiii–xxvii. Edited by Kraig Adler. Athens, Ohio: Society for the Study of Amphibians and Reptiles, 1976.

Wright, Mabel Osgood. *Birdcraft: A Field Book of Two Hundred Song, Game, and Water Birds*. New York: Macmillan and Co., 1895.

Wright, Mabel Osgood. *Gray Lady and the Birds: Stories of the Bird Year for Home and School*. New York: The Macmillan Company, 1907.

Wright, Mabel Osgood, with Elliott Coues. *Citizen Bird: Scenes from Bird-Life in Plain English for Beginners*. New York: The Macmillan Company, 1897.

Wyckoff, William C. "A Perilous Fossil Hunt: Professor Marsh's Last Trip to the Bad Lands." *The New York Tribune*, 22 December 1874, pp. 3–4.

Young, Robert M. *Darwin's Metaphor: Nature's Place in Victorian Culture*. Cambridge, U.K.: Cambridge University Press, 1985.

# Index